Chemical Process Industries

Volume 1

Inorganic Chemicals and Allied Industries

Chemical Process Industries

Volume 1

Inorganic Chemicals and Allied Industries

W Smith

R Chapman

CBS

CBS Publishers & Distributors Pvt Ltd

New Delhi • Bengaluru • Chennai • Kochi • Kolkata • Mumbai
Bhopal • Bhubaneswar • Hyderabad • Jharkhand • Nagpur • Patna • Pune
• Uttarakhand • Dhaka (Bangladesh)

Chemical Process Industries

Volume 1

Inorganic Chemicals and Allied Industries

ISBN: 978-81-239-2845-6

Copyright © Publisher

First Edition: 2016

Reprint: 2019

Published by Satish Kumar Jain and produced by Varun Jain for

CBS Publishers & Distributors Pvt Ltd

4819/XI Prahlad Street, 24 Ansari Road, Daryaganj, New Delhi 110 002, India.

Ph: 23289259, 23266861, 23266867   Website: www.cbspd.com

Fax: 011-23243014   e-mail: delhi@cbspd.com; cbspubs@airtelmail.in.

Corporate Office: 204 FIE, Industrial Area, Patparganj, Delhi 110 092

Ph: 4934 4934   Fax: 4934 4935   e-mail: publishing@cbspd.com; publicity@cbspd.com

Branches

- **Bengaluru:** Seema House 2975, 17th Cross, K.R. Road, Banasankari 2nd Stage, Bengaluru 560 070, Karnataka, India.
  Ph: +91-80-26771678/79   Fax: +91-80-26771680   e-mail: bangalore@cbspd.com
- **Chennai:** 7, Subbaraya Street, Shenoy Nagar, Chennai 600 030, Tamil Nadu, India.
  Ph: +91-44-26680620, 26681266   Fax: +91-44-42032115   e-mail: chennai@cbspd.com
- **Kochi:** 42/1325, 1326, Power House Road, Opp. KSEB Power House, Ernakulam 682 018, Kochi, Kerala, India.
  Ph: +91-484-4059061-65   Fax: +91-484-4059065   e-mail: kochi@cbspd.com
- **Kolkata:** 6/B, Ground Floor, Rameswar Shaw Road, Kolkata-700 014, West Bengal, India.
  Ph: +91-33-22891126, 22891127, 22891128   e-mail: kolkata@cbspd.com
- **Mumbai:** 83-C, Dr E Moses Road, Worli, Mumbai-400018, Maharashtra, India.
  Ph: +91-22-24902340/41   Fax: +91-22-24902342   e-mail: mumbai@cbspd.com

Representatives

• Bhopal	0-8319310552	• Bhubaneswar	0-9911037372	• Hyderabad	0-9885175004
• Jharkhand	0-9811541605	• Nagpur	0-9021734563	• Patna	0-9334159340
• Pune	0-9623451994	• Uttarakhand	0-9716462459	• Dhaka (Bangladesh)	01912-003485

Printed at India Binding House, Noida, UP, India

Preface

The encyclopaedic range of the subject of chemical process industries has been condensed into a handy, easy-to-use reference/textbook, presented in a concise, compact and lucid manner.

Products of the chemical process industry are used in all areas of everyday life. The raising of food plants and animals requires chemical fertilisers, insecticides, food supplements, and disinfectants. Clothing utilises many synthetic fibres and dyes. Transportation depends upon gasoline and other fuels. Written communication uses paper and printing ink and electronic communication requires many chemically processed insulators and conductors. The nation's health is maintained by drugs and pharmaceuticals, soaps and detergents, insecticides and disinfectants — all products of the chemical process industry. In addition, many chemicals never reach the consumer in their original form but are sold within the industry for further processing or use in the production of other chemicals for consumer use. It is often said that the chemical industry is its own best customer.

Modern chemical processes are often extremely complex operations involving hundred of pieces of equipment. Without a systematic approach, it would be impossible to analyse an existing process or to design a new process. Therefore, chemical processes are broken down into individual steps that recur in many other processes.

To the practicing engineer, particularly the chemical engineer, the process flow sheet is the key instrument for defining, refining, and documenting a chemical process. The process flow diagram is the authorised process blueprint, the framework for specifications used in equipment designation and design; it is the single, authoritative document employed to define, construct, and operate the chemical process. This type of diagram is also used in other processes and industries.

In all chemical processes it is necessary to know such process data as flow rates, compositions, pressures and temperatures, so that the operator and production engineer can tell that the process is functioning properly. In the typical chemical process many instruments are used to measure, indicate and record the necessary process data. It is often desirable to use an automatic control, so that not only does the instrument measure and record a variable, but it also maintains the variable at a predetermined value.

Many chemical engineers design and operate large-scale and complex chemical production facilities supplying diverse chemical products to society. In performing these functions, a chemical engineer will likely assume a number of roles during a career. The engineer may become involved in raw materials extraction, intermediate materials processing, or production of pure chemical substances; in each activity, the minimisation and management of waste streams will have important economic and environmental consequences. Chemical engineers are involved in the production of bulk and speciality chemicals, petrochemicals, integrated circuits, pulp and paper, consumer products, minerals and pharmaceuticals.

This volume dealing with the inorganic aspects of chemical industries is the first volume of the reference/textbook on chemical process industries.

Each chapter covers an important aspect of chemical processing industries with an accurate, up-to-date account of each topic.

Chapter 1 deals with chemical processing industries: a review. Chapter 2 concentrates on water conditioning and environmental protection. Various physical, chemical water quality parameters along with treatment of water are discussed in detail. Chemical process industries consume more than a third of energy used by all manufacturing industries, chemical engineer should be familiar with the broad technical aspects of the production of power, cold and heat. Considering this chapter 3 is devoted to energy, fuels, air conditioning and refrigeration. Chapter 4 discusses coal chemicals, its types and gases produced from coal. Chapter 5 deals with fuel gases and discusses natural, producer and synthetic gases. Chapter 6 focuses on industrial gases which have performed varied and essential functions in our economy, as some are raw materials for other manufacture chemicals. Carbon is essential to all known living systems, and without it life as we know it could not exist. Keeping this in mind chapter 7 is devoted to industrial carbon and its various allotropic forms are discussed. Chapter 8 concentrates on ceramic industries such as clay, refractory, pottery and their manufacturing aspects. Chapter 9 focuses on portland cements, calcium and magnesium compounds. Chapter 10 discusses glass industries along with their properties and manufacturing aspects. Chapter 11 deals with salt and miscellaneous sodium compounds and chapter 12 discusses chlor-alkali industries such as soda ash, caustic soda and chlorine. Chapter 13 is devoted to electrolytic industries, as energy in the form of electricity causes chemical reactions to take place in electrolytic industries. Chapter 14 focuses on electrothermal industries which include abrasives, silicon carbide, aluminium oxide and calcium carbide. The use of artificial fertilisers, phosphoric acid, phosphate salts and their derivatives has increased recently. Considering this chapter 15 concentrates on phosphorus industries. Chapter 16 and 17 are devoted to potassium and nitrogen industries respectively. Chapter 18 deals with sulphur and sulphuric acid which are the basic raw materials of various chemical process and allied industries. Extraction, production, properties and uses of sulphur, sulphuric acid, sulphur dioxide are discussed in detail. Chapter 19 discusses hydrochloric acid and miscellaneous inorganic chemicals along with their manufacturing aspects and uses.

The simple, lucid and readable style of this book will help the students to comprehend the concepts and information presented, and ensures an easy grasp of the key points which are vividly illustrated by diagrams, tables, and index.

This invaluable model reference/textbook is specifically intended for B.E./B. Tech. institutions which offer courses in chemical engineering, and biotechnology. The researchers in analytical and instrumental techniques, chemists and industrialists will also find it highly useful and informative, as its primary aim is to provide an exhaustive coverage of principles and concepts which form the basis of chemical process industries.

It has been prepared with meticulous care, aiming at making the book error-free. Constructive suggestions are always welcome from users of this book.

W Smith

R Chapman

Contents at a Glance

Contents

Chapter 1

Chemical Process Industries: A Review

INTRODUCTION

Products of the chemical process industry are used in all areas of everyday life. The raising of food plants and animals requires chemical fertilisers, insecticides, food supplements, and disinfectants. Clothing utilises many synthetic fibres and dyes. Transportation depends upon gasoline and other fuels. Written communication uses paper and printing ink and electronic communication requires many chemically processed insulators and conductors. The nation's health is maintained by drugs and pharmaceuticals, soaps and detergents, insecticides and disinfectants — all products of the chemical process industry. In addition, many chemicals never reach the consumer in their original form but are sold within the industry for further processing or use in the production of other chemicals for consumer use. It is often said that the chemical industry is its own best customer.

Any definition or description of the chemical process industry is bound to be incomplete. Most processes in the chemical industry involve a chemical change. The term 'chemical change' should be interpreted to include not only chemical reactions but also physico-chemical changes, such as the separation and purification of the components of a mixture. Purely mechanical changes are usually not considered part of the chemical process, unless they are essential to later chemical changes. As an example, the manufacture of the plastic polythene, using ethylene produced from petroleum or natural gas, involves a chemical process. On the other hand, the moulding and fabrication of the resulting plastic resin into final shapes for consumer products would not be considered part of the chemical process.

SYSTEMATIC ANALYSIS OF CHEMICAL PROCESSES

Early chemical processes usually involved a few simple steps. Materials were processed in small batches. Process improvements evolved slowly from experience gained by the operators. As the demand for chemical products increased and the chemical industry grew, processes had to be developed to produce large quantities at the lowest possible cost. Improvements based on experience were no longer sufficient; it became necessary to build new plants for products never before produced. Systematic analysis of chemical processes elucidated many underlying principles which could be used in the synthesis of new processes.

Modern chemical processes are often extremely complex operations involving hundred of pieces of equipment. Without a systematic approach, it would be impossible to analyse an existing process or to design a new process. Therefore, chemical processes are broken down into individual steps that recur in many other processes. The general principles of these steps have been carefully studied and are frequently well understood. If the principles are known, it is possible to design the step to do the best possible job. The typical chemical process is analysed with the following interdependent considerations:

1. Mass and energy balances.
2. Thermochemistry.
3. Unit operations.
4. Plant equipment.
5. Ancillery equipment.
6. Process flow diagrams.
7. Instrumentation and control.
8. Economics.

Mass and Energy Balances

Here some of the fundamental principles that engineers and scientists employ in performing design calculations and predicting the performance of plant equipment are discussed.

Conservation of mass

The conservation law for mass can be applied to any process or system. The conservation law for mass finds its major application in performing pollution prevention audits. A pollution prevention assessment is a systematic, planned procedure with the objective of identifying ways to reduce or eliminate waste.

Conservation of energy

A presentation of the conservation law for energy would be incomplete without a brief review of some introductory thermodynamic principles.

Thermodynamics is defined as the science that deals with the relationships between the various forms of energy.

Thermochemistry

Consider now the energy effects associated with a chemical reaction. To introduce this subject, the reader is reminded that engineers and applied scientists are rarely concerned with the magnitude or amount of energy in a system; their primary concern is with changes in the amount of energy. It has been found in measuring energy changes for systems that the enthalpy is the most convenient term to work with. There are many different types of enthalpy effects; these include sensible heat, latent heat, and heat of reaction.

The heat of reaction is defined as the enthalpy change of a system undergoing chemical reaction. If the reactants and products are at the same temperature and in their standard states, the heat of reaction is termed the standard heat of reaction. For engineering purposes, the standard state of a chemical may be taken as the pure chemical at 1 atm pressure.

Chemical kinetics

Chemical kinetics involves the study of reaction rates and the variables that affect these rates. It is a topic that is critical for the analysis of reacting systems. The objective here is to develop a working understanding of this subject that will permit us to apply chemical kinetics principles in the area of pollution prevention.

Ideal gas law

The ideal gas law was derived from experiments in which the effects of pressure and temperature on gaseous volumes were measured over a moderate range of temperatures and pressures. As a general rule, this law works best when the molecules of the gas are far apart, that is, when the pressure is low and the temperature is high.

Under these conditions, the gas is said to behave ideally. For engineering calculations the ideal gas law is almost always assumed to be valid, since it generally works well for the temperature and pressure ranges used in most applications.

Phase equilibrium

The term phase, for a pure substance, indicates a state of matter—that is, solid, liquid, or gas. For mixtures, however, a more stringent connotation must be used, since a totally liquid or solid system may contain more than one phase. A phase is characterised by uniformity or homogeneity; the same composition and properties must exist throughout the phase region. At most temperatures and pressures, a pure substance normally exists as a single phase. At certain temperatures and pressures, two or perhaps even three phases can coexist in equilibrium.

Unit Operations

In the chemical and other physical processing industries, including the food and biological processing industries, there are many similarities in the manner in which the entering feed materials are modified or processed into final products. These seemingly different chemical, physical, or biological processes can be broken down into a series of separate and distinct steps called unit operations. These unit operations are common to all types of diverse process industries.

For example, the unit operation distillation is used to purify or separate alcohol in the beverage industry and hydrocarbons in the petroleum industry. Drying of grain and other foods is similar to drying of lumber, filtered precipitates, and rayon yarn. The unit operation absorption occurs in absorption of oxygen from air in a fermentation process or in a sewage treatment plant, and in absorption of hydrogen gas in a process for liquid hydrogenation of oil.

Evaporation of salt solutions in the chemical industry is similar to evaporation of sugar solutions in the food industry. Settling and sedimentation of suspended solids are similar in the sewage and the mining industries. Flow of liquid hydrocarbons in the petroleum refinery and flow of milk in a dairy plant are carried out in a similar fashion. Thus, the most efficient method of organising the subject matter of unit operations is based on two facts: (i) although the number of individual processes is great, each one may be separated into a series of steps, called operations, each of which in turn appears in process after process; (ii) the individual operations have common techniques and are based on the same scientific principles. The unit operation concept, therefore, is this: By systematically studying these operations themselves— operations that clearly cross industry and process lines—the treatment of all processes is unified and simplified.

Plant Equipment

The plant equipment provides details on a number of commonly used process units: reactors, heat exchangers, columns of various types (distillation, absorption, adsorption, evaporation, extraction), dryers, and grinders. The purpose of each unit or operation and the many configurations in which the units can be found are also discussed.

Chemical reactors

The reactor is often the heart of a chemical process. It is the place in the process where raw materials are usually converted into products, and reactor design is therefore a vital step in the overall design of the process.

Heat exchangers

The transfer of heat to and from process fluids is an essential part of most chemical processes. The most commonly used type of heat transfer equipment is the shell and tube heat exchanger.

Mass transfer equipment

Distillation

Distillation is probably the most widely used separation process in the chemical industry. Its applications range from the rectification of alcohol, which has been practised since antiquity, to the fractionation of crude oil. The separation of liquid mixtures by distillation is based on differences in volatility between the components. The greater the relative volatilities, the easier the separation. Vapour flows up a column and liquid flows countercurrently down the column. The vapour and liquid are brought into contact on plates, or packing. Part of the condensate from the condenser is returned to the top of the column to provide liquid flow above the feed point (reflux), and part of the liquid from the base of the column is vapourised in the reboiler and returned to provide the vapour flow.

Adsorption

In the adsorption process, one or more components in a mixture are preferentially removed from the mixture by a solid (referred to as the adsorbent). Adsorption is influenced by the surface area of the adsorbent, the nature of the solvent being adsorbed, the pH of the operating system (liquid application), and the temperature of operation. These are important parameters to be aware of when designing or evaluating an adsorption process.

Absorption

The process of absorption conventionally refers to the intimate contacting of a mixture of gases with a liquid so that part of one or more of the constituents of the gas will dissolve in the liquid. The contact usually takes place in some type of packed column.

Evaporation

The processing industry has given the operations involving heat transfer to a boiling liquid the general name evaporation. The most common application is the removal of water from a processing stream. Evaporation is used in the food, chemical, and petrochemical industries, and it usually results in an increase in the concentration of a certain species.

Extraction

Extraction (sometimes called leaching) encompasses liquid-liquid as well as liquid-solid systems. Liquid-liquid extraction involves the transfer of solutes from one liquid phase into another liquid solvent; it is normally conducted in a mixer-settler, plate and agitated-tower contracting equipment, or packed or spray towers. Liquid-solid extraction, in which a liquid solvent is passed over a solid phase to remove some solute, is carried out in fixed-bed, moving-bed or agitated-solid columns.

Drying

Drying generally involves the removal from solids of relatively small amounts of water or organic liquids, whereas evaporation removes larger amounts. Drying removes the liquid as a vapour by warm gas (usually air) currents. Drying can be accomplished on a batch or continuous basis.

Ancillary Equipment

This section considers devices for transporting gases and liquids to, from, or between units of process equipment. Some of these devices are simply conduits for the moving of material (pipes, ducts, fittings, stacks); others control the flow of material (valves); still others provide the mechanical driving force for the flow (fans, pumps, and compressors). This also covers storage facilities, holding tanks, materials-handling devices and techniques, and utilities (e.g. gas, steam, water) along with air, water, and solid waste control equipment.

Process Diagrams

To the practicing engineer, particularly the chemical engineer, the process flow sheet is the key instrument for defining, refining, and documenting a chemical process. The process flow diagram is the authorised process blueprint, the framework for specifications used in equipment designation and design; it is the single, authoritative document employed to define, construct, and operate the chemical process. This type of diagram is also used in other processes and industries.

Instrumentation and Control

In all chemical processes it is necessary to know such process data as flow rates, compositions, pressures and temperatures, so that the operator and production engineer can tell that the process is functioning properly. In the typical chemical process many instruments are used to measure, indicate and record the necessary process data. It is often desirable to use an automatic control, so that not only does the instrument measure and record a variable, but it also maintains the variable at a predetermined value. If the variable begins to change from the proper value, the automatic control initiates corrective action to return the variable to its proper value. For example, the temperature in a chemical reactor must often be controlled to give the desired yield of products. If the reaction is endothermic, the reactor might have a steam jacket to supply heat. The temperature of the materials in the reactor is measured by a temperature-sensing instrument, such as a thermocouple. The signal is transmitted to a paper-chart recorder to give the operator a record of the temperature variations with time. The temperature signal is also sent to the controller, which has been set by the operator to maintain the desired temperature. If the temperature of the reactor begins to fall below the desired value, the controller opens the steam valve to supply more heat to the heating jacket on the reactor. Conversely, if the reactor temperature is too high, the controller closes the steam valve. The steam valve may be in any position between fully open and fully closed.

Automatic control may be applied to almost any process variable that can be measured. Control of a variable is often maintained by the measurement of another more easily measured variable. Automatic control reduces the number of human operators. It can give faster and more accurate control than a human operator.

Economics

No matter how efficiently a process operates to produce a final product of high purity, the process is a failure if the product cannot be sold at a profit. Before a chemical plant is built, a thorough market analysis is made to determine how much of the product can be sold and at what price. Often it is possible to sell more of the product if the price is lower. Present and future competition from other producers must be carefully evaluated. As the plant is designed, many economic analyses are made to determine the least expensive design which will produce the desired quantity of product at a minimum price. If the cost of manufacture is sufficiently below the selling price to give an attractive profit, the plant will be built. If the product is successful and profitable, a competitor may find the market attractive and enter it with perhaps a somewhat better product produced at a lower price by an improved process. It is then necessary for the older producer to improve his process and product, or he will be forced out of the market.

WHAT IS CHEMICAL ENGINEERING?

Chemical engineering is a synthesis of chemistry and engineering. Although the profession grew out of industrial chemistry, today it is based as much on physical principles as on chemical fundamentals. There are many definitions of chemical engineering. Many of them are either too specific or too vague. One popular definition—'a chemical engineer carries out on a large scale reactions developed in the laboratory by the chemist', is true as far as it goes but it does not show the broad scope of problems encountered by the chemical engineer. The unit operations are not obviously a part of this definition. The typical chemist has little knowledge of any of the unit operations in theory and practice. A facetious definition of chemical engineering makes a point, 'a chemical engineer is one who talks engineering in the presence of chemists, chemistry in the presence of engineers and politics in the presence of both'. A unique characteristic of the chemical engineer is that he can talk to and understand both chemists and other engineers. He must understand the research findings of the chemist and he must be able to use engineering fundamentals to design and operate a new chemical process based on these research findings. It is this versatility that has made the chemical engineer indispensable in the chemical process industry.

Chemical engineering is unique among engineering disciplines. It requires a thorough understanding of chemistry, physics and mathematics, whereas the other fields of engineering usually do not require a thorough understanding of chemistry. For this reason chemical engineering will probably always remain somewhat apart from both chemistry and the other branches of engineering.

As the chemical industry grew in size and complexity, the work of the chemical engineer required an ever-growing background of education and experience. Today a chemical engineer may work in one of many fields within the profession of chemical engineering. Some of the fields require special training beyond an undergraduate curriculum.

ROLES AND RESPONSIBILITIES OF CHEMICAL ENGINEERS

Many chemical engineers design and operate large-scale and complex chemical production facilities supplying diverse chemical products to society. In performing these functions, a chemical engineer will likely assume a number of roles during a career. The engineer may become involved in raw materials extraction, intermediate materials processing, or production of pure chemical substances; in each activity, the minimisation and management of waste streams will have important economic and environmental consequences. Chemical engineers are involved in the production of bulk and speciality chemicals, petrochemicals, integrated circuits, pulp and paper, consumer products, minerals and pharmaceuticals. Chemical engineers also find employment in research, consulting organisations and educational institutions. The engineer may perform functions such as process and production engineering, process design, process control, technical sales and marketing, community relations and management.

As engineers assume such diverse roles, it is increasingly important that they be aware of their responsibilities to the general public, colleagues and employers, the environment, and also to their profession. One of the central roles of chemical engineers is to design and operate chemical processes yielding chemical products that meet customer specifications and that are profitable. Another important role is to maintain safe conditions for operating personnel and for residents in the immediate vicinity of a production facility. Finally, chemical process designs need to be protective of the environment and of human health. Environmental issues must be considered not only within the context of chemical production but also during other stages of a chemical's life cycle, such as transportation, use by customers, recycling activities and ultimate disposal. The types of procedures that will be used in designing processes that minimise environmental impacts, and the responsibilities of chemical engineers to reduce pollution generation within chemical processes, and briefly notes some of the other professional responsibilities of chemical engineers, i.e. issues dealing with engineering ethics.

Chemical Process Safety

A major objective for chemical process design is the inclusion of safeguards that minimise the number and severity of accidental releases of toxic chemicals and the incidence of fires and explosions. A number of chemical plant accidents have occurred in the relatively recent past illustrating the importance of integrating safety into process designs. These accidents resulted in loss of life, permanent disability, and the destruction of chemical plant, process equipment and neighbouring residences. The most famous accidents occurred in Flixborough, England (1974) and Bhopal, India (1984).

In incidents such as this, loss of life and injuries are tragic, and economic consequences are severe. Engineers have a special role to play in preventing such incidents. Part of an engineer's professional responsibility is to design processes and products that are as safe as possible. Traditionally, this has meant identifying hazards, evaluating their severity and then applying several layers of protection as a means of mitigating the risk of an accident. This approach can be very effective and has resulted in significant improvement of the safety performance of chemical processes. However, the layer of protection approach has disadvantages that place limitations on its effectiveness: (i) the layers are expensive to build and maintain, and (ii) the hazard remains and there is always a finite risk that an accident will happen despite the layers of protection.

Inherently safer design is a fundamentally different approach to chemical process safety. Instead of working with existing hazards in a chemical process and adding layers of protection, the engineer is challenged to reconsider the design and eliminate or reduce the source of the hazard within the process. Approaches to the design of inherently safer processes have been grouped into the four categories listed below. This list contains a short checklist of questions related to inherently safer processes.

1. Minimise: Use smaller quantities of hazardous substances.
    (a) Have all in-process inventories of hazardous materials in storage tanks been minimised?
    (b) Are all of the proposed in-process storage tanks really needed?

(c) Can other types of unit operations or equipment reduce material inventories (for example, continuous in-line mixers in place of mixing vessels)?

2. Substitute: Use a less hazardous material in place of a more hazardous substance.
   (a) Is it possible to completely eliminate hazardous raw materials, process intermediates, or by-products by using an alternative process or chemistry?
   (b) Is it possible to substitute less hazardous raw materials or to substitute noncombustible for flammable solvents?

3. Moderate: Use less hazardous conditions or facilities which minimise the impacts of a release of a hazardous material or energy.
   (a) Can the supply pressure of raw materials be limited to less than the working pressure of the vessels they are delivered to?
   (b) Can reaction conditions (temperature, pressure) be made less severe by using a catalyst, or by using a better catalyst?

4. Simplify: Design facilities which eliminate unnecessary complexity and make operating errors less likely, and which are forgiving of errors that are made.
   (a) Can equipment be sufficiently designed to totally contain the maximum pressure generated, even if the 'worst credible event' occurs?

It is useful to recognise analogies between chemical process safety and the design of processes that minimise environmental impacts.

A new generation of inherently safer processes relies on designs that reduce hazards, rather than providing protection from hazards. Similarly, new generations of processes that minimise environmental impact do not rely on treating wastes, but instead are designed so that they do not generate wastes.

Environmental Protection

When the method for managing environmental performance is to treat wastes, the process is designed, wastes are generated, and treatment technologies are deployed. The design method for meeting environmental objectives is sequential. In contrast, if the primary design, rather than the design of peripheral waste treatment units, is to be modified to meet environmental objectives, a key question to answer is 'at what stage in the design should environmental considerations be considered'?

Designs for new processes and retrofitting of existing procedures are multistep procedures. The first step is the definition of a primitive problem, such as identifying the chemical to be produced and the annual quantity. This is followed by a process creation step that includes choosing reaction chemistry, the use of design heuristics to identify process equipment and operating conditions, development of a base case flowsheet, and process simulation. The third step is a more detailed process synthesis of separation trains and a heat/power integration analysis. What follows is a detailed design and simulation of the flowsheet, profitability analysis, and optimisation. The final steps include a plantwide control liability assessment, startup assessment, and reliability and safety analysis.

As part of their professional responsibilities, engineers should, through their designs, continuously improve the environmental performance of chemical processes. Recently the chemical manufacturers association (CMA, now American Chemistry Council) has adopted the pollution prevention code of management practice, which outlines tangible steps along a path to continuous reductions in the amounts of all contaminants released to air, water, and soil.

Engineering Ethics

Process safety and environmental protection are not the only responsibilities of professional engineers. Engineers also have responsibilities to clients, to colleagues, and to the profession. The American Institute of Chemical Engineers has assembled a code of ethics that highlights the main issues in the area of professional conduct. This code can be found at AIChE website (http://www.aiche.org/membership/ethics.htm). Some of the responses dealt with putting health, safety, and environmental issues ahead of profits; placing self-respect as professionals above loyalty to companies; working within organisations versus whistleblowing to promote ethical behaviour; and taking career risks in order to get a company to do the right thing.

Chapter 2

Water Conditioning and Environmental Protection

INTRODUCTION

Water is one of the abundantly available substances in nature. It is an essential constituent of all animal and vegetable matter and forms about 75 per cent of the matter of earth's crust. It is also an essential ingredient of animal and plant life. Water is distributed in nature in different forms, such as rain water, river water, spring water and mineral water.

Water is an essential ingredient of animal and plant life. Generally the municipal water is used for drinking purposes and other domestic purposes in cities and towns and hence water conditioning and waste-water treatment have long been essential practical functions of municipalities. The importance of suitably preparing water for chemical industry is now well recognised and various processes for purification or conditioning of water obtained from natural sources have actually been proposed.

The treatment of water to which it is subjected, depends on the purpose for which the treated water has to be applied. The choice generally depends on the use, whether for power generation, heating, cooling or actual incorporation in a product or its manufacturing process. The quality or quantity of surface as well as groundwater are very important in the location of the chemical plant. Rivers, lakes, streams, rains, etc. are the various natural sources of water which supply abundant water containing large number of impurities in most cases. The impurities present in water vary greatly from one place to another. Treatment of water obtained from different sources is, therefore, essential so that a part or whole of the treated water may be used safely for municipal purpose, laundry purpose, boiler purpose or other industrial purpose.

PHYSICAL AND CHEMICAL WATER QUALITY PARAMETERS

The availability of a water supply adequate in terms of both quantity and quality is essential to human existence. Early people recognised the importance of water from a quantity viewpoint. Civilisation developed around water bodies that could support agriculture and transportation as well as provide drinking water. Recognition of the importance of water quality developed more slowly. Early humans could judge water quality only through the physical senses of sight, taste and smell. Not until the biological, chemical and medical sciences developed, were methods available to measure water quality and to determine its effects on human health and well-being.

Physical parameters define those characteristics of water that respond to the senses of sight, touch, taste or smell. Suspended solids, turbidity, colour, taste and odour and temperature fall into this category.

The development of the science of water chemistry roughly paralleled that of water microbiology. Many of the chemicals used in industrial processes and agricultrure have been identified in water. However, the effort to identify other chemical compounds which may already be found in trace quantities in many water supplies and to determine their effect on human health was only recently begun.

Suspended Solids

Solids can be dispersed in water in both suspended and dissolved forms. Although some dissolved solids may be perceived by the physical senses, they fall more appropriately under the category of chemical parameters and will be discussed more fully in a later section.

Solids suspended in water may consist of inorganic or organic particles or of immiscible liquids. Inorganic solids such as clay, silt and other soil constituents are common in surface water. Organic material such as plant fibres and biological solids (algal cells, bacteria, etc.) are also common constituents of surface waters. These materials are often natural contaminants resulting from the erosive action of water flowing over surfaces. Because of the filtering capacity of the soil, suspended material is seldom a constituent of groundwater. Other suspended material may result from human use of the water. Domestic waste-water usually contains large quantities of suspended solids that are mostly organic in nature. Industrial use of water may result in a wide variety of suspended impurities of either organic or inorganic nature. Immiscible liquids such as oils and greases are often constituents of waste-water.

Suspended material may be objectionable in water for several reasons. It is aesthetically displeasing and provides adsorption sites for chemical and biological agents. Suspended organic solids may be degraded biologically, resulting in objectionable by-products. Biologically active (live) suspended solids may include disease-causing organisms as well as organisms such as toxin-producing strains of algae.

Turbidity

A direct measurement of suspended solids is not usually performed on samples from natural bodies of water or on potable (drinkable) water supplies. The nature of the solids in these waters and the secondary effects they produce are more important than the actual quantity. For such waters a test for turbidity is commonly used. Turbidity is a measure of the extent to which light is either absorbed or scattered by suspended material in water. Because absorption and scattering are influenced by both size and surface characteristics of the suspended material, turbidity is not a direct quantitative measurement of suspended solids. For example, one small pebble in a glass of water would produce virtually no turbidity. If this pebble were crushed into thousands of particles of colloidal size, a measurable turbidity would result, even though the mass of solids had not changed. Most turbidity in surface waters results from the erosion of colloidal material such as clay, silt, rock fragments and metal oxides from the soil. Vegetable fibres and micro-organisms may also contribute to turbidity. Household and industrial waste-waters may contain a wide variety of turbidity-producing material. Soaps, detergents and emulsifying agents produce stable colloids that result in turbidity. Although turbidity measurements are not commonly run on waste-water, discharges of waste-waters may increase the turbidity of natural bodies of water.

Colour

Pure water is colourless, but water in nature is often coloured by foreign substances. Water whose colour is partly due to suspended matter is said to have apparent colour. Colour contributed by dissolved solids that remain after removal of suspended matter is known as true colour.

Industrial wastes from textile and dyeing operations, pulp and paper production, food processing, chemical production and mining, refining and slaughterhouse operations may add substantial colouration to water in receiving streams.

Taste and Odour

The terms taste and odour are themselves definitive of this parameter. Because the sensations of taste and smell are closely related and often confused, a wide variety of tastes and odours may be attributed to water by consumers. Substances that produce an odour in water will almost invariably impart a taste as well. The converse is not true, as there are many mineral substances that produce taste but no odour.

Temperature

Temperature is not used to evaluate directly either potable water or waste-water. It is, however, one of the most important parameters in natural surface-water systems. The temperature of surface waters governs to a large extent the biological species present and their rates of activity. Temperature has an effect on most chemical reactions that occur in natural water systems. Temperature also has a pronounced effect on the solubilities of gases in water.

Chemical Water Quality Parameters

Water has been called the universal solvent and chemical parameters are related to the solvent capabilities of water. Total dissolved solids, alkalinity, hardness, fluorides, metals, organics and nutrients are chemical parameters of concern in water quality management.

Total Dissolved Solids

The material remaining in the water after filtration for the suspended-solids analysis is considered to be dissolved. This material is left as a solid residue upon evaporation of the water and constitutes a part of total solids.

Alkalinity

Alkalinity is defined as the quantity of ions in water that will react to neutralise hydrogen ions. Alkalinity is thus a measure of the ability of water to neutralise acids.

In large quantities, alkalinity imparts a bitter taste to water. The principal objection to alkaline water, however, is the reactions that can occur between alkalinity and certain cations in the water. The resultant precipitate can foul pipes and other water-systems appurtenances.

Alkalinity measurements are made by titrating the water with an acid and determining the hydrogen equivalent. Alkalinity is then expressed as milligrams per litre of $CaCO_3$.

Hardness

Hardness is defined as the concentration of multivalent metallic cations in solution. At supersaturated conditions, the hardness cations will react with anions in the water to form a solid precipitate. Hardness is classified as carbonate hardness and noncarbonate hardness, depending upon the anion with which it associates.

The hardness that is equivalent to the alkalinity is termed carbonate hardness, with any remaining hardness being called noncarbonate hardness.

Carbonate hardness is sensitive to heat and precipitates readily at high temperatures.

$$Ca(HCO_3)_2 \xrightarrow{\Delta} CaCO_3 + CO_2 + H_2O \qquad ... (2.1)$$

$$Mg(HCO_3)_2 \xrightarrow{\Delta} Mg(OH)_2 + 2CO_2 \qquad ... (2.2)$$

Sources

The multivalent metallic ions most abundant in natural waters are calcium and magnesium. Others may include iron and manganese in their reduced states (Fe^{2+}, Mn^{2+}), strontium (Sr^{2+}) and aluminium (Al^{3+}). The latter are usually found in much smaller quantities than calcium and magnesium and for all practical purposes, hardness may be represented by the sum of the calcium and magnesium ions.

Impacts

Soap consumption by hard waters represents an economic loss to the water user. Sodium soaps react with multivalent metallic cations to form a precipitate, thereby losing their surfactant properties. A typical divalent cation reaction is:

$$2NaCO_2C_{17}H_{33} + cation^{2+} \longrightarrow cation^{2+} (CO_2C_{17}H_{33})_2 + 2Na^+ \quad ... (2.3)$$

Soap   Precipitate

Lathering does not occur until all of the hardness ions are precipitated, at which point the water has been 'softened' by the soap. The precipitate formed by hardness and soap adheres to surfaces of tubs, sinks and dishwashers and may stain clothing, dishes and other items. Residues of the hardness soap precipitate may remain in the pores, so that skin may feel rough and uncomfortable. In recent years these problems have been largely alleviated by the development of soaps and detergents that do not react with hardness.

Boiler scale, the result of the carbonate hardness precipitate may cause considerable economic loss through fouling of water heaters and hot water pipes. Changes in pH in the water distribution systems may also result in deposits of precipitates. Bicarbonates begin to convert to the less soluble carbonates at pH values above 9.0.

Magnesium hardness, particularly associated with the sulphate ion, has a laxative effect on persons unaccustomed to it. Magnesium concentrations of less than 50 mg/l are desirable in potable waters, although many public water supplies exceed this amount.

Calcium hardness presents no public health problem. In fact, hard water is apparently beneficial to the human cardiovascular system.

Measurement

Hardness can be measured by using spectrophotometric techniques or chemical titration to determine the quantity of calcium and magnesium ions in a given sample. Hardness can be measured directly by titration with ethylenediamine tetraacetic acid (EDTA) using eriochrome black T (EBT) as an indicator. The EBT reacts with the divalent metallic cations, forming a complex that is red in colour. The EDTA replaces the EBT in the complex and when the replacement is complete, the solution changes from red to blue. If 0.01 M EDTA is used, 1.0 ml of the titrant measures 1.0 mg of hardness as $CaCO_3$.

Fluoride

Generally associated in nature with a few types of sedimentary or igneous rocks, fluoride is seldom found in appreciable quantities in surface waters and appears in groundwater in only a few geographical regions. Fluoride is toxic to humans and other animals in large quantities, while small concentrations can be beneficial. Concentrations of approximately 1.0 mg/l in drinking water help to prevent dental cavities in children. During formation of permanent teeth, fluoride combines chemically with tooth enamel, resulting in harder, stronger teeth that are more resistant to decay. Fluoride is often added to drinking water supplies if sufficient quantities for good dental formation are not naturally present. Excessive intakes of fluoride can result in discolouration of teeth. Noticeable discolouration, called mottling, is relatively common when fluoride concentrations in drinking water exceed 2.0 mg/l, but is rare when concentrations are less than 1.5 mg/l. Adult teeth are not affected by fluoride, although both the benefits and liabilities of fluoride during tooth-formation years carry over into adulthood.

Excessive dosages of fluoride can also result in bone fluorosis and other skeletal abnormalities. Concentrations of less than 5 mg/l in drinking water are not likely to cause bone fluorosis or related problems and some water supplies are known to have somewhat higher fluoride concentrations with no discernible problem other than severe mottling of teeth. On the assumption that people drink more water in warmer climates, EPA drinking-water standards base upper limits for fluoride on ambient temperatures.

Metals

All metals are soluble to some extent in water. While excessive amounts of any metal may present health hazards, only those metals that are harmful in relatively small amounts are commonly labelled toxic; other metals fall into the nontoxic group. Sources of metals in natural waters include dissolution from natural deposits and discharges of domestic, industrial or agricultural waste-waters. Measurement of metals in water is usually made by atomic absorption spectrophotometry.

Nontoxic metals

In addition to the hardness ions, calcium and magnesium, other nontoxic metals commonly found in water include sodium, iron, manganese, aluminium, copper and zinc. Sodium, by far the most common nontoxic metal found in natural waters, is abundant in the earth's crust and is highly reactive with other elements. The salts of sodium are soluble in water. Excessive concentrations cause a bitter taste in water and are a health hazard to cardiac and kidney patients. Sodium is also corrosive to metal surfaces and, in large concentrations, is toxic to plants. Iron and manganese quite frequently occur together and present no health hazards at concentrations normally found in natural waters.

Toxic metals

As noted earlier, toxic metals are harmful to humans and other organisms in small quantities. Toxic metals that may be dissolved in water include arsenic, barium, cadmium, chromium, lead, mercury and silver. Cumulative toxins such as arsenic, cadmium, lead and mercury are particularly hazardous. These metals are concentrated by the food chain, thereby posing the greatest danger to organisms near the top of the chain.

Organics

Many organic materials are soluble in water. Organics in natural water systems may come from natural sources or may result from human activities. Most natural organics consist of the decay products of organic solids, while synthetic organics are usually the result of waste-water discharges or agricultural practices. Dissolved organics in water are usually divided into two broad categories: biodegradable and nonbiodegradable (refractory).

Biodegradable organics

Biodegradable material consists of organics that can be utilised for food by naturally occurring micro-organisms within a reasonable length of time. In dissolved form, these materials usually consist of starches, fats, proteins, alcohols, acids, aldehydes and esters. They may be the end product of the initial microbial decomposition of plant or animal tissue or they may result from domestic or industrial waste-water discharges. Although some of these materials can cause colour, taste and odour problems, the principal problem associated with biodegradable organics is a secondary effect resulting from the action of micro-organisms on these substances.

Nonbiodegradable organics

Some organic materials are resistant to biological degradation. Tannic and lignic acids, cellulose and phenols are often found in natural water systems. These constituents of woody plants biodegrade so slowly that they are usually considered refractory. Molecules with exceptionally strong bonds (some of the polysaccharides) and ringed structures (benzene) are essentially nonbiodegradable. An example is the detergent compound alkyl benzene sulphonate (ABS) which, with its benzene ring, does not biodegrade. Being a surfactant, ABS causes frothing and foaming in waste-water treatment plants and increases turbidity by stabilising colloidal suspensions. This problem was largely alleviated when detergent manufacturers switched to a linear alkyl sulphonate (LAS) compound, which is biodegradable. Many of the organics associated with petroleum and with its refining and processing also contain benzene and are essentially nonbiodegradable. Some organics are nonbiodegradable because they are toxic to organisms.

These include the organic pesticides, some industrial chemicals and hydrocarbon compounds that have combined with chlorine.

Measurement of nonbiodegradable organics is usually by the chemical oxygen demand (COD) test. Nonbiodegradable organics may also be estimated from a total organic carbon (TOC) analysis. Both COD and TOC measure the biodegradable fraction of the organics, so the BOD must be subtracted from the COD or TOC to quantify the nonbiodegradable organics.

Specific organic compounds can be identified and quantified through analysis by gas chromatography.

Nutrients

Nutrients are elements essential to the growth and reproduction of plants and animals and aquatic species depend on the surrounding water to provide their nutrients. Although a wide variety of minerals and trace elements can be classified as nutrients, those required in most abundance by aquatic species are carbon, nitrogen and phosphorus. Carbon is readily available from many sources. Carbon dioxide from the atmosphere, alkalinity and decay products of organic matter all supply carbon to the aquatic system. In most cases, nitrogen and phosphorus are the nutrients that are the limiting factors in aquatic plant growth.

Nitrogen

Nitrogen gas (N_2) is the primary component of the earth's atmosphere and is extremely stable. It will react with oxygen under high-energy conditions (electrical discharges or flame incineration) to form nitrogen oxides. Although a few biological species are able to oxidise nitrogen gas, nitrogen in the aquatic environment is derived primarily from sources other than atmospheric nitrogen.

Nitrogen is a constituent of proteins, chlorophyll and many other biological compounds. Upon the death of plants or animals, complex organic matter is broken down to simple forms by bacterial decomposition. Proteins, for instance, are converted to amino acids and further reduced to ammonia (NH_3).

If oxygen is present, the ammonia is oxidised to nitrite (NO_2^-) and then to nitrate (NO_3^-). The nitrate can then be reconstituted into living organic matter by photosynthetic plants.

Other sources of nitrogen in aquatic systems include animal wastes, chemical (particularly chemical fertilisers) and waste-water discharges.

Nitrogen from these sources may be discharged directly into streams or may enter waterways through surface runoff or groundwater discharge. Nitrogen compounds can be oxidised to nitrate by soil bacteria and may be carried into the groundwater by percolating water. Once in the aquifer, nitrates move freely with the groundwater flow. Groundwater contamination by nitrogen from animal feedlots and septic-tank drain fields has been recorded in numerous instances.

Tests for nitrogen forms in water commonly include analysis for ammonia (including both ammonia and ammonium), nitrate and organic nitrogen. The results of the analyses are usually expressed as milligrams per litre of the particular species as nitrogen. Tests for ammonium and organic nitrogen are more common on waste-water and other polluted waters, while the test for nitrate is the most common on clean-water samples and treated waste-waters.

Phosphorus

Phosphorus appears exclusively as phosphate (PO_4^{3-}) in aquatic environments. There are several forms of phosphate, however, including orthophosphate, condensed phosphates (pyro, meta and polyphosphates) and organically bound phosphates. These may be insoluble or particulate form or may be constituents of plant or animal tissue. Like nitrogen, phosphates pass through the cycles of decomposition and photosynthesis.

Phosphate is a constituent of soils and is used extensively in fertiliser to replace and/or supplement natural quantities on agricultural lands. Phosphate is also a constituent of animal waste and may become incorporated into the soil in grazing and feeding areas. Runoff from agricultural areas is a major contributor to phosphate in surface waters. The tendency for phosphate to adsorb to soil particles limits its movement in soil moisture and groundwater, but results in its transport into surface waters by erosion.

Municipal waste-water is another major source of phosphate in surface water. Condensed phosphates are used extensively as builders in detergents and organic phosphates are constituents of body waste and food residue. Other sources include industrial waste in which phosphate compounds are used for such purposes as boiler water conditioning.

While phosphates are not toxic and do not represent a direct health threat to human or other organisms, they do represent a serious indirect threat to water quality.

Phosphates are measured colourimetrically. Orthophosphates can be measured directly, while condensed forms must be converted to orthophosphate by acid hydrolysation and organic phosphates must be converted to orthophosphates by acid digestion. Results of the analysis are reported as milligrams per litre of phosphate as phosphorus. Careful handling of samples prior to analysis is crucial. For example, acid-washed glass bottles should be used for sampling, as bottles washed in phosphate detergent may contaminate samples.

BIOLOGICAL WATER QUALITY PARAMETERS

Water may serve as a medium in which literally thousands of biological species spend part, if not all, of their life cycles. Aquatic organisms range in size and complexity from the smallest single-cell micro-organism to the largest fish. All members of the biological community are, to some extent, water quality parameters, because their presence or absence may indicate in general terms the characteristics of a given body of water. As an example, the general quality of water in a trout stream would be expected to exceed that of a stream in which the predominant species of fish is carp. Similarly, abundant algal populations are associated with a water rich in nutrients. Based on their known tolerance for a given pollutant, certain organisms can be used as indicators of the presence of pollutants.

Pathogens

From the perspective of human use and consumption, the most important biological organisms in water are pathogens, those organisms capable of infecting or of transmitting diseases to humans. These organisms are not native to aquatic systems and usually require an animal host for growth and reproduction. They can, however, be transported by natural water systems, thus becoming a temporary member of the aquatic community. Many species of pathogens are able to survive in water and maintain their infectious capabilities for significant periods of time. These waterborne pathogens include species of bacteria, viruses, protozoa and helminths (parasitic worms).

Bacteria

Bacteria are considered to be single-celled plants because of their cell structure and the way they take in food. They utilise soluble food taken in through a rigid cell wall. But unlike green plants that use photosynthesis, bacteria do not produce their own food. One of the most important factors affecting the growth and reproduction of bacteria is temperature. At low temperatures, bacteria grow and reproduce slowly.

Algae

Algae are microscopic plants that contain photosynthetic pigments, such as chlorophyll. They are autotrophic organisms that support themselves by converting inorganic materials into organic matter using energy from the sun. During the process of photosynthesis, they take in carbon dioxide from the air and give off oxygen. A basic characteristic of these simple plants is their lack of roots, stems and leaves. Free-floating algae are also called phytoplankton. (Plankton are tiny floating plants or animals that live in either fresh or salt waters).

Protozoa

Protozoa are the simplest of animal species. These single-celled microscopic animals consume solid particles, bacteria and algae for food. They are, in turn, ingested as food by higher-level multicellular animals. Floating freely in water, these zooplankton, as they are sometimes called, are a vital part of the natural aquatic food chain. They are also of significance in biological waste-water treatment systems.

Viruses

Viruses are the smallest biological structures known to contain all the genetic information necessary for their own reproduction. So small that they can only be 'seen' with the aid of an electron microscope, viruses are obligate parasites that require a host in which to live. Symptoms associated with waterborne viral infections usually involve disorders of the nervous system rather than of the gastrointestinal tract. Waterborne viral pathogens are known to cause poliomyelitis and infectious hepatitis and several other viruses are known to be or suspected of being, waterborne.

Helminths

The life cycles of helminths, or parasitic worms, often involve two or more animal hosts, one of which can be human and water contamination may result from human or animal waste that contains helminths. Contamination may also be via aquatic species of other hosts, such as snails or insects. While aquatic systems can be the vehicle for transmitting helminthal pathogens, modern water-treatment methods are very effective in destroying these organisms.

Thus, helminths pose hazards primarily to those persons who come into direct contact with untreated water. Sewage plant operators, swimmers in recreational lakes polluted by sewage or storm water runoff from cattle feedlots and farm labourers employed in agricultural irrigation operations are at particular risk.

ENGINEERING SYSTEMS FOR WATER PURIFICATION

An adequate supply of pure water is absolutely essential to human existence. The development of effective water-treatment methods has virtually eliminated major waterborne epidemics in developed countries. This is not to suggest, however, that the problem of waterborne diseases has been eliminated.

Developing nations, where treated water is not available to all the population, still experience occasional epidemics of cholera and typhoid, as well as many outbreaks of less severe disease.

Chemical contamination of water supplies has become a concern in more recent times. Industrial facilities in developed countries produce and use literally thousands of chemical compounds. Along with an abundant array of household and agricultural chemicals, these materials often find their way into water supplies. While some of these chemical compounds are known toxicants, mutagens or carcinogens, the health effects of many others are not presently known. It is ironic that the high standard of living that allows industrialised nations to provide biologically pure water to the majority of the population also results in the discharge of chemical waste that may eventually have more deleterious effects on human health than the domestic waste that helped spread the plagues of past centuries.

The following section gives an overview of modern water treatment processes, while the remaining sections of the chapter contain a detailed description of the individual processes.

Water Treatment Processes

The processes selected for the treatment of potable water depend on the quality of the raw water supply. Most groundwaters are clear and pathogen-free and do not contain significant amounts of organic materials. Such waters may often be used in potable systems with a minimal dose of chlorine to prevent contamination in the distribution system. Other groundwaters may contain large quantities of dissolved solids or gases. When these include excessive amounts of iron, manganese or hardness, chemical and physical treatment processes may be required. Treatment systems commonly used to prepare potable water from groundwater are shown in Fig. 2.1.

Surface waters often contain a wider variety of contaminants than groundwater and treatment processes may be more complex. Most surface waters contain turbidity in excess of drinking-water standards. Although fast-moving streams may carry larger material in suspension, most of the solids will be colloidal in size and will require chemical coagulation for removal. Depending on the geology of the watershed, hardness may or may not be a problem in surface waters. If low levels of colour and other organic material are present, adsorption onto surface-active material, a process not significant in natural water systems, may be necessary. A wide variety of micro-organisms, some of which may be pathogenic, are also common constituents of surface waters. Treatment systems commonly used in treating surface waters are shown in Fig. 2.2.

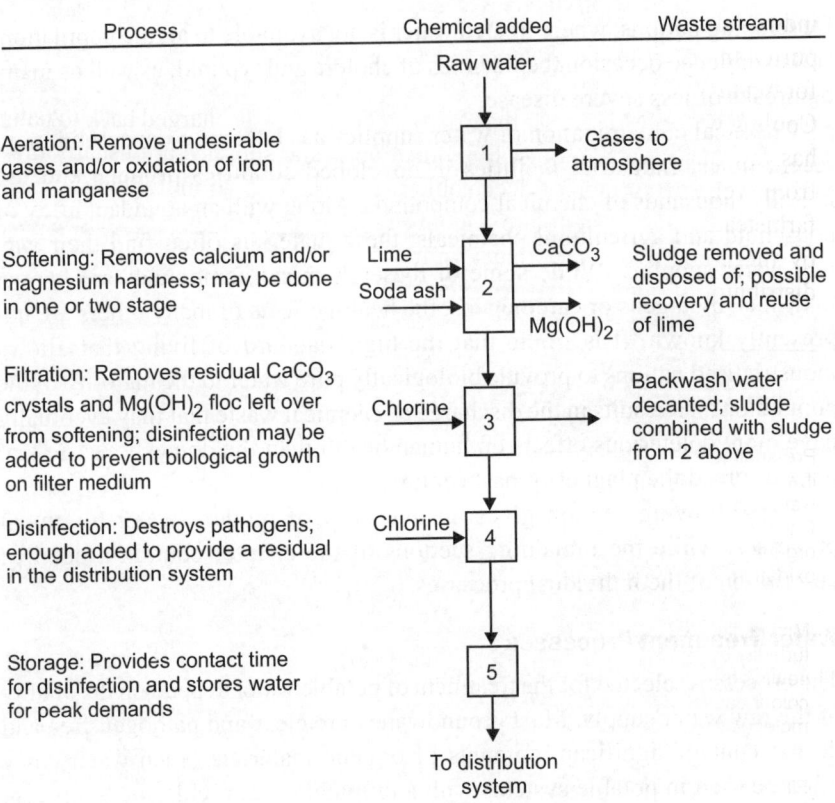

Process	Chemical added	Waste stream

Raw water

Aeration: Remove undesirable gases and/or oxidation of iron and manganese — 1 — Gases to atmosphere

Softening: Removes calcium and/or magnesium hardness; may be done in one or two stage — Lime, Soda ash — 2 — $CaCO_3$, $Mg(OH)_2$ — Sludge removed and disposed of; possible recovery and reuse of lime

Filtration: Removes residual $CaCO_3$ crystals and $Mg(OH)_2$ floc left over from softening; disinfection may be added to prevent biological growth on filter medium — Chlorine — 3 — Backwash water decanted; sludge combined with sludge from 2 above

Disinfection: Destroys pathogens; enough added to provide a residual in the distribution system — Chlorine — 4

Storage: Provides contact time for disinfection and stores water for peak demands — 5

To distribution system

Fig. 2.1. Typical plant treating hard groundwater.

Water treatment processes: theory and application

It is generally convenient to group human use of water into two broad categories depending upon the location of the use relative to the source. In-place use of water includes navigation, recreation, wildlife propagation and the dilution, assimilation and transportation of waste-water. Although hydroelectric power generation requires brief diversion of water through turbine penstocks, this use is also considered an in-place use.

For irrigation and industrial use and for individual and public domestic supplies, water must be withdrawn from streams, lakes or aquifers in the natural hydrologic cycle. The pollutants most deleterious to crops (inorganic salts and metals) are difficult and expensive to remove. The vast quantity of irrigation water used and the low margin of profit associated with farming virtually preclude any treatment of this water. Water not suited for irrigation is simply abandoned and available capital is used instead to secure an alternate source of acceptable quality. Many industries with needs for small amounts of essentially potable water obtain their supplies from public systems. Some industrial water supplies, such as boiler feed water, may require a chemical purity an order of magnitude greater than potable water. Engineering design for treatment of other types of industrial water supplies may also be necessary. Cooling water, particularly that used only once and discharged back to nature, has few quality constraints. Individual domestic supplies are usually drawn from wells or springs of acceptable quality and serve individual homes or farmsteads. Such systems are seldom engineered but are installed and operated by the home owners, perhaps with the advice of the well-driller and the distributor of home water-treatment units.

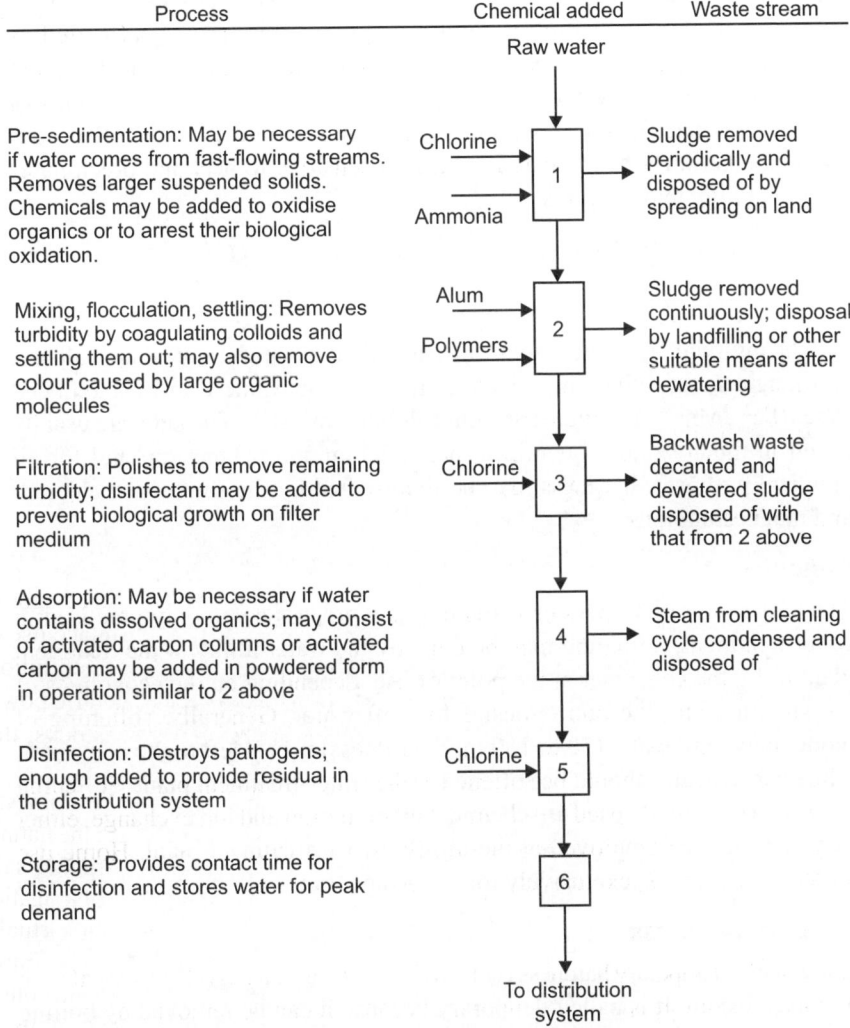

Fig. 2.2. Typical plant treating turbid surface water with organics.

Public water supplies, while only a fraction of the total water use, require by far the largest amount of effort expended by environmental engineers in the water-treatment field.

Water Conditioning and Softening

Softening of water is the removal of bivalent calcium and magnesium ions (Ca^{+2}, Mg^{+2}). These ions come from the dissolved compounds of calcium and magnesium. Their presence is known as hardness of water.

Hardness is defined as the concentration of multivalent metallic cations in solution. At supersaturated conditions, the hardness cations will react with anions in the water to form a solid precipitate. Hardness is classified as carbonate hardness and noncarbonate hardness, depending upon the anion with which it associates. The hardness that is equivalent to the alkalinity is termed carbonate hardness with any remaining hardness being called non-carbonate hardness. Carbonate hardness is sensitive to heat and precipitates readily at high temperatures.

$$Ca(HCO_3)_2 \xrightarrow{\Delta} CaCO_3 + CO_2 + H_2O \qquad \text{... (2.4)}$$

$$Mg(HCO_3)_2 \xrightarrow{\Delta} Mg(OH)_2 + 2CO_2 \qquad \text{... (2.5)}$$

The multivalent metallic ions most abundant in natural waters are calcium and magnesium. Others may include iron and manganese in their reduced states (Fe^{2+}, Mn^{2+}), strontium (Sr^{2+}) and aluminium (Al^{3+}). The latter are usually found in much smaller quantities than calcium and magnesium and for all practical purposes, hardness may be represented by the sum of the calcium and magnesium ions.

Softening

The reduction of hardness or softening, is a process commonly practised in water treatment. Softening may be done by the water utility at the treatment plant or by the consumer at the point of use, depending on the economics of the situation and the public desire for soft water. Generally, softening of moderately hard water (50 to 150 mg/l hardness) is best left to the consumer, while harder water should be softened at the water-treatment plant. Softening processes commonly used are chemical precipitation and ion exchange, either of which may be employed at the utility owned treatment plant. Home use softeners are almost exclusively ion-exchange units.

Types of hardness

Carbonate or temporary hardness is mainly due to bicarbonates (HCO_3^-) of calcium and magnesium. It is called temporary because it can be removed by boiling the water, which converts some of the bicarbonates into insoluble carbonates.

$$Ca(HCO_3)_2 \longrightarrow CaCO_3\downarrow + H_2O + CO_2\uparrow$$

Noncarbonate or permanent hardness is caused by soluble compounds other than bicarbonates, such as sulphates, nitrates and chlorides of calcium and magnesium. These compounds are more stable than bicarbonates; they are not removed by boiling the water. Calcium sulphate (gypsum) and magnesium sulphate (epsom) are the common causes of noncarbonate hardness.

Problems caused by hardness

Hardness is undesireable for several reasons. For example, hardness is:
1. A nuisance in laundering, due to wastage of soap and collection of dirty precipitate on fibres.
2. A nuisance in bathing.
3. A source of a dirty ring in the tubs and sinks.
4. Responsible for a residue on washed objects like cars and utensils.
5. Responsible for deposits on faucets and shower heads.
6. Responsible for forming a carbonate scale inside the steam boilers.

Impacts

Soap consumption by hard waters represents an economic loss to the water user. Sodium soaps react with multivalent metallic cations to form a precipitate, thereby losing their surfactant properties.

A typical divalent cation reaction is:

$$\underset{\text{Soap}}{2NaCO_2C_{17}H_{33}} + \text{cation}^{2+} \rightarrow \underset{\text{Precipitate}}{\text{cation}^{2+}(CO_2C_{17}H_{33})_2} + 2Na^+ \quad ...(2.6)$$

Lathering does not occur until all of the hardness ions are precipitated, at which point the water has been 'softened' by the soap. The precipitate formed by hardness and soap adheres to surfaces of tubs, sinks and dishwashers and may stain clothing, dishes and other items. Residues of the hardness soap precipitate may remain in the pores, so that skin may feel rough and un-comfortable. In recent years these problems have been largely alleviated by the development of soaps and detergents that do not react with hardness. Boiler scale, the result of the carbonate hardness precipitate may cause considerable economic loss through fouling of water heaters and hot water pipes.

Changes in pH in the water distribution systems may also result in deposits of precipitates. Bicarbonates begin to convert to the less soluble carbonates at pH values above 9.0.

Magnesium hardness, particularly associated with the sulphate ion, has a laxative effect on persons unaccustomed to it. Magnesium concentrations of less than 50 mg/l are desirable in potable waters, although many public water supplies exceed this amount. Calcium hardness presents no public health problem. In fact, hard water is apparently beneficial to the human cardio-vascular system.

Measurement

Hardness can be measured by using spectrophotometric techniques or chemical titration to determine the quantity of calcium and magnesium ions in a given sample. Hardness can be measured directly by titration with ethylenediamine tetraacetic acid (EDTA) using eriochrome black T (EBT) as an indicator. The EBT reacts with the divalent metallic cations, forming a complex that is red in colour. The EDTA replaces the EBT in the complex and when the replacement is complete, the solution changes from red to blue. If 0.01M EDTA is used. 1.0 ml of the titrant measures 1.0 mg of hardness as $CaCO_3$.

Softening Methods

The water-softening methods can be classified as chemical precipitation and nonchemical precipitation methods.

Chemical precipitation methods

Lime or lime-soda ash softening method

In this method, soluble calcium and magnesium compounds are converted to insoluble calcium carbonate ($CaCO_3$) and magnesium hydroxide [$Mg(OH)_2$], respectively. For this purpose, lime [quick lime, CaO or slaked lime, $Ca(OH)_2$] and soda ash/sodium carbonate (Na_2CO_3) are added to the water in the coagulation or flocculation basins. Lime is added after the alum and soda ash is applied after the lime. This sequence is important for proper reactions. Alum needs to react with turbidity and precipitate it out before reacting with lime. After the alum and lime reaction, soda ash is added to react with permanent hardness (to prevent its reaction with alum or lime). Lime and soda ash should never be feed through a common line because they will react and plug up the line. Lime–soda ash reactions occur during the flocculation to form a part of the floc. Insoluble calcium carbonate and magnesium hydroxide settle out in the sedimentation basins along with the turbidity.

Lime removes all the carbonate hardness and noncarbonate magnesium hardness. It forms insoluble $CaCO_3$ and $Mg(OH)_2$. The following chemical reactions show the removal of these hardnesses:

$$Ca(HCO_3)_2 + Ca(OH)_2 \longrightarrow 2CaCO_3\downarrow + 2H_2O$$

$$Mg(HCO_3)_2 + 2Ca(OH)_2 \longrightarrow Mg(OH)_2\downarrow + 2CaCO_3\downarrow + 2H_2O$$

$$MgSO_4 + Ca(OH)_2 \longrightarrow Mg(OH)_2\downarrow + CaSO_4$$

These reactions show that removal of magnesium carbonate hardness, requires twice the amount of lime than the removal of calcium carbonate hardness and magnesium noncarbonate hardness removal produces the equivalent amount of calcium noncarbonate hardness, which needs soda ash for its removal. Soda ash removes the noncarbonate calcium hardness.

$$CaSO_4 + Na_2CO_3 \longrightarrow CaCO_3\downarrow + Na_2SO_4$$

Removal of calcium chlorides and nitrates is similar, except that the products are sodium chloride and sodium nitrate. Due to a high amount of lime use, magnesium hardness, removal needs pH above 10.6, whereas calcium is removed above pH 9.4. These equations are used to determine the lime and soda ash doses corresponding to the degree of hardness removal.

Terminology of lime and lime-soda ash softening treatment based on the removal of various degrees of hardness

1. Partial lime softening uses a small amount of lime to remove the desired amount of calcium carbonate hardness.
2. Lime softening is done with lime only. It is applied when water has only high carbonate hardness. This process requires pH of 9.6 to 9.8.
3. Excess lime softening is the use of excess lime to remove high magnesium and calcium hardness. It needs pH above 10.6.
4. Lime-soda ash softening is the use of lime and some soda ash. This process is used for waters with high calcium carbonate hardness, low magnesium hardness and only some of the noncarbonate calcium hardness.
5. Excess lime-soda ash softening uses excess lime and soda ash to remove high calcium and magnesium carbonate hardness and high noncarbonate hardness.

Dose calculation

It is important to know how much of each chemical is needed to remove the desired amount of hardness. To calculate the hardness removal, determine the total alkalinity, total hardness, calcium hardness and possibly the CO_2 contents of the water. As discussed before, total alkalinity is equal to carbonate hardness and noncarbonate hardness is the difference between total hardness and alkalinity.

Lime also reacts with coagulants, carbon dioxide, iron and manganese. Therefore, an excess amount of lime is needed. This can be determined by jar testing that takes into consideration all the reactants. The jar test simplifies the calculations.

Permutit or zeolite process

In this process, water is softened through a natural or artificial zeolite. Permutit is an artificial zeolite, called as hydrate of sodium aluminium orthosilicate and can be obtained in the form of a coarse sand by fusing together sodium carbonate (Na_2CO_3), alumina (Al_2O_3) and silica (SiO_2). Zeolites, known as green sand are used for water softening, but artificial zeolite, known as permutit is more common and it has general formula.

$$Na_2O \cdot Al_2O_3 \cdot nSiO_2 \cdot xH_2O \ (n = 5\text{--}13, x = 3\text{--}4).$$

Permutit or zeolite is insoluble in water, but can act as base exchanger when brought in contact with water containing cations. The zeoloite or permutit is placed in a suitable column as shown in the Fig. 2.3 and hard water containing Ca^{++} and Mg^{++} ions is allowed to percolate through it. This process removes both temporary and permanent hardness.

$$Na_2Z + Ca(HCO_3)_2 \rightarrow 2NaHCO_3 + CaZ \qquad Na_2Z + MgSO_4 \rightarrow Na_2SO_4 + MgZ$$

$$Na_2Z + Mg(HCO_3)_2 \rightarrow 2NaHCO_3 + MgZ \qquad Na_2Z + CaCl_2 \rightarrow 2NaCl + CaZ$$

$$Na_2Z + CaSO_4 \rightarrow Na_2SO_4 + CaZ \qquad Na_2Z + MgCl_2 \rightarrow 2NaCl + MgZ$$

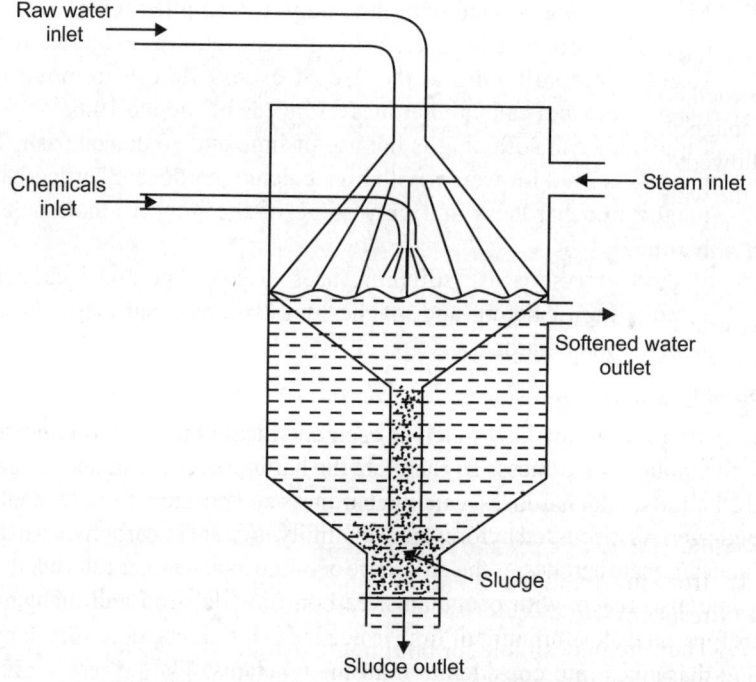

Fig. 2.3. Continuous type hot lime soda softener.

The permutit or zeolite is represented as Na_2Z and after base exchange it is converted into CaZ and MgZ. Sodium present in the zeolite is replaced by divalent Ca^{++} and Mg^{++} ions present in hard water. Water is thus softened as sodium salts do not cause hardness. After some time of use, the whole of zeolite gets exhausted. The zeolite may be regenerated by treatment for some hours with 10 per cent solution of sodium chloride, when the sodium salt of zeolite is formed. The soluble chloride of Ca and Mg passing into the solution can thus be washed away.

$$CaZ + 2NaCl \longrightarrow CaCl_2 + Na_2Z$$

The permutit or zeolite process has become a commercial success, because calcium and magnesium zeolite formed by passing hard water through the bed of sodium zeolite can be easily recovered into sodium zeolite as discussed above. Soft water obtained by this method is used mostly for laundry purpose. Water softened by this process cannot be used for boiler purpose.

Nonprecipitation methods

Membrane softening

Membrane filtration, such as reverse osmosis, removes hardness without producing any residual solids. Currently, membrane filtration is used only for small operations. In the future, it may replace lime-soda ash softening treatment.

Ion exchange softening

This method uses zeolites (Z) or ion exchange resins that exchange calcium and magnesium ions for sodium or hydrogen ions, receptively. If hard water is allowed to stand in the sodium zeolite, Ca^{++} and Mg^{++} ions replace Na^+ ions and the water is softened.

Iron and manganese in water

The presence of other metal ions such as iron (Fe^{+2}) and manganese (Mn^{+2}) also cause hardness, but the amount is insignificant. They are commonly present in water with low pH and no dissolved oxygen. Both iron and manganese can be present in all groundwaters. Water containing more than 0.3 mg/l of iron and 0.05 mg/l of manganese is aesthetically objectionable. Soluble forms of iron and manganese are ferrous (Fe^{+2}) and manganous (Mn^{+2}) compounds. They are removed as insoluble ferric (Fe^{+3}) and manganic (Mn^{+3}) compounds.

Problems caused by iron and manganese

1. Iron and manganese stain clothes and enamel yellow and black, respectively.
2. They are undesirable for bottling, laundries, paper mills, tanning and ice manufacturing.
3. Iron deposits in the distribution systems cause red water complaints and inaccurate meter readings.

These metals can be removed by precipitation and non-precipitation methods similar to calcium and magnesium removal. In all the precipitation methods, they form ferric oxide (Fe_2O_3) and manganic oxide (Mn_2O_3). Ferric oxide is commonly called rust.

Chemical precipitation

1. Lime-soda ash treatment removes iron and manganese while removing calcium and magnesium. They are removed above pH 9.4. They precipitate out as Fe_2O_3 and Mn_2O_3. When hydrated (wet), they are ferric hydroxide $[Fe(OH)_3]$ and manganic hydroxide $[Mn(OH)_3]$.
2. Aeration is the adding of oxygen into water. It is done by passing air through the water in aeration towers, by cascading or mechanical aerating. Oxygen reacts with iron and manganese compounds to form ferric and manganic oxides. For effective iron removal, pH should be around 7.5. Manganese oxidises slower than iron and requires a higher pH.
3. Chlorination is the process by which chlorine also removes iron and manganese.
4. Chlorine dioxide and ozone treatment use the disinfectants chlorine dioxide and ozone to remove iron and manganese.

Non-precipitation method

The zeolite treatment is a non-precipitation treatment. Removal of iron and manganese by the cation exchanger method is similar to and simultaneous with the removal of calcium and magnesium. Water should not be aerated before the zeolite treatment to avoid any accumulation of ferric hydroxide in the zeolite bed.

Stabilisation

Complete removal of hardness cannot be accomplished by chemical precipitation. Under conditions normally prevailing in water-treatment plants, up to 40 mg/l $CaCO_3$ and 10 mg/l $Mg(OH)_2$ usually remain in the softened water. Precipitation of the supersaturated solution of $CaCO_3$ will continue slowly, however, resulting in deposits in water lines and storage facilities. It is, therefore, necessary to stabilise the water by converting the supersaturated $CaCO_3$ back to the soluble form: $Ca^{2+} + 2(HCO_3)^-$.

Stabilisation can be accomplished by the addition of anyone of several acids. Using sulphuric acid as an example:

$$2CaCO_3 + H_2SO_4 \longrightarrow 2Ca^{2+} + 2(HCO_3)^- + SO_4^{2-} \qquad \text{... (2.7)}$$

$$Mg(OH)_2 + H_2SO_4 \longrightarrow Mg^{2+} SO_4^{2-} + 2H_2O \qquad \text{... (2.8)}$$

The most common practice, however, is to make the conversion with carbon dioxide:

$$CaCO_3 + CO_2 + H_2O \longrightarrow Ca^{2+} + 2(HCO_3)^- \qquad \text{... (2.9)}$$

$$Mg(OH)_2 + 2CO_2 \longrightarrow Mg^{2+} + 2(HCO_3) \qquad \text{... (2.10)}$$

This process is generally called recarbonation.

If the pH has been raised to facilitate the precipitation of magnesium, it will be necessary to neutralise the excess hydroxyl ions prior to stabilisation.

This necessitates a two-stage treatment process. Typical reactions are:

With sulphuric acid

$$Ca^{2+} + 2OH^- + H_2SO_4 \longrightarrow Ca^{2+} + SO_4^{2-} + 2H_2O \qquad ...(2.11)$$

$$2Na^+ + 2OH^- + H_2SO_4 \longrightarrow 2Na^+ + SO_4^{2-} + 2H_2O \qquad ...(2.12)$$

With carbon dioxide

$$Ca^{2+} + 2OH^- + 2CO_2 \longrightarrow CaCO_3 + H_2O \qquad ...(2.13)$$

$$2Na^+ + 2OH^{2-} + CO_2 \longrightarrow 2Na^+ + CO_3^{2-} + H_2O \qquad ...(2.14)$$

The pH must be lowered to approximately 9.5 before significant stabilisation occurs.

Softening Problems and Possible Solutions

Refer to Table 2.1 for some common problems and possible solutions.

Table 2.1. Softening problems and their solutions.

Problems	Possible causes	Possible solutions
Soda ash does not effectively remove hardness	Hardness is gone up in the raw by water	Run hardness test; determine right dose jar test; apply it
	Lime and soda ash react together before they react with the hardness	Check feed lines because they might be feeding too closely. Feed lime first and soda ash later
	Feeder does not feed correctly	Check feeder belt speed; look for any obstruction in feeding system and correct it
Soda ash line plugs up	Mixing of lime slurry with soda ash solution	Check lines and any other possible mixing of these two chemicals. They react and produce calcium carbonate. Calcium carbonate will deposit in the lines
Soda ash solution is milky	In some way, there is a mixing of and gritty	Check soda ash bin for any contamination some lime with soda ash in the storage bin with lime
Lime is not effectively softening	Lime dose is too high. It will be indicated by the light and flakey floc and slightly milky water due to undissolved calcium hydroxide which causes high calcium hardness	Check lime dose by jar testing and reduce it as required

(Contd ...)

Problems	Possible causes	Possible solutions
	Coagulant (alum) dose is too low	Run jar test; determine right dose of coagulant; and feed correct. Alum will react with lime and reduce hardness
	Quick lime (CaO) is not slaking properly. It is indicated by grit still slaking in the grit drum, which means an improper grade of lime	Ask supplier to provide proper grade of lime
During winter, softening sludge is heavy, gritty, and like white sand	An insufficient alum dose. Especially during winter, an improper dose of alum can cause sandlike calcium carbonate sludge formation that separates from the rest of the lighter sludge. It is easily visible as a bottom layer in a graduated cylinder while running the settlability test. If not removed, this sludge can lock up scrapers	Run jar test and determine optimum alum dose until all the sludge has a uniform density. It does not stratify in the jar test

Aeration

Aeration occupies a significant place in waste-water quality management and is an important factor in the purification of polluted water. Gas transfer is a physical phenomena in which gas molecules are exchanged between a liquid and a gas at a gas-liquid interface. This physical phenomenon of gas molecules exchanged between the liquid and gas at the liquid-gas interface may also be accompanied by biological, biochemical, biophysical and chemical action. These results are often the primary purpose of the gas transfer operation and methods of achieving the desired results may vary. Principal objectives of aeration, however, usually add or remove gases or volatile substances to water or carry-out both objectives simultaneously. In the biological process, aerators function to transfer the required oxygen and include sufficient mixing to maintain uniform dispersed oxygen throughout the basin and keep biological solids in suspension in aerobic basins and the activated sludge process. For high rate organic loadings, the power required may be determined by oxygen transfer requirements rather than mixing.

Aeration is one of the important unit operation of gas transfer. The aim of the aeration is to create extensive, new and self-renewing interfaces between air and water, to keep interfacial films from building up in thickness.

Objectives

Aeration of water is done to accomplish the following objectives:
1. It removes tastes and odours caused by gases due to organic decomposition.
2. It increases the dissolved oxygen content of the water.
3. It removes hydrogen sulphide and hence odour due to this is also removed.
4. It decreases the carbon dioxide content of water and thereby reduces its corrosiveness and raises its pH value.
5. It converts iron and manganese from their soluble states to their insoluble states, so that these can be precipitated and removed.
6. Due to agitation of water during aeration, bacteria may be killed to some extent.
7. It is also used for mixing chemicals with water, as in the Aeromix process and in the use of diffused compressed air.

Coagulation, Flocculation and Filtration

Coagulation and flocculation convert non-settlable turbidity particles into settlable form for their effective removal by gravity. After presedimentation, these particles are mostly colloidal type. Colloidal turbidity particles are too small (1–100 nm) to settle by gravity. They stay suspended and cause turbidity. Mostly, they are negatively charged. Their removal is accomplished by using substances that make them clump together to form large and heavy particles known as floc that will settle. These substances are known as coagulants. A coagulant is an electrolyte that provides cations (positively charged ions) to precipitate out the negatively charged colloidal turbidity particles.

As a rule, the higher the charge on the cation, the more effective is the coagulant. Therefore, commonly used coagulants are aluminium and ferric compounds that provide Al^{+3} and Fe^{+3} cations, respectively. Some other substances are often used to facilitate the coagulants; those are known as coagulant aids. This treatment phase is the second barrier to remove turbidity, waterborne pathogens and other contaminants. It consists of three parts: rapid mixing, coagulation and flocculation.

Coagulation

Coagulation is the precipitation of the colloidal turbidity particles, coagulants and coagulant aids.

Steps of coagulation

Coagulation occurs in three steps. First, Al^{+3} or Fe^{+3} ions attract a considerable number of negative colloidal turbidity particles. Second, due to aggregation, they form small clumps, called micro-floc. Third, micro-floc, due to its positive charge, still attracts negative ions such as alkalinity (OH^- from lime) and floc compounds precipitate due to their low solubility.

Factor affecting coagulation

Coagulant

Different sources of water need different coagulants, such as the following: (i) filter alum, (ii) activated alum, (iii) black alum (iv) ferric sulphate, (v) ferrous sulphate, (vi) chlorinated copper, and (vii) sodium aluminate.

Coagulant aids

These help the coagulation by creating better coagulation conditions, such as proper pH, alkalinity and particulate nuclei. Some of them act as secondary coagulants: (i) pH adjusting coagulant aids, (ii) non-pH affecting coagulant aids, and (iii) coagulating aids acting as secondary coagulants.

pH

Effectiveness of a coagulant is generally pH dependent. Different water requires different coagulants based on its pH. Water with a colour will coagulate better at low pH (4.4–6) with alum.

Alkalinity

It is needed to provide anions, such as (OH^-) for forming insoluble compounds to precipitate them out. It could be naturally present in the water or needed to be added as hydroxides, carbonates or bicarbonates, as coagulant aids. Generally, 1 part alum uses 0.5 part alkalinity for proper coagulation.

Temperature

The higher the temperature, the faster the reaction and the more effective is the coagulation. Winter temperature will slow down the reaction rate, which can be helped by an extended detention time. Mostly, it is naturally provided due to lower water demand in winter.

Time

Proper mixing and detention times are important.

Velocity

The higher velocity causes the shearing or breaking of floc particles and lower velocity will let them settle in the flocculation basins. Velocity around 1 ft/sec in the flocculation basins should be maintained.

Zeta potential

It is the charge at the boundary of the colloidal turbidity particle and the surrounding water. The higher the charge, the more is the repulsion between the turbidity particles, less the coagulation and *vice versa*. Higher zeta potential requires the higher coagulant dose. An effective coagulation is aimed at reducing zeta potential charge to almost 0. Selection of a proper coagulant and a coagulant aid for a water supply is important. A jar test for water should be run to determine which coagulant and coagulant aid are economical and most effective.

Flocculation

Flocculation is the clumping of microfloc particles to form large particles called floc. It is achieved by the gentle mixing of coagulated water, in tanks known as flocculation basins to allow further clumping of the coagulated matter and turbidity particles, to form large floc particles. Flocculation basins have slow mixing mechanical paddles, known as flocculators and baffles to provide adequate mixing and low velocity. The hardness removal is also achieved at this phase of treatment by using lime and soda ash. If chlorine dioxide is used for predisinfection, then chlorite removal can also be done here by using ferrous ions. For the proper chemical reactions, these chemicals should be applied in the following sequence: ferrous ions, alum, lime and then soda ash.

Filtration

Filtration is the mechanical removal of turbidity particles by passing the water through a porous medium, which is either a granular bed or a membrane. Filtration's purpose is to remove all the turbidity particles carried over from the sedimentation phase, thus producing a sparkling clear water with almost zero turbidity.

Thus, filtration is a fundamental unit operation that separates suspended particle matter from water. Although industrial applications of this operation vary significantly, all filtration equipment operate by passing the solution or suspension through a porous membrane or medium, upon which the solid particles are retained on the medium's surface or within the pores of the medium, while the fluid, referred to as the filtrate, passes through.

In a very general sense, the operation is performed for one or both of the following reasons. It can be used for the recovery of valuable products (either the suspended solids or the fluid) or it may be applied to purify the liquid stream, thereby improving product quality or both. Examples of various processes that rely on filtration include adsorption, chromatography, operations involving the flow of suspensions through packed columns, ion exchange and various reactor engineering applications. In petroleum engineering, filtration principles are applied to the displacement of oil with gas (i.e. liquid–liquid separations), in the separation of water and miscible solvents (including solutions of surface-active agents) and in reservoir flow applications.

Membrane filtration

This process is the passing of pre-treated water under pressure through a membrane to remove specific sized particles. A membrane is a very thin paperlike structure. Membranes can achieve the degree of treatment comparable to a conventional treatment plant. Membrane treatment is one of the best treatment technologies to meet the present and expected Safety for Drinking Water Act (SDWA) challenges. It is capable of removing most of the regulated contaminants.

Membrane structure

Membranes are either hollow fine fibre (HFF) or spiral wound (SW) structures formed of cellulose acetate and synthetic materials, such as polypropylene or polyfuron.

Mechanism of particle removal

There are three basic mechanisms: sieving, selective diffusion and charge repulsion:
1. Sieving: Each membrane has a uniform and specific pore size. All particles bigger than the pore size are sieved out/rejected. The smallest rejected particle is slightly bigger than the pore size. The smallest molecular weight that will be strained out is known as a cut-off molecular weight (COMW).
2. Selective diffusion: Passing only selected dissolved particles through the membrane is selective diffusion. For selective diffusion, a membrane needs to be semipermeable, which means it will allow only certain chemicals to diffuse through and all others will be rejected (e.g. reverse osmosis membranes).
3. Charge repulsion: Filtration uses a direct electric current; anions go to the anode and cations go to the cathode, e.g. electrodialysis membranes. Electrodialysis is dialysis aided by electrodes. Dialysis is the separation of dissolved and suspended substances by a membrane.

Membrane treatment

Membrane filtration uses an appropriate membrane system that takes into account source water quality and treatment requirements. Currently, there are three commonly used membrane systems: microfiltration, ultrafiltration and reverse osmosis. The first two are the most feasible alternatives to conventional water treatment to meet the requirements of the enhanced surface water treatment, the total coliform and the disinfectants and disinfection by-products rules. Here are some guidelines to select a membrane system for different source water:

1. Microfiltration: Microfiltration is used for surface water where conventional treatment requires pre-treatment, sedimentation and filtration.
2. Ultrafiltration: Ultrafiltration is used for surface water that needs the removal of very small particles such as viruses and dissolved organics.
3. Reverse osmosis: Reverse osmosis is suitable when water has a very high concentration of dissolved substances such as chlorides, nitrates and fluorides, in addition to other contaminants (e.g. salty water).

Ion Exchange and Carbon Adsorption

Ion exchange and carbon adsorption are unrelated technologies and often have different objectives. They are, however, often times used in compliment to achieve high water quality attributes.

Ion exchange is a reversible chemical reaction wherein an ion (an atom or molecule that has lost or gained an electron and thus acquired an electrical charge) from solution is exchanged for a similarly charged ion attached to an immobile solid particle. These solid ion exchange particles are either naturally occurring inorganic zeolites or synthetically produced organic resins. The synthetic organic resins are the predominant type used today because their characteristics can be tailored to specific applications. An organic ion exchange resin is composed of high-molecular-weight polyelectrolytes that can exchange their mobile ions for ions of similar charge from the surrounding medium. Each resin has a distinct number of mobile ion sites that set the maximum quantity of exchanges per unit of resin. The industry application most familiar with ion exchange technology is metal plating. Most plating process water is used to cleanse the surface of the parts after each process bath. To maintain quality standards, the level of dissolved solids in the rinse water must be regulated. Freshwater added to the rinse tank accomplishes this purpose and the overflow water is treated to remove pollutants and then discharged. As the metal salts, acids and bases used in metal finishing are primarily inorganic compounds, they are ionised in water and could be removed by contact with ion exchange resins.

In a water deionisation process, the resins exchange hydrogen ions (H^+) for the positively charged ions (such as nickel, copper and sodium) and hydroxyl ions (OH^-) for negatively charged sulphates, chromates and chlorides.

Because the quantity of H^+ and OH^- ions is balanced, the result of the ion exchange treatment is relatively pure, neutral water. Ion exchange technology is applied in many other industry sectors, including the petroleum and chemical industries, as well as general waste-water treatment applications. The technology is most often compared to reverse osmosis, since both technologies are often aimed at similar objectives. In this regard, in addition to discussing ion exchange as a technology, we will also review some of the operational trade-offs and economics of the two processes in this chapter.

The process of ion exchange is uniquely suited to the removal of ionic species from water supplies for several reasons. First, ionic impurities may be present in rather low concentrations. Second, modern ion exchange resins have high capacities and can remove unwanted ions preferentially. Third, modern ion-exchange resins are stable and readily regenerated, thereby allowing their reuse. Other advantages ion exchange offers are: (i) the process and equipment are a proven technology. Designs are well developed into pre-engineered units that are rugged and reliable, with well-established, applications; (ii) fully manual to completely automatic, units are available; (iii) there are many models of ion exchange systems on the market which keep costs competitive; (iv) temperature effects over a fairly wide range (from 0°C to 35°C) are negligible; and (v) the technology is excellent for both small and large installations, from home water softeners to large utility/industrial applications.

Ion exchange is a well-known method for softening or for demineralising water. Although softening could be useful in some instances, the most likely application for ion exchange in waste-water treatment is for demineralisation. Many ion exchange materials are subject to fouling by organic matter. It is possible that treatment of secondary effluent for suspended-solids removal and possibly soluble organic removal will be required before carrying out ion exchange. Many natural materials and, more importantly, certain synthetic materials have the ability to exchange ions from an aqueous solution for ions in the material itself.

Cation-exchange resins can, for example, replace cations in solution with hydrogen ions. Similarly, anion-exchange resins can either replace anions in solution with hydroxyl ions or absorb the acids produced from the cation-exchange treatment. A combination of these cation-exchange and anion-exchange treatments results in a high degree of demineralisation.

Working of ion exchange

Ion exchangers are materials that can exchange one ion for another, hold it temporarily and then release it to a regenerant solution. In a typical demineraliser, this is accomplished in the following manner:

The influent water is passed through a hydrogen cation-exchange resin which converts the influent salt (e.g. sodium sulphate) to the corresponding acid (e.g. sulphuric acid) by exchanging an equivalent number of hydrogen (H^+) ions for the metallic cations (Ca^{+2}, Mg^{+2}, Na^+).

These acids are then removed by passing the effluent through an alkali regenerated anion-exchange resin which replaces the anions in solution (Cl^-, SO_4^-, NO_3^-) with an equivalent number of hydroxide ions. The hydrogen ions and hydroxide ions neutralise each other to form an equivalent amount of pure water. During regeneration, the reverse reaction takes place. The cation resin is regenerated with either sulphuric or hydrochloric acid and the anion resin is regenerated with sodium hydroxide. Figure 2.4 illustrates a basic scheme for ion exchange demineralisation.

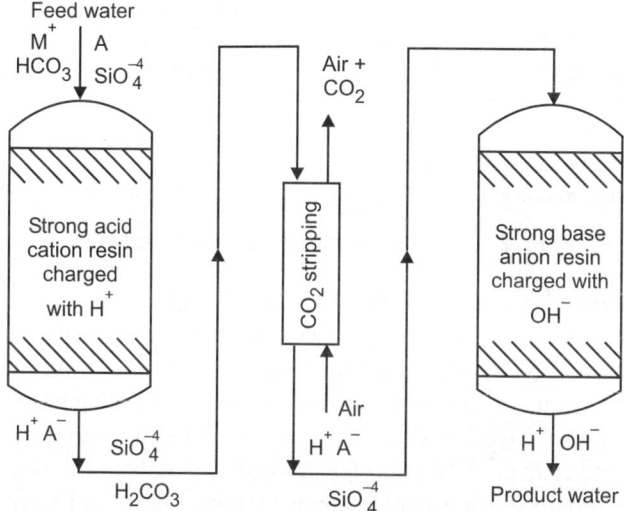

Fig. 2.4. Ion exchange demineralisation scheme.

There are various arrangements or equipment possible but in all cases, except in mixed-bed demineralisation, the water should first pass through a cation exchanger. In mixed demineralisation, the two exchange materials (that is, the cation-exchange resin and the anion-exchange resin) are placed in one shell instead of two separate shells.

In operation, the two types of exchange materials are thoroughly mixed so that we have, in effect, a number of multiple demineralisers in series. Higher quality water is obtained from a mixed-bed unit than from a two-bed system (Fig. 2.5 for an example).

To be suitable for industrial use, an ion exchange resin must exhibit durable physical and chemical characteristics such as—functional groups, solubility, bead size, resistance to fracture etc.

Types of resins

As noted earlier, ion-exchange materials are grouped into four specific classifications depending on the functional groups attached; strong-acid cation, strong-base anion, weak-acid cation or weak-base anion. In addition to these, we also have inert resins that do not have chemical properties.

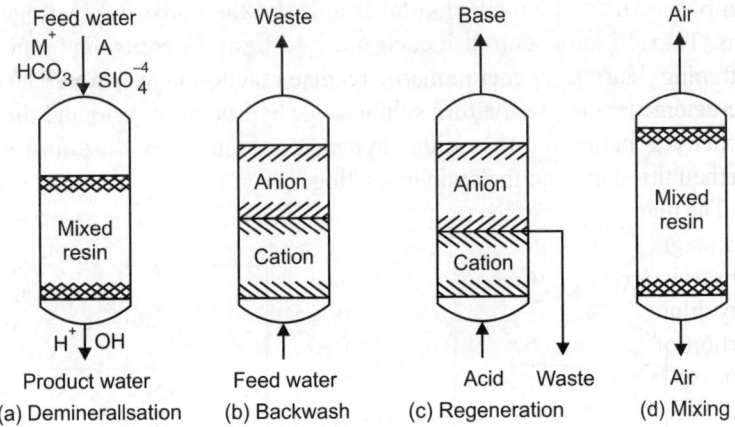

Fig. 2.5. Mixed resin demineralisation scheme.

Ion exchange softening (sodium zeolite softening)

This is one of the ion-exchange processes used in water purification. In this process, sodium ions from the solid phase are exchanged with the hardness ions from the aqueous phase. Consider a bed of ion-exchange resin having sodium as the exchangeable ion, with water containing calcium and magnesium hardness allowed to percolate through this bed. Let us denote the ion exchange resinous material as RNa, where R stands for resin matrix and Na is its mobile exchange ion. The hard water will exchange Ca and Mg ions rapidly, so that water at the effluent will be almost completely softened. Calcium and magnesium salts will be converted into corresponding sodium salts.

The reaction will proceed toward the right-hand side to its completion until the bed gets completely exhausted or saturated with Ca and Mg ions. In order to reverse the equilibrium so that the reaction proceeds toward the left-hand side, the concentration of sodium ions has to be increased. This increase in sodium ions is accomplished by using a brine solution of sufficient strength so that the total sodium ions present in the brine are more than the total equivalent of Ca and Mg in the exhausted bed. This reverse reaction is carried out in order to bring the exhausted resin back to its sodium form. This process is known as regeneration. When the softener with the fresh resin in sodium form is put in service, the sodium ions in the surface layer of the bed are immediately exchanged with calcium and magnesium, thereby producing soft water with very little residual hardness in the effluent. As the process continues, the resin bed keeps exchanging its sodium ions with calcium and magnesium ions until the hardness concentration increases rapidly and the softening run is ended.

This softening process can be extended to a point where the hardness coming in and going out is the same. When this condition is reached, the bed is completely exhausted and does not have any further capacity to exchange ions. This capacity is called the total breakthrough capacity. In practice, the softening process is never extended to reach this stage as it is ended at some pre-determined effluent hardness, much lower than the influent hardness. This capacity is called the operating exchange capacity. After the resin bed has reached this capacity, the resin bed is regenerated with a brine solution.

The regeneration of the resin bed is never complete. Some traces of calcium and magnesium remain in the bed and are present in the lower-bed level. In the service run, sodium ions exchanged from the top layers of the bed form a very dilute regenerant solution which passes through the resin bed to the lower portion of the bed. This solution tends to leach some of the hardness ions not removed by previous regeneration.

These hardness ions appear in the effluent water as leakage. Hardness leakage is also dependent on the raw water characteristics. If the Na/Ca ratio and calcium hardness are very high in the raw water, leakage of the hardness ions will be higher. Most industrial applications of ion exchange use fixed-bed column systems, the basic component of which is the resin column (Fig. 2.6). The column design must:

1. Contain and support the ion exchange resin.
2. Uniformly distribute the service and regeneration flow through the resin bed.
3. Provide space to fluidise the resin during backwash.
4. Include the piping, valves and instruments needed to regulate flow of feed, regenerant and backwash solutions.

After the feed solution is processed to the extent that the resin becomes exhausted and cannot accomplish any further ion exchange, the resin must be regenerated. In normal column operation, for a cation system being convened first to the hydrogen then to the sodium form, regeneration employs the following basic steps:

1. The column is backwashed to remove suspended solids collected by the bed during the service cycle and to eliminate channels that may have formed during this cycle. The back wash flow fluidises the bed, releases trapped particles and reorients the resin particles according to size. During backwash the larger, denser particles will accumulate at the base and the particle size will decrease moving up the column. This distribution yields a good hydraulic flow pattern and resistance to fouling by suspended solids.
2. The resin bed is brought in contact with the regenerant solution. In the case of the cation resin, acid elutes the collected ions and converts

the bed to the hydrogen form. A slow water rinse then removes any residual acid.

3. The bed is brought in contact with a sodium hydroxide solution to convert the resin to the sodium form. Again, a slow water rinse is used to remove residual caustic. The slow rinse pushes the last of the regenerant through the column.

4. The resin bed is subjected to a fast rinse that removes the last traces of the regenerant solution and ensures good flow characteristics.

5. The column is returned to service.

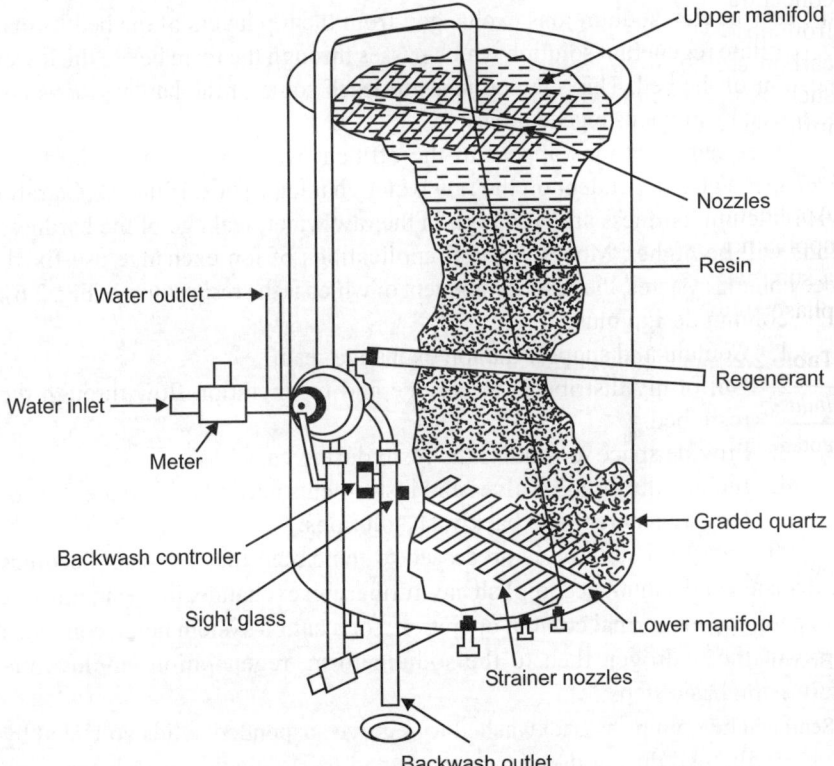

Fig. 2.6. Ion exchange unit.

For resins that experience significant swelling or shrinkage during regeneration, a second backwash should be performed after regeneration to eliminate channelling or resin compression. Regeneration of a fixed-bed column usually requires between 1 and 2 hours. Frequency depends on the volume of resin in the exchange columns and the quantity of heavy metals and other ionised compounds in the waste-water.

Carbon adsorption in water teatment

Activated carbon is a crude form of graphite, the substance used for pencil leads. It differs from graphite by having a random imperfect structure which is highly porous over a broad range of pore sizes from visible cracks and crevices to molecular dimensions. The graphite structure gives the carbon it's very large surface area which allows the carbon to adsorb a wide range of compounds. Activated carbon can have a surface of greater than 1000 m^2/g. This means 5 grams of activated carbon can have the surface area of a football field.

Adsorption is the process by which liquid or gaseous molecules are concentrated on a solid surface, in this case activated carbon. This is different from absorption, where molecules are taken up by a liquid or gas. Activated carbon can made from many substances, containing a high carbon content such as coal, wood and coconut shells. The raw material has a very large influence on the characteristics and performance activated carbon.

Applications

Applications of carbon adsorption go far beyond conventional water treatment applications which we will discuss in a general sense shortly. Table 2.2 provides a summary of the key applications of carbon adsorption systems for liquid phase applications.

Table 2.2. Liquid phase applications of carbon adsorption.

Industry	Description	Use
Potable water	Granular activated carbons (GAC) installed in rapid	Removal of dissolved organic treatment contaminants, gravity filters control of taste and odour problems
Soft drinks	Potable water treatment, sterilisation with chlorine	Chlorine removal and adsorption of dissolved organic materials
Brewing	Potable water treatment	Removal of trihalomethanes (THM) and phenolics
Semi-conductors	Ultra-high purity water	Total organic carbon (TOC) reduction
Gold recovery	Operation of carbon in leach, carbon in pulp, and heap leach circuits	Recovery of gold from tailings dissolved in sodium cyanide
Petrochemical	Recycling of steam condensate for boiler feed water	Removal of oil and hydrocarbon contamination
Groundwater	Industrial contamination of ground-water reserves	Reduction of total organic halogens (TOX) and

(Contd ...)

Industry	Description	Use
		adsorbable organic halogens (AOX) including chloroform, tetrachloroethylene and trichloroethylene
Industrial waste-water	Process effluent treatment to meet environmental discharge standards	Reduction of total organic halogens (TOX), biological oxygen demand (BOD), and chemical oxygen demand (COD)
Swimming pools	Ozone injection for removal of organic	Removal of residual ozone and contaminants control of chloramine levels

The most common application of carbon adsorption in municipal water treatment is in the removal of taste and odour compounds.

Figure 2.7 provides an example of a process flow diagram for a municipal water treatment plant. In this example water is pumped from the river into a flotation unit, which is used for the removal of suspended solids such as algae and particulate matter.

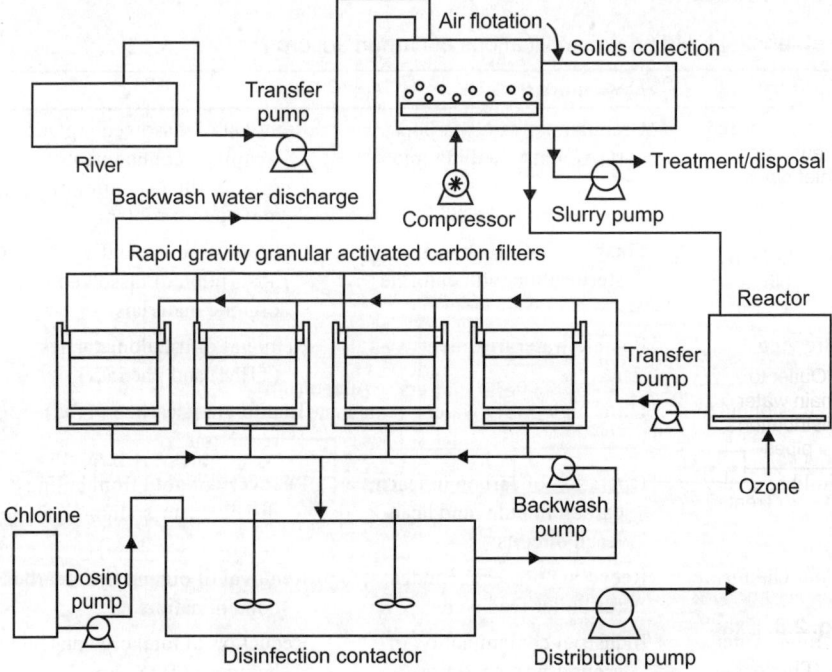

Fig. 2.7. Process flow sheet for municipal water plant.

Dissolved air is injected under pressure into the basin. This action creates micro bubbles which become attached to the suspended solids, causing them to float. This results in a layer of suspended solids on the surface of the water, which is removed using a mechanical skimming technique.

The next step in the process involves the production of ozone by-passing high tension, high frequency electrical discharges through air in specially designed units. Ozone is injected into the water to provide bactericidal action and to break down the natural humic compounds that are the cause of taste and odour problems.

The water then passes through a rapid gravity filtration system filled with activated carbon (GAC), which adsorbs the compounds resulting from the ozone treatment. Following adsorption, the water is disinfected for supply to the distribution network. Understand that treatment plants are unique, in many ways like oil refineries, i.e. design basis can be substantially different depending on the nature of the water being treated.

Figure 2.8 provides another example of a municipal water treatment facility using PAC. Again the plant is used for the removal of taste and odour compounds.

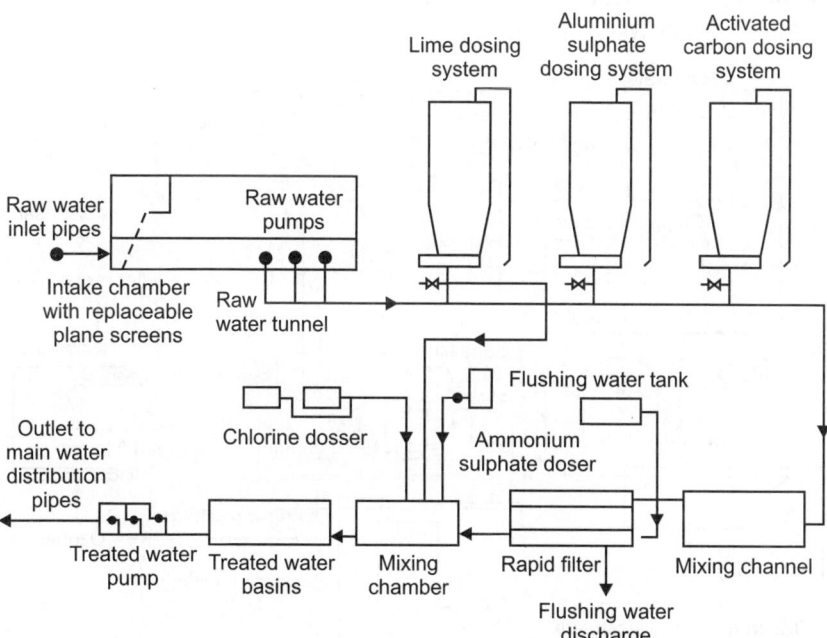

Fig. 2.8. Example of a municipal water treatment plant for taste and odour control.

There are regions where the treatment of water is intended for potable purposes is not necessary at all times during the year. The presence of taste, odour and naturally occurring toxins largely depends on the biological action in areas where lake or reservoir water supply is common. In these situations it is more cost effective to use intermittent dosing of activated carbon into the water during those times of the year where it is needed.

The use of PAC is preferred in these cases, mainly because no costly fixed bed filtration equipment is required. The PAC can be dosed directly to existing flocculant tanks at a prescribed rate to achieve the level of pollutant removal required. Shown in Fig. 2.8, following the dosing of PAC the activated carbon is removed as part of the flocculation process or it can be filtered out by mechanical means. The final stage of water treatment is disinfection, whereupon the water is pumped to the distribution network.

Non-potable water treatment is also well within the economical applications of liquid phase adsorption systems. There, in fact, are so many unique examples of process water treatment throughout the chemical industry that we could go on for days discussing specify systems.

Figure 2.9 shows a process diagram for the removal of creosote and pesticides from the liquid phase in a timber treatment facility. A storage dosing tank is used for smoothing the flow from where the water is pumped into a chemical dosing system for pH adjustment.

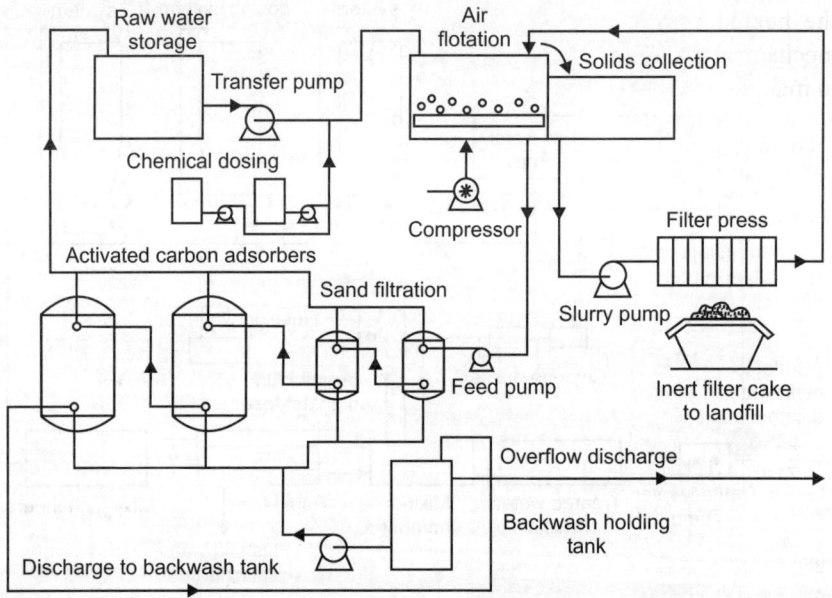

Fig. 2.9. Example of a process water treatment facility.

Then, ferric sulphate is added to form a precipitate with suspended solids, which is subsequently flocculated by the addition of polyelectrolyte.

The water is then pumped through series operated sand filters, which provide the final stage of suspended solids removal and protect the granular activated carbon (GAC) filters from particulate contamination. Series operated GAC filters are then used to remove the dissolved creosote and pesticides from the water.

To achieve compliance with specifications levels, water should be sampled and analysed after leaving the first GAC filter. The second GAC filter normally serves as a guard bed.

A final example of application and process layout is shown in Fig. 2.10. In this example the process relies on activated carbon to remove colour bodies from a recycled glucose intermediary prior to use in the production of confectionary. The glucose containing the colour taint must be mildly heated (to about 70°C), so that the normally solid product becomes less viscous and easier to pump. The syrup is pumped through a series of high efficiency filters (mechanical type) that remove entrained particulate matter and crystallised sugar formed during the heating process. Filtered syrup is then passed through columns containing GAC using a high residence time (a variation is simply the addition of PAC on an as needed basis—this obviously has cost advantages for batch operated systems). During these stages the colour bodies are physically adsorbed by the activated carbon. When PAC is added to the process, the heated syrup is agitated. Following agitation, the syrup undergoes mechanical filtration to remove entrained PAC prior to the glucose being used to manufacture the confectionary.

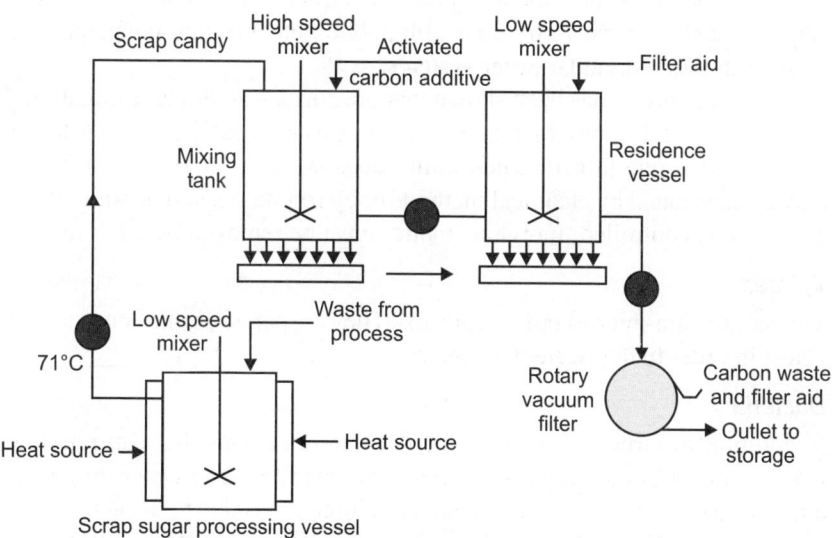

Fig. 2.10. Example of a decolourisation treatment facility.

This is a good example of a pollution prevention technology, because the reprocessing of waste in this manner allows it to be suitable for reuse as a saleable product after further use.

This technique is adaptable in diverse applications such as pharmaceutical processes, chemical intermediaries manufacturing and soft drink production. These examples help to illustrate the versatility of activated carbon in standard water treatment applications. Another application which merits a distinct discussion is goundwater remediation.

Water Sterilisation Technologies

This section provides with a very basic overview of the principles and technologies associated with water purification or more specifically, sterilisation. In very simplistic terms, there are two general classes of technologies, namely those based on chemical methods and those based on non-chemical technologies. The major application is in purifying water for human consumption purposes.

Waterborne diseases

Untreated waters contain a number of harmful pollutants which give the water colour, taste and odour. These pollutants include viruses, bacteria, organic materials and soluble inorganic compounds and these must be removed or rendered harmless before the water can be used again. A breakdown of the documented outbreaks identifies acute gastroenteritis, hepatitis shigellosis, ciardiasis, chemical poisoning, typhoid fever and salmonellosis. Sources of contaminated water can be traced to semipublic water systems, municipal water systems and to individual water systems.

In cell culture, it has been shown that one virion can produce infection. In the human host, because of acquired resistance and a variety of other factors, the one virion/one infection possibility does not exist. Viruses and bacteria may be eliminated by chemical methods or by irradiation and organic poisons may also be controlled. Inorganic matter must be removed by other means.

Viruses

Viruses are ultra-microscopic organisms. They are parasites; they need to infest a host in order to duplicate themselves.

Bacteria

In addition to viruses, bacteria (microscopic organisms that can reproduce without a host in the proper conditions) are also found in water. In general, damage to the human body from bacterial infection is due to the action of the toxins they produce. Bacteria found in water are derived from contact with air, soil, living and decaying plants and animals and animal excrements.

Survival of viruses

A variety of factors is responsible for the survival of viruses in water bodies. The survival of enteric viruses under laboratory conditions and in estuaries varies from a few hours to up to 200 days. Survival in winter is superior to that at summer temperatures.

Treatment options

Primary treatment of municipal waste involving settling and retention removes very few viruses. Sedimentation effects some removal. Virus removal of up to 90 per cent (which is a minimal removal efficiency) has been observed after the activated sludge step. Further physical-chemical treatment can result in large reductions of virus titer, coagulation being one of the most effective treatments achieving as much as 99.9 per cent removal of virus suspended in water.

Non-conventional treatment methods

Electromagnetic (EM) waves

Electromagnetic radiation is the propagation of energy through space by means of electric and magnetic fields that vary in time. Electromagnetic radiation may be specified in terms of frequency, vacuum wavelength or photon energy. For water purification, EM waves up to the low end of the UV band will result in heating the water.

Sound

A sound wave is an alteration in pressure, stress, particle displacement, particle velocity or a combination of these that is propagated in an elastic medium. Sound waves, therefore, require a medium for transmission; that is, they may not be transmitted in a vacuum.

Electron beams

The electron is the lightest stable elementary particle of matter known and carries a unit of negative charge. Electrons are emitted from the cathode by a number of mechanisms: (i) thermionic emission, (ii) shottky emission, (iii) high field emission, (iv) photoemission, and (v) secondary emission.

Electromagnetism

In a high-gradient magnetic separator, the force on a magnetised particle depends on the intensity of the magnetising field and on the gradient of the field. When a particle is magnetised by an applied magnetic field, the particle develops an equal number of north and south poles. Hence, in a uniform field, a dipolar particle experiences a torque, but not a attractive force.

Direct and alternating currents

Electrolytic treatment is achieved when two different metal strips are dipped in water and a direct current is applied from a rectifier. The higher the voltage the greater the force pushing electrons across the gap between the electrodes. If the water is pure, very few electrons cross the path between the electrodes. Impurities increase conductivity, hence decreasing the required voltage. Additionally, chemical reactions occur at both the cathode and the anode. The major reaction taking place at the cathode is the decomposition of water with the evolution of hydrogen gas.

Ozonation

Ozone has been used continuously for nearly 90 years in municipal water treatment and the disinfection of water supplies. This practice began in France, then extended to Germany, Holland, Switzerland and other European countries and in recent years to Canada and other developing countries. Ozone is a strong oxidising substance with bactericidal properties similar to those of chlorine. In test conditions it was shown that the destruction of bacteria was between 600 and 3000 times more rapid by ozone than by chlorine. Further, the bactericidal action of ozone is relatively unaffected by changes in pH while chlorine efficacy is strongly dependent on the pH of the water.

Ozone's high reactivity and instability as well as serious obstacles in producing concentrations in excess of 6 per cent preclude central production and distribution with its associated economies of scale.

Ultraviolet radiation

One way of implementing the UV disinfection process at existing activated sludge plants involves suspending the UV lights (in the form of low pressure mercury arc UV lamps with associated reflectors) above the secondary clarifiers. The effluent is exposed to the UV radiation as it rises over the wire in a thin film.

Electron beam

The idea of using ionising radiation to disinfect water is not new. Ionising radiations can be produced by various radio-active sources (radioisotopes), by X-ray and particle emissions from accelerators and by high-energy electrons. The advances in reliable, relatively low-cost devices for producing high-energy electrons are more significant.

Disinfection by chlorination

Disinfection has received increased attention over the past several years from regulatory agencies through the establishment and enforcement of rigid bacteriological effluent standards. In upgrading existing waste-water treatment facilities, the need for improved disinfection as well as the elimination of odour problems are frequently encountered. Adequate and reliable disinfection is essential in ensuring that waste-water treatment plants are both environmentally safe and aesthetically acceptable to the public. Chlorine is the most widely used disinfectant in water and waste-water treatment. It is used to destroy pathogens, control nuisance micro-organisms and for oxidation. As an oxidant, chlorine is used in iron and manganese removal, for destruction of taste and odour compounds and in the elimination of ammonia nitrogen. It is, however, a highly toxic substance and recently concerns have been raised over handling practices and possible residual effects of chlorination.

Disinfection with interhalogens and halogen mixtures

The interhalogen compounds are the bromine- and iodine-base materials. It is the larger, more positive halogen that is the reactive portion of the interhalogen molecule during the disinfection process. Although only used on a limited basis at present, there are members of this class that show great promise as environmentally safe disinfectants.

Engineering Systems for Waste-water Treatment and Disposal

All industrial operations produce some waste-water which must be returned to the environment. Waste-waters can be classified as: (i) domestic waste-waters, (ii) process waste-waters, and (iii) cooling waste-waters. Domestic waste-waters are produced by plant workers, shower facilities and cafeterias. Process waste-waters result from spills, leaks and product washing. Cooling waste-waters are the result of various cooling processes and can be once-pass systems or multiple-recycle cooling systems. Once-pass cooling systems employ large volumes of cooling waters that are used once and returned to the environment. Multiple-recycle cooling systems have various types of cooling towers to return excess heat to the environment and require periodic blow down to prevent excess build-up of salts.

Domestic waste-waters are generally handled by the normal sanitary sewerage system to prevent the spread of pathogenic micro-organisms which might cause disease. Normally, process waste-waters do not pose the potential for pathogenic micro-organisms, but they do pose potential damage to the environment through either direct or indirect chemical reactions.

Some process wastes are readily biodegraded and create an immediate oxygen demand. Other process wastes are toxic and represent a direct health hazard to biological life. Cooling waste-waters are the least dangerous, but they can contain process waste-waters as a result of leaks in the cooling systems. Recycle cooling systems tend to concentrate both inorganic and organic contaminants to a point at which damage can be created.

Waste-water characteristics

Waste-waters are usually classified as industrial waste-water or municipal waste-water. Industrial waste-water with characteristics compatible with municipal waste-water is often discharged to the municipal sewers. Many industrial waste-waters require pre-treatment to remove non-compatible substances prior to discharge into the municipal system. Characteristics of industrial waste-water vary greatly from industry to industry and consequently, treatment processes for industrial waste-water also vary, although many of the processes used to treat municipal waste-water are also used in industrial waste-water treatment. Water collected in municipal waste-water systems, having been put to a wide variety of uses, contains a wide variety of contaminants. A list of contaminants commonly found in municipal waste-water along with their sources and their environmental consequences is given in Table 2.3.

Table 2.3. Important waste-water contaminants.

Contaminant	Source	Environmental significance
Suspended solids anaerobic conditions	Domestic use, industrial wastes, erosion by infiltration/inflow	Cause sludge deposits and in aquatic environment
Biodegradable organics	Domestic and industrial waste	Cause biological degradation, which may use up oxygen in receiving water and result in undesirable conditions
Pathogens	Domestic waste	Transmit communicable diseases
Nutrients	Domestic and industrial waste	May cause eutrophication
Refractory organics	Industrial waste	May cause taste and odour problems, may be toxic or carcinogenic
Heavy metals	Industrial waste, mining, etc.	Are toxic, may interfere with effluent reuse
Dissolved inorganic solids	Increases above level in water supply by domestic and/or industrial use	May interfere with effluent reuse

Terminology in waste-water treatment

The terminology used in waste-water treatment is often confusing to the uninitiated person. Terms such as unit operations, unit processes, reactors, systems and primary, secondary and tertiary treatment frequently appear in the literature and their usage is not always consistent. The meanings of these terms, as used in this text, are discussed here. Methods used for treating municipal waste-waters are often referred to as either unit operations or unit processes. Generally, unit operations involve contaminant removal by physical forces, while unit processes involve biological and/or chemical reactions.

The term reactor refers to the vessel or containment structure, along with all of its appurtenances, in which the unit operation or unit process takes place. Although unit operations and processes are natural phenomena, they may be initiated, enhanced or otherwise controlled by altering the environment in the reactor. Reactor design is a very important aspect of waste-water treatment and requires a thorough understanding of the unit processes and unit operations involved. A waste-water-treatment system is composed of a combination of unit operations and unit processes designed to reduce certain constituents of waste-water to an acceptable level. Many different combinations are possible. Although practically all waste-water treatment systems are unique in some respects, a general grouping of unit operations and unit processes according to target contaminants has evolved over the years. Unit operations and processes commonly used in waste-water treatment are listed in Table 2.4 and are arranged according to conventional grouping. Actually, only a few waste-water treatment methods fall completely into one category. Thus the usefulness of this classification system is somewhat compromised.

Table 2.4. Unit operations, unit processes and systems for waste-water treatment.

Contaminant	Unit operation, unit process or treatment system
Suspended solids	Sedimentation
	Screening and comminution
	Filtration variations
	Flotation
	Chemical-polymer addition
	Coagulation/sedimentation
	Land treatment systems
Biodegradable organics	Activated-sludge variations
	Fixed-film: trickling filters
	Fixed-film: rotating biological contactors
	Lagoon and oxidation pond variations
	Intermittent sand filtration
	Land treatment systems
	Physical-chemical systems
Pathogens	Chlorination
	Hypochlorination
	Ozonation
	Land treatment systems

(Contd ...)

Contaminant	Unit operation, unit process or treatment system
Nutrients	
Nitrogen	Suspended-growth nitrification and denitrification variations
	Fixed-film nitrification and denitrification variations
	Ammonia stripping
	Ion exchange
	Breakpoint chlorination
	Land treatment systems
Phosphorus	Metal-salt addition
	Lime coagulation/sedimentation
	Biological-chemical phosphorus removal
	Land treatment systems
Refractory organics	Carbon adsorption
	Tertiary ozonation
	Land treatment systems
Heavy metals	Chemical precipitation
	Ion exchange
	Land treatment systems
Dissolved inorganic solids	Ion exchange
	Reverse osmosis
	Electrodialysis

Municipal waste-water treatment systems are often divided into primary, secondary and tertiary subsystems. The purpose of primary treatment is to remove solid materials from the incoming waste-water. Large debris may be removed by screens or may be reduced in size by grinding devices. Inorganic solids are removed in grit channels and much of the organic suspended solids is removed by sedimentation. A typical primary treatment system (Fig. 2.11) should remove approximately one-half of the suspended solids in the incoming waste-water. The BOD associated with these solids accounts for about 30 per cent of the influent BOD.

Secondary treatment usually consists of biological conversion of dissolved and colloidal organics into biomass that can subsequently be removed by sedimentation. Contact between micro-organisms and the organics is optimised by suspending the biomass in the waste-water or by passing the waste-water over a film of biomass attached to solid surfaces.

The most common suspended biomass system is the activated-sludge process shown in Fig. 2.12a. Recirculating a portion of the biomass maintains a large number of organisms in contact with the waste-water and speeds up the conversion process. The classical attached-biomass system is the trickling filter shown in Fig. 2.12b.

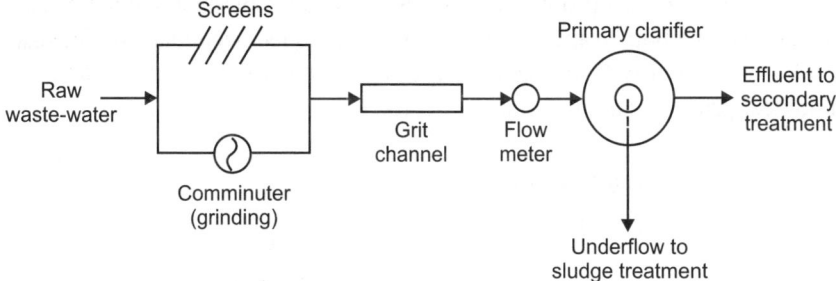

Fig. 2.11. Typical primary treatment system.

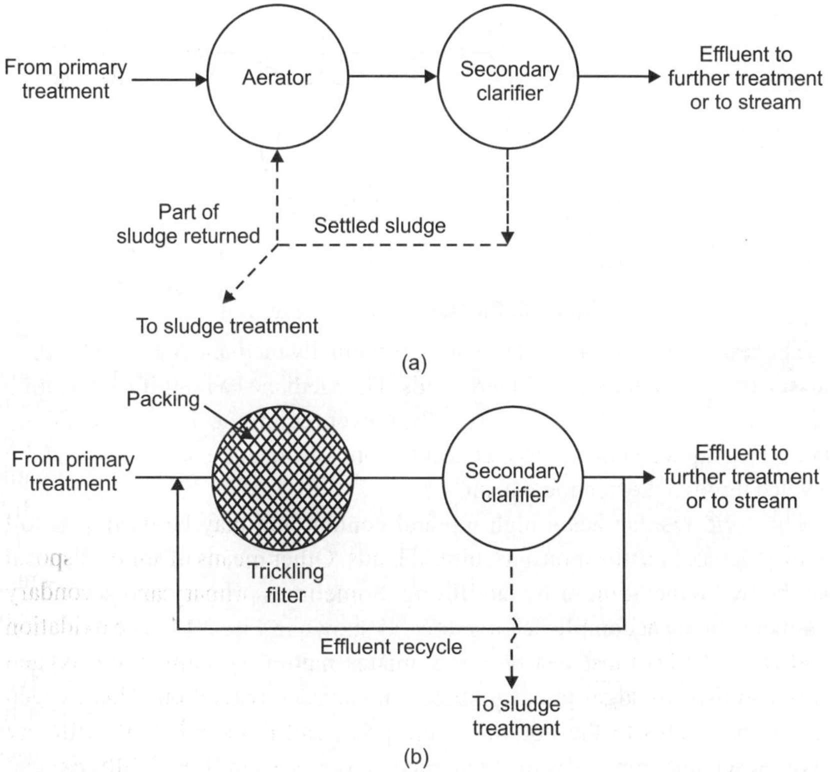

Fig. 2.12. Secondary treatment system: (a) activated sludge system; and (b) trickling filter system.

Stones or other solid media are used to increase the surface area for biofilm growth. Mature biofilms peel off the surface and are washed out to the settling basin with the liquid underflow. Part of the liquid effluent may be recycled through the system for additional treatment and to maintain optimal hydraulic flow rates.

Secondary systems produce excess biomass that is biodegradable through endogenous catabolism and by other micro-organisms. Secondary sludges are usually combined with primary sludge for further treatment by anaerobic biological processes as shown in Fig. 2.13.

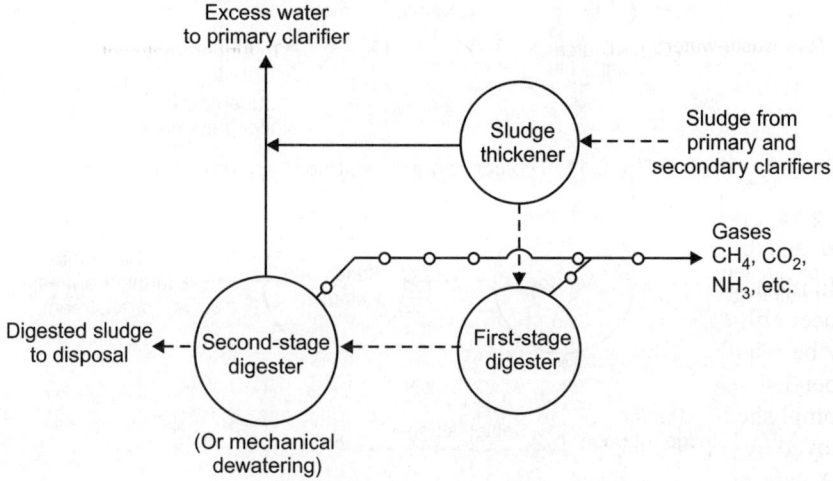

Fig. 2.13. Sludge treatment system.

The results are gaseous end products, principally methane (CH_4) and carbon dioxide (CO_2) and liquids and inert solids. The methane has significant heating value and may be used to meet part of the power requirements of the treatment plant. The liquids contain large concentrations of organic compounds and are recycled through the treatment plant.

The solid residue has a high mineral content and may be used as a soil conditioner and fertiliser on agricultural lands. Other means of solids disposal may be by incineration or by landfilling. Sometimes primary and secondary treatment can be accomplished together, as shown in Fig. 2.14. The oxidation pond (Fig. 2.14a) most nearly approximates natural systems, with oxygen being supplied by algal photosynthesis and surface reaeration. This oxygen seldom penetrates to the bottom of the pond and the solids that settle are decomposed anaerobically. In the aerated lagoon system (Fig. 2.14b) oxygen is supplied by mechanical aeration and the entire depth of the pond is aerobic. Decomposition of the biomass occurs by aerobic endogenous catabolism. The small quantity of excess sludge that is produced is retained in the bottom sediments.

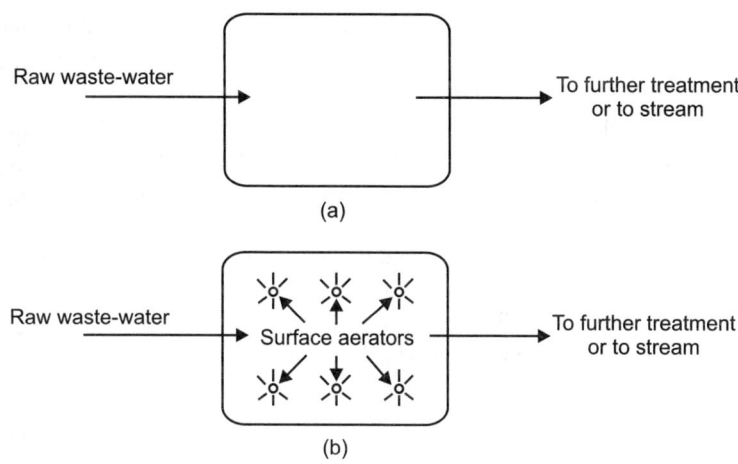

Fig. 2.14. Primary and secondary water treatment in combination: (a) oxidation pond, and (b) aerated lagoon.

In most cases, secondary treatment of municipal waste-water is sufficient to meet effluent standards. In some instances, however, additional treatment may be required. Tertiary treatment most often involves further removal of suspended solids and/or the removal of nutrients. Solids removal may be accomplished by filtration and phosphorus and nitrogen compounds may be removed by combinations of physical, chemical and biological processes.

A careful inspection of Figs 2.11 through 2.14 leads to an interesting observation. The removal processes in waste-water treatment are essentially concentrating or thickening, processes. Suspended solids are removed as sludges and dissolved solids are converted to suspended solids and subsequently become removable sludges. Hammer states that primary and secondary treatment, followed by sludge thickening, may concentrate organic material represented by 250 mg/l of suspended solids and 200 mg/l BOD in 375 litres of municipal waste-water (the average per capita contribution) to 2 litres of sludge containing 50,000 mg/l of solids. Most of the objectionable material initially in the waste-water is concentrated in the sludges and must be disposed of in a safe and environmentally acceptable manner. Vesilind notes that a majority of the expenses, effort and problems of waste-water treatment and disposal are associated with the sludges.

Design of waste-water-treatment systems is an important part of an environmental engineer's work. A thorough understanding of the unit operations and processes is necessary before the reactors can be designed.

Advanced Waste-water Treatment

The quality of effluent provided by secondary treatment may not always be sufficient to meet discharge requirements. This is often the case when large quantities of effluent are discharged into small streams or when delicate ecosystems are encountered. In these instances, additional treatment to polish the effluent from secondary systems will be required or an alternative method of waste-water disposal must be found.

Additional treatment, usually referred to as tertiary treatment, often involves the removal of nitrogen and phosphorus compounds, plant nutrients associated with eutrophication. Further treatment may be required to remove additional suspended solids, dissolved inorganic salts and refractory organics. Combinations of the above processes can be used to restore waste-water to potable quality, although at considerable expense. Referred to as reclamation, this complete treatment of waste-water can seldom be justified except in water-scarce areas where some form of reuse is mandated.

The term advanced treatment is frequently used to encompass any or all of the above treatment techniques and this term would seem to imply that advanced treatment follows conventional secondary treatment. This is not always the case, as some unit operations or unit processes in secondary or even primary treatment may be replaced by advanced-treatment systems. Advanced-treatment processes and operations are described in the following section of this chapter. Because treatment systems are selected to meet discharge or reuse criteria with respect to specific parameters, the discussion is arranged according to treatment objectives.

Nitrogen and phosphorus control

Nitrogen compounds often move within the environment as they change form. Most of the problems caused by nitrogen compounds occur when they enter groundwater or surface water bodies. Nitrogen reaches fresh surface water through precipitation, dustfall, surface runoff, subsurface groundwater entry and the discharge of waste-water effluents. There are also blue-green algae and some bacteria which are able to fix nitrogen from nitrogen gas in the atmosphere.

Nitrification

Nitrification is the first of two stages in biological nitrogen removal in which ammonia-nitrogen is biologically oxidised to nitrate, a less objectable form which does not exert an oxygen demand on the receiving water.

In the first step of nitrification, ammonium ions are oxidised to nitrite ions according to the following reaction in which 58 to 84 kcal/mole of ammonium is released:

$$NH_4 + 1.5O_2 \rightarrow H^+ + H_2O + NO_2$$

The bacteria responsible for this oxidation are usually nitrosomonas, although sometimes nitrosococcus can be involved. These bacteria are aerobic autotrophs. Autotrophs, unlike heterotrophs, which obtain their energy from the oxidation of carbonaceous (organic) matter, get their energy for growth from the oxidation of inorganic nitrogenous matter and use inorganic carbon rather than organic carbon, as heterotrophs do, for cell synthesis. Being aerobic, these autotrophs require the presence of oxygen to convert the nitrogen into a usable form.

In the second step of nitrification, the nitrite ion is further oxidised by nitrobacter bacteria to nitrate releasing only 15–21 kcal/mole of nitrite oxidised.

Nitrobacter bacteria are aerobic autotrophs also. The energy freed by the nitrification reactions is used by the bacteria for growth. Since the nitrosomonas obtain more energy than the nitrobacter bacteria per mole of nitrogen oxidised, their mass in any nitrification system is greater.

The nitrobacter bacteria require three times the substrate needed by nitrosomonas to get the same energy and therefore their population is 1/3 that of the nitrosomonas.

Denitrification

Denitrification is the second and final stage in the biological removal of nitrogen. With denitrification, nitrates are reduced to nitrogen gas.

When methanol is used as a source of carbonaceous matter, the reaction for denitrification is:

$$5CH_3OH + 6H^+ + 6NO_3^- \rightarrow 5CO_2 + 3N_2 + 13H_2O$$

It is also possible for nitrites to be converted directly to nitrogen gas. The bacteria responsible for this transformation are heterotrophs, which derive their energy from organic chemicals through the reduction of nitrate or nitrite. These bacteria include pseudomonas, achromobacter, bacillus and micrococcus, which are facultative, meaning that they can survive with or without the presence of oxygen. The bacteria prefer oxidising organic matter with oxygen rather than by reducing nitrite or nitrate. Therefore, anaerobic (no oxygen) conditions must be maintained in a denitritication system.

With both nitrification and denitritication of wastes, nitrogen removals of 70–90 per cent can be obtained. While such removal will serve to reduce or prevent most algal growth and eutrophication, many blue-green algal blooms cannot be affected since these algae can fix nitrogen gas for their synthesis from the atmosphere.

Biological nitrification methods

Biological nitrification can be achieved by several means and through various add on treatment and upgrading methods for new and existing treatment systems.

Domestic waste effluent has been one of the main contributors of nitrogenous compounds to our environment. Nitrogen in such wastes usually exists as organic nitrogen or as free ammonia. The nitrate and nitrite concentrations are generally small in raw wastes in relation to the other forms.

The most significant of the above compounds is ammonia since it can lower the dissolved oxygen of a receiving stream by nitrification. The ammonia content is derived from urea and to a lesser extent proteins. The organic-nitrogens are in the form of purines, pyrimidines, proteins, urea and amino acids. Much of the organic-nitrogen is transformed to ammonia through hydrolysis before it reaches the waste-water treatment plant. Conversion of organic-nitrogen to ammonia continues to occur within a conventional treatment plant due to the actions of heterotrophic bacteria.

After passing through conventional biological treatment, the secondary sanitary effluent will have a typical nitrogen content of 20–50 mg/l, indicating that nitrogen just passes through such systems. However, biological nitrification does not remove nitrogen better than any conventional treatment.

Biological nitrification can be achieved in conventional carbonaceous removal systems which have been modified to combine nitrification and by the addition of a separate tertiary system. Generally if the BODs/total kjeldahl nitrogen (TKN) ratio is less than 3, a separate nitrification system must be added. If the BODs/TKN ratio is greater than 5, a combined system should work. There is no special recommendation for wastes with ratios between 3 and 5 at this time. There are varied opinions as to the merits of both methods.

Combined carbon and nitrogen removal systems

The first nitrification processes developed were combined systems made by modifying extended aeration systems. Combining operations is advantageous in terms of cost for existing carbonaceous systems which can be upgraded to include nitrification. Combined carbon and nitrogen removal systems have a high proportion of influent organic loading relative to the ammonia-nitrogen concentration. As a result, the population of nitrifiers is small compared to heterotrophs. In addition, the conditions required for the carbon oxidising heterotrophs and nitrifying autotrophs are different and therefore operating parameters must be carefully controlled in combined systems.

Combined systems should be based on the sludge growth rate or solids retention time. This generally means an additional oxygen supply, longer mean cell residence times (about 10 days) and operating temperatures of 21° to 22°C.

Two-stage carbon and nitrogen removal systems

Physical separation of carbon and nitrogen removal functions can improve the control and efficiency of nitrification in certain cases. By reducing the

BODs load in the first stage significantly to the influent ammonia concentration, more nitrifiers can be established. The value of separating the heterotrophic and autotrophic populations is realised in the reduced residence time required— 6 days total versus 10 days in combined systems.

Plug-flow systems are the favoured activated sludge method for obtaining nitrification. Lower effluent ammonia concentrations can be achieved than in completely mixed units.

Trickling filters

Trickling filter systems are the major type of attached growth system used to perform nitrification. Due to larger land requirements relative to activated sludge systems, trickling filters are often used for sanitary treatment in smaller cities of less than 10,000 people and less populated areas. Because of the relative stability of trickling filters compared to other biological systems, they are used for treating high-strength industrial wastes too. A diagram of a typical trickling filter is shown in Fig. 2.15.

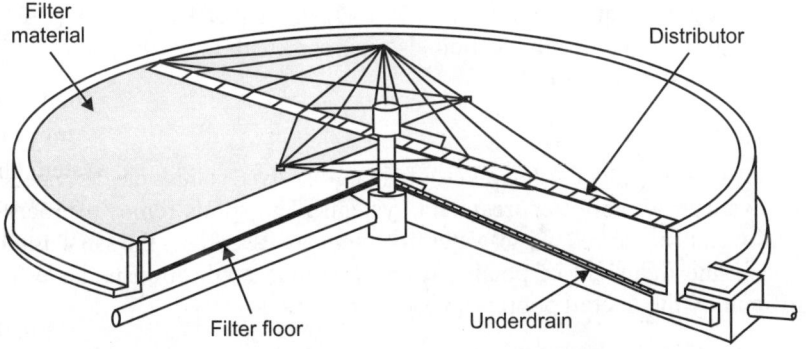

Fig. 2.15. Cutaway view of a trickling filter.

Trickling filters are usually circular in form. Waste effluent is distributed by rotary sprays over the media and is collected underneath by an underdrain system.

Media to which organisms attach can be made of rock, plastic or redwood. The liquid waste percolates down through the media and the substrate and inorganic and organic waste matter is assimilated by the organisms attached to the media. Aerobic degradation takes place on the outer portions of the biological film which develops on the media. As the mass of organisms becomes thicker, anaerobic conditions occur near the media surface. The surface organisms then die and are washed off periodically (sloughing).

Effluent peaks can be taken care of by designing the system with additional surface area and increasing recirculation rates during low-flow periods to keep the media from drying out. Clarification normally is not needed following trickling filters since the solids are maintained within the units. Trickling filters are the homes of a varied assortment of organisms including: aerobic, anaerobic or facultative bacteria; fungi, during low pHs; algae; protozoans, worms, insect larvae and snails, which in turn feed on the bacteria. Due to the unstable characteristics of the slime, a kinetic theory for the biological activities has not yet been developed. Conclusions regarding nitrification in such systems are based on empirical results.

Nitrification and carbonaceous oxidation can occur simultaneously in trickling filters. It is better to use a media with a lower specific surface and higher voids, such as a maximum of 35 sq. ft/cu. ft, to prevent clogging in combined systems. When nitrification is separate from carbonaceous oxidation, plugging is less of a problem for the nitrification system and application of a media with a high specific surface (up to 67 sq.ft/cu.ft) is okay and reduces space requirements. In two-stage systems where organic carbon and nitrogen activities occur separately, increases in nitrification have been proportional to increases in surface areas. The surface area requirements for two-stage systems increases greatly at temperatures of 7° to 11°C than at 13° to 19°C. Surface area requirements for nitrification also increase with the degree of ammonia nitrogen reduction desired.

Solids removal

Removal of suspended solids and sometimes dissolved solids, may he necessary in advanced waste-water treatment systems. The solids removal processes employed in advanced waste-water treatment are essentially the same as those used in the treatment of potable water, although application is made more difficult by the overall poorer quality of the waste-water.

Suspended solids removal

As an advanced treatment process, suspended-solids removal implies the removal of particles and flocs too small or too lightweight to be removed in gravity settling operations. These solids may be carried over from the secondary clarifier or from tertiary systems in which solids were precipitated.

Several methods are available for removing residual suspended solids from waste-water. Removal by centrifugation, air flotation, mechanical micro-screening and granular-media filtration have all been used successfully. In current practice, granular media filtration is the most commonly used process. Basically, the same principles apply to filtration of particles from potable water apply to the removal of residual solids in waste-water.

Dissolved solids removal

Both secondary treatment and nutrient removal decrease the dissolved organic solids content of waste-water. Neither process, however, completely removes all dissolved organic constituents and neither process removes significant amounts of inorganic dissolved solids. Further treatment will be required where substantial reductions in the total dissolved solids of waste-water must be made.

Ion exchange, microporous membrane filtration, adsorption and chemical oxidation can be used to decrease the dissolved solids content of water. These processes, were developed to prepare potable water from a poor-quality raw water. Their use can be adopted to advanced waste-water treatment if a high level of pre-treatment is provided. The removal of suspended solids is necessary prior to any of the processes. Removal of the dissolved organic material (by activated carbon adsorption) is necessary prior to microporous membrane filtration to prevent the larger organic molecules from plugging the micropores.

Advanced waste-water treatment for dissolved solids removal is complicated and expensive. Treatment of municipal waste-water by these processes can be justified only when reuse of the waste-water is anticipated.

Treating the Sludge

The similarity between a water treatment facility and any manufacturing operation, whether it be a rubber producing plant or an auto-making facility or an iron and steel plant, are the reliance upon combinations of unit processes and unit operations that work in harmony to produce a high quality product. But that is where the similarity ends. A normal manufacturing operation aims not only to produce a high quality product, but efficient businesses strive to do so by eliminating or minimising their wasteful by-products — simply because those by-products have little to no market value and add cost to production. If our cost of production is higher, then profit margins are lower. That in fact is the basis for pollution prevention and waste minimisation practices of modern day industry.

A water treatment facility differs in this regard because the primary objective is to produce high quality water by removing or destroying as much of the contaminants as possible. We cannot produce high quality water without generating the wasteful by-product, sludge, very often in large quantities. Water treatment plants are simply pollution control technologies, whether they are applied to industrial applications or municipal. That does not mean that pollution prevention practices are not appropriate for water treatment plants — they most certainly are and can minimise solid waste generation. But understand that we cannot eliminate the wasteful by-product of sludge as one might try and do if we had manufacturing facility and we identified another technology to make our product and eliminate a wastestream generated by the older technology.

Sludge or solid waste is unavoidably produced in the treatment of water containing suspended solids. There are, however different technologies that we can select among that will indeed concentrate these solids and thereby reduce the volumes that we ultimately must dispose of. In addition, some sludge can be stabilised and treated, which can impart a low, but none-the-less marketable value to this waste. These technologies and practices do indeed constitute pollution prevention and waste minimisation programmes within water treatment plant operations and they can have a very significant and positive impact on the overall costs of the operation. This brings us to a collection of technologies that focus on: (i) sludge concentration, (ii) sludge stabilisation, and (iii) sludge handling and disposal. Some technologies fall into the category of pollution prevention, while others are within the normal arena of solid waste management and disposal.

Note that pollution prevention or P2 technologies, as in other industry sectors, are not necessarily the preferred choices. Specific technology selection quite often depends on localised conditions. By this, we mean the properties of the sludge, the volumes handled and the comparative costs between technologies and or practices. In a very general sense, pollution prevention technologies are only appropriate when they are financially attractive for an operation. Like any other engineering project, the investment into a technology that falls within the pollution prevention area must have financial attractiveness. An alternative way of stating this is that there are indeed situations where more conventional methods resulting in large volumes of sludge are more cost effective than a leading-edge technology that minimises or reduces sludge volumes. The financial attractiveness of an investment needs to be assessed on a case by case basis.

This section focuses on sludge processing and post-processing technologies. Where appropriate, we will point out which technologies may be considered P2.

When we think of sludge, what automatically comes to mind is sewerage. Water carriage systems of sewerage provide a simple and economical means for removing offensive and potentially dangerous wastes from household and industry. The solution and suspension of solids in the transporting of water produces sewage. Thus, the role of solids and sludge removal at sewage treatment plants is apparent. Sludge removal is complicated by the fact that some of the waste matters go into solution while others are colloidal or become finely divided in their flow through the sewage system. Ordinarily, less than half of such waste remains in suspension in a size or condition that can be separated by being strained out, skimmed off or settled out. The remainder must then be precipitated out by chemical means, filtered mechanically or be subjected to biological treatment whereby they are either removed from the water or changed in character as to be rendered innocuous.

Sewage contains mineral and organic matter in suspension (coarse and fine suspended matter), in colloidal state (very finely dispersed matter) and in solution. Living organisms, notably bacteria and protozoa, find sewage to be an abundant source of food and their lives activities result in the decomposition of sewage.

Sewage becomes offensive due to its own instability together with the objectionable concentration of suspended materials. In addition, the potential presence of disease producing organisms makes sewage dangerous. Removal or stabilisation of sewage matters may be accomplished in treatment works by a number of different methods or by a suitable combination of these methods.

While sewage sludge is rich in nutrients and organic matter, offering the potential for applications as a biosolid or it has a heating value making it suitable for incineration, many industrial sludge are often unsuitable for reuse. A more common practice with industrial sludge is to try and identify a reclaim value, i.e. if the sludge can be concentrated sufficiently there may be portion of this waste which is reclaimable or may enter into a recycling market.

Stabilisation and conditioning

Pre-stage basics

Before sludge undergoes treatment such as dewatering or thickening, it must be stored and pretreated. Sludge storage is an important, integral part of every waste-water sludge treatment and disposal system. Sludge storage provides many benefits including equalisation of sludge flow to downstream processes, allowing sludge accumulation during times of non-operation of sludge-processing facilities and allowing a uniform feed rate that enhances thickening, conditioning and dewatering operations.

Sludge is stored within waste-water treatment process tankage, sludge treatment process systems or separately in specially designed tanks. Sludge can be stored on a short-term or a long-term basis. Small treatment plants, where storage time may vary from several to 24 hours, may store sludge in waste-water clarification basins or sludge-thickening tanks. Larger plants often use aerobic digester, facultative lagoons and other processes with long detention times to store sludge. The pretreatment of sludge is often necessary before dewatering or thickening can take place. It includes degritting and grinding. Sludge degritting involves the installation of grit removal and processing facilities at the head works where raw waste-water first enters the treatment plant. As a result, there is reduced wear on influent pumping systems and primary sludge pumping, piping and thickening systems. Sludge grinding involves shearing of large sludge solids into smaller particles. This method is used to prevent problems with operation of downstream processes.

Sludge-pumping systems play an important part in waste-water treatment plants, particularly those operations experiencing average flows of greater than 1 million gallons per day (mgd). There are different types of pumps within this process. Typical advantages of kinetic pumps for sludge transport include lower purchase cost, lower maintenance cost due to wear, less space used and availability of both dry-well and submersible pumps. Advantages of positive displacement pumps include improved process control and pumping capability at high pressure and low flow.

Sludge cake storage (where a cake is the dewatered solid part of sludge) provides similar benefits for downstream disposal alternatives, like composting and incineration, to sludge storage which is used for thickening and dewatering. Storage of sludge cakes increases operational reliability, evens outflow fluctuations and allows accumulation when downstream operations are not in service.

Before any of the sludge can proceed to dewatering or thickening processes, it must be conditioned. Sludge conditioning involves chemical or thermal treatment to improve the efficiency of the downstream processes. Chemical conditioning involves use of inorganic chemicals or organic polyelectrolytes or both. The most commonly used inorganic chemicals are ferric chloride and lime. Organic polymers are used for both sludge-thickening and dewatering processes. Their advantage over inorganics is that polymers do not greatly increase the amount of sludge production: 1 kg of inorganic chemicals added will produce 1 kg of extra sludge. The disadvantage of polymers is their relatively high cost. There are several important factors that affect conditioning of sludge. They include: sludge characteristics, sludge handling and sludge coagulation and flocculation. The fundamental purpose of sludge conditioning is to cause the aggregation of fine solids by coagulation and inorganic chemicals, flocculation with organic polymers or both. A critical design parameter in conditioning is dosage. Selection of the right dosage of a chemical conditioner is critical for good performance. The dosage affects the solids content of sludge cakes as well as solids capture rate and solids disposal cost. Dosage is determined from pilot studies, bench tests and on-line tests. In the following sections we will cover the basics of sludge stabilisation and then conditioning.

Chemical stabilisation

Chemical stabilisation is a process whereby the sludge matrix is treated with chemicals in different ways to stabilise the sludge solids. Two common methods employed are lime stabilisation and the use of chlorine.

The lime stabilisation process can be used to treat raw primary, waste activated, septage and anaerobically digested sludge. The process involves mixing a large enough quantity of lime with the sludge to increase the pH of the mixture to 12 or more. This normally reduces bacterial hazards and odour to a negligible value, improves vacuum filter performance and provides satisfactory means of stabilising the sludge prior to ultimate disposal.

Stabilisation by chlorine addition has been developed and is marketed under the registered trade name Purifax. The chemical conditioning of sludge with chlorine varies greatly from the more traditional methods of biological digestion or heat conditioning. First, the reaction is almost instantaneous. Second, there is very little volatile solids reduction in the sludge. There is some breakdown of organic material and formation of carbon dioxide and nitrogen; however, most of the conditioning is by the substitution or addition of chlorine to the organic compound to form new compounds that are biologically inert.

Stabilisation via aerobic digestion

Aerobic digestion is an extension of the activated sludge aeration process whereby waste primary and secondary sludge are continually aerated for long periods of time. In aerobic digestion the micro-organisms extend into the endogenous respiration phase. This is a phase where materials previously stored by the cell are oxidised, with a reduction in the biologically degradable organic matter. This organic matter, from the sludge cells is oxidised to carbon dioxide; water and ammonia.

The ammonia is further converted to nitrates as the digestion process proceeds. Eventually, the oxygen uptake rate levels off and the sludge matter is reduced to inorganic matter and relatively stable volatile solids.

The primary advantage of aerobic digestion is that it produces a biologically stable end product suitable for subsequent treatment in a variety of processes. Volatile solids reductions similar to anaerobic digestion are possible. Some parameters affecting the aerobic digestion process are:
1. The rate of sludge oxidation.
2. Sludge temperature.
3. System oxygen requirements.
4. Sludge loading rate.
5. Sludge age.
6. Sludge solids characteristics.

Sludge conditioning by thermal methods

There are two basic processes for thermal treatment of sludge. One, wet air oxidation, is the flameless oxidation of sludge at temperatures of 450° to 550°F and pressures of about 1200 psig. The other type, heat treatment, is similar but carried out at temperatures of 350° to 400°F and pressures of 150 to 300 psig. Wet air oxidation (WAO) reduces the sludge to an ash and heat treatment improves the dewaterability of the sludge. The lower temperature and pressure heat treatment is more widely used than the oxidation process.

When the organic sludge is heated, heat causes water to escape from the sludge. Thermal treatment systems release water that is bound within the cell structure of the sludge and thereby improves the dewatering and thickening characteristics of the sludge. The oxidation process further reduces the sludge to ash by wet incineration (oxidation). Sludge is ground to a controlled particle size and pumped to a pressure of about 300 psi. Compressed air is added to the sludge (wet air oxidation only), the mixture is brought to a temperature of about 350°F by heat exchange with treated sludge and direct steam injection and then is processed (cooked) in the reactor at the desired temperature and pressure. The hot treated sludge is cooled by heat exchange with the incoming sludge. The treated sludge is settled from the supernatant before the dewatering step. Gases released at the separation step are passed through a catalytic after burner at 650° to 705°F or deodourised by other means. In some cases these gases have been returned through the diffused air system in the aeration basins for deodourisation. An advantage of thermal treatment is that a more readily dewaterable sludge is produced than with chemical conditioning.

Sludge pasteurisation process

This process is really sludge disinfection. Its aim is the destruction or inactivation of pathogenic organisms in the sludge. Destruction is defined as the physical disruption or disintegration of a pathogenic organism, while inactivation is defined as the removal of a pathogen's ability to infect.

Sludge dewatering operations

Another term for dewatering the sludge is sludge thickening. The objective is to concentrate the sludge and quite frankly—make it as dry as economically possible for post processing and disposal purposes. There are both mechanical and thermal techniques for achieving this. Among the mechanical processes used to dewater sludge are belt filter presses and drum filters (vacuum technologies), pressure filter presses and centrifugation.

Vacuum filtration

The vacuum filter for dewatering sludge is a drum over which is laid the filtering medium consisting of a cloth of cotton, wool, nylon, dynel, fibre glass or plastic, or a stainless steel mesh or a double layer of stainless steel coil springs.

Volume reduction

As title implies, we will now focus our attention to those technologies aimed at reducing the volume of the final form of the sludge. Dewatering or thickening technologies can only bring us so far in concentrating the form of the waste. Ultimately, we must find ways of either disposing of this waste or in using it. Of immediate concern is how we can reduce the volume of so-called dry sludge, at solids contents ranging anywhere from 30 to 60 per cent, even further.

Incineration

Incineration of sludge has gained popularity throughout the world, especially at large plants. It has the advantages of economy, freedom of odour, independence of weather and the great reduction in the volume and weight of end product to be disposed of. There is a minimum size of sewage treatment plant below which incineration is not economical. There must be enough sludge to necessitate reasonable use of costly equipment. One of the difficulties in operating an incinerator is variations in tonnage and moisture of sludge handled.

There are two major incinerator technologies used in this process. They are: (i) the multiple hearth incinerator, and (ii) the fluidised-bed incinerator. An incinerator is usually part of a sludge treatment system which includes sludge thickening, macerations, dewatering (such as vacuum filter, centrifuge, or filter press), an incinerator feed system, air pollution control devices, ash handling facilities and the related automatic controls. The operation of the incinerator cannot be isolated from these other system components. Of particular importance is the operation of the thickening and dewatering processes because the moisture content of the sludge is the primary variable affecting the incinerator fuel consumption.

Incineration may be thought of as the complete destruction of materials by heat to their inert constituents. This material that is being destroyed is the waste product (i.e. the sludge). Sewer sludge as sludge cake normally contains from 55 to 85 per cent moisture. It cannot burn until the moisture content has been reduced to no more than 30 per cent. The purpose of incineration is to reduce the sludge cake to its minimum volume, as sterile ash. There are three objectives incineration must accomplish:

1. Dry the sludge cake.
2. Destroy the volatile content by burning.
3. Produce a sterile residue or ash.

Chapter 3

Energy, Fuels, Air Conditioning and Refrigeration

INTRODUCTION

Since the chemical process industries consume more than a third of the energy used by all the manufacturing industries, chemical engineers should be familiar with the broad technical aspects of the production of power, cold, and heat. They should also be prepared to work with power, refrigeration, and air conditioning engineers for the proper coordination of the production of these essential tools and their use in chemical processes in order to attain the cheapest manufacturing costs. Frequently, the cost of power, particularly if it is to be used electrochemically, is the deciding factor in choosing the location of a factory. Process industries under the direction of chemical engineers are in most instances outstanding consumers of steam for evaporation, heating, and drying. Consequently, these industries need large quantities of steam, usually in the form of low-pressure or exhaust steam from turbines. Occasionally, however, certain exothermic reactions, as in the contact sulphuric acid process, can be employed to generate steam for use. If only electricity is desired from a steam power plant, naturally the turbines are run condensing; if, on the other hand, both steam and power are needed, as in the chemical process industries, it is economical to lead the high-pressure steam directly from the boilers through noncondensing turbines, obtaining exhaust steam from these prime movers to supply the heat necessary for drying, evaporation, and endothermic chemical reactions throughout the plant.

ENERGY

Energy is a scalar physical quantity that describes the amount of work that can be performed by a force. Energy is an attribute of objects and systems that is subject to a conservation law. Several different forms of energy exist to explain all known natural phenomena. These forms include (but are not limited to) kinetic, potential, thermal, gravitational, sound, light, elastic, and electromagnetic energy. The forms of energy are often named after a related force.

72

Any form of energy can be transformed into another form, but the total energy always remains the same. This principle, the conservation of energy, was first postulated in the early 19th century, and applies to any isolated system.

According to Noether's theorem, the conservation of energy is a consequence of the fact that the laws of physics do not change over time.

Although the total energy of a system does not change with time, its value may depend on the frame of reference. For example, a seated passenger in a moving airplane has zero kinetic energy relative to the airplane, but non-zero kinetic energy relative to the earth.

Energy is subject to a strict global conservation law; that is, whenever one measures (or calculates) the total energy of a system of particles whose interactions do not depend explicitly on time, it is found that the total energy of the system always remains constant.

1. The total energy of a system can be subdivided and classified in various ways. For example, it is sometimes convenient to distinguish potential energy (which is a function of coordinates only) from kinetic energy (which is a function of coordinate time derivatives only). It may also be convenient to distinguish gravitational energy, electric energy, thermal energy, and other forms. These classifications overlap; for instance thermal energy usually consists partly of kinetic and partly of potential energy.

2. The transfer of energy can take various forms; familiar examples include work, heat flow, and advection.

In classical physics energy is considered a scalar quantity, the canonical conjugate to time. In special relativity energy is also a scalar (although not a Lorentz scalar but a time component of the energy-momentum 4-vector). In other words, energy is invariant with respect to rotations of space, but not invariant with respect to rotations of space-time (= boosts).

Because energy is strictly conserved and is also locally conserved (wherever it can be defined), it is important to remember that by definition of energy the transfer of energy between the system and adjacent regions is work.

Chemical Energy

Chemical energy is the energy due to associations of atoms in molecules and various other kinds of aggregates of matter. It may be defined as a work done by electric forces during rearrangement of mutual positions of electric charges, electrons and protons, in the process of aggregation. So, basically it is electrostatic potential energy of electric charges. If the chemical energy of a system decreases during a chemical reaction, the difference is transferred to the surroundings in some form (often heat or light); on the other hand if the chemical energy of a system increases as a result of a chemical reaction the difference then is supplied by the surroundings (usually again in form of heat or light). For example, when two hydrogen atoms react to form a dihydrogen molecule, the chemical energy decreases by 724 zJ (the bond energy of the H—H bond); when the electron is completely removed from a hydrogen atom, forming a hydrogen ion (in the gas phase), the chemical energy increases by 2.18 aJ (the ionisation energy of hydrogen). It is common to quote the changes in chemical energy for one mole of the substance in question: typical values for the change in molar chemical energy during a chemical reaction range from tens to hundreds of kilojoules per mole. Table 3.1 listed the examples of the interconversion of energy.

Table 3.1. The examples of the interconversion of energy.

Chemical energy is converted	
Into	By
Mechanical energy	Muscle
Thermal energy	Fire
Electric energy	Fuel cell
Electromagnetic radiation	Glowworms
Chemical energy	Chemical reaction

The chemical energy as defined above is also referred to by chemists as the internal energy, U: technically, this is measured by keeping the volume of the system constant. However, most practical chemistry is performed at constant pressure and, if the volume (V) changes during the reaction (e.g. a gas is given off), a correction must be applied to take account of the work done by or on the atmosphere to obtain the enthalpy, H:

$$\Delta H = \Delta U + p\Delta V$$

A second correction, for the change in entropy, S, must also be performed to determine whether a chemical reaction will take place or not, giving the Gibbs free energy, G:

$$\Delta G = \Delta H - T\Delta S$$

These corrections are sometimes negligible, but often not (especially in reactions involving gases).

Since the industrial revolution, the burning of coal, oil, natural gas or products derived from them has been a socially significant transformation of chemical energy into other forms of energy, the energy consumption (one should really speak of energy transformation) of a society or country is often quoted in reference to the average energy released by the combustion of these fossil fuels:

1 ton of coal equivalent (TCE) = 29 GJ

1 ton of oil equivalent (TOE) = 41.87 GJ

RENEWABLE ENERGY

Renewable energy is energy generated from natural resources—such as sunlight, wind, rain, tides and geothermal heat—which are renewable (naturally replenished). In 2006, about 18 per cent of global final energy consumption came from renewables, with 13 per cent coming from traditional biomass, such as wood-burning. Hydroelectricity was the next largest renewable source, providing 3 per cent (15 per cent of global electricity generation), followed by solar hot water/heating, which contributed 1.3 per cent. Modern technologies, such as geothermal energy, wind power, solar power, and ocean energy together provided some 0.8 per cent of final energy consumption.

Climate change concerns coupled with high oil prices, peak oil and increasing government support are driving increasing renewable energy legislation, incentives and commercialisation.

Wind power is growing at the rate of 30 per cent annually, with a worldwide installed capacity of over 100 GW, and is widely used in several European countries and the United States. The manufacturing output of the photovoltaics industry reached more than 2000 MW in 2006, and photovoltaic (PV) power stations are particularly popular in Germany and Spain. Solar thermal power stations operate in the USA and Spain, and the largest of these is the 354 MW SEGS power plant in the Mojave desert. The world's largest geothermal power installation is The Geysers in California, with a rated capacity of 750 MW. Brazil has one of the largest renewable energy programmes in the world, involving production of ethanol fuel from sugar cane, and ethanol now provides 18 per cent of the country's automotive fuel. Ethanol fuel is also widely available in the USA.

While there are many large-scale renewable energy projects and production, renewable technologies are also suited to small off-grid applications, sometimes in rural and remote areas, where energy is often crucial in human development. Kenya has the world's highest household solar ownership rate with roughly 30,000 small (20–100 watt) solar power systems sold per year.

Some renewable energy technologies are criticised for being intermittent or unsightly, yet the market is growing for many forms of renewable energy. In response to the G8's call on the IEA for 'guidance on how to achieve a clean, clever and competitive energy future', the IEA reported that the replacement of current technology with renewable energy could help reduce CO_2 emissions by 50 per cent by 2050.

Main Renewable Energy Technologies

The majority of renewable energy technologies are powered by the sun. The earth-atmosphere system is in equilibrium such that heat radiation into space is equal to incoming solar radiation, the resulting level of energy within the earth-atmosphere system can roughly be described as the earth's climate. The hydrosphere (water) absorbs a major fraction of the incoming radiation. Most radiation is absorbed at low latitudes around the equator, but this energy is dissipated around the globe in the form of winds and ocean currents. Wave motion may play a role in the process of transferring mechanical energy between the atmosphere and the ocean through wind stress. Solar energy is also responsible for the distribution of precipitation which is tapped by hydroelectric projects, and for the growth of plants used to create biofuels.

Wind power

Airflows can be used to run wind turbines. Modern wind turbines range from around 600 kW to 5 MW of rated power, although turbines with rated output of 1.5–3 MW have become the most common for commercial use; the power output of a turbine is a function of the cube of the wind speed, so as wind speed increases, power output increases dramatically. Areas where winds are stronger and more constant, such as offshore and high altitude sites, are preferred locations for wind farms.

Since wind speed is not constant, a wind farm's annual energy production is never as much as the sum of the generator nameplate ratings multiplied by the total hours in a year. The ratio of actual productivity in a year to this theoretical maximum is called the capacity factor. Typical capacity factors are 20–40 per cent, with values at the upper end of the range in particularly favourable sites. For example, a 1 MW turbine with a capacity factor of 35 per cent will not produce 8760 MWh in a year, but only $0.35 \times 24 \times 365 = 3066$ MWh, averaging to 0.35 MW.

Globally, the long-term technical potential of wind energy is believed to be five times total current global energy production, or 40 times current electricity demand. This could require large amounts of land to be used for wind turbines, particularly in areas of higher wind resources. Offshore resources experience mean wind speeds of 90 per cent greater than that of land, so offshore resources could contribute substantially more energy. This number could also increase with higher altitude ground-based or airborne wind turbines. Wind power is renewable and produces no greenhouse gases during operation, such as carbon dioxide and methane.

Water power

Energy in water (in the form of kinetic energy, temperature differences or salinity gradients) can be harnessed and used. Since water is about 800 times denser than air, even a slow flowing stream of water, or moderate sea swell, can yield considerable amounts of energy.

There are many forms of water energy:

1. Hydroelectric energy is a term usually reserved for large-scale hydroelectric dams.

2. Micro hydro systems are hydroelectric power installations that typically produce up to 100 kW of power. They are often used in water rich areas as a remote area power supply (RAPS). There are many of these installations around the world, including several delivering around 50 kW in the Solomon Islands.

3. Damless hydro systems derive kinetic energy from rivers and oceans without using a dam.

4. Ocean energy describes all the technologies to harness energy from the ocean and the sea.

5. Marine current power, similar to tidal stream power, uses the kinetic energy of marine currents.

6. Ocean thermal energy conversion (OTEC) uses the temperature difference between the warmer surface of the ocean and the colder lower recesses. To this end, it employs a cyclic heat engine.

7. Tidal power captures energy from the tides. Two different principles for generating energy from the tides are used at the moment:

   (a) Tidal motion in the vertical direction: Tides come in, raise water levels in a basin, and tides roll out. Around low tide, the water in the basin is discharged through a turbine, exploiting the stored potential energy.

   (b) Tidal motion in the horizontal direction or tidal stream power. Using tidal stream generators, like wind turbines but then in a tidal stream. Due to the high density of water, about eight hundred times the density of air, tidal currents can have a lot of kinetic energy. Several commercial prototypes have been build, and more are in development.

8. Wave power uses the energy in waves. Wave power machines usually take the form of floating or neutrally buoyant structures which move relative to one another or to a fixed point. Wave power has now reached commercialisation.

9. Saline gradient power, or osmotic power, is the energy retrieved from the difference in the salt concentration between seawater and river water.

10. Vortex power is generated by placing obstacles in rivers in order to cause the formation of vortices which can then be tapped for energy.

11. Deep lake water cooling, although not technically an energy generation method, can save a lot of energy in summer. It uses submerged pipes as a heat sink for climate control systems. Lake-bottom water is a year-round local constant of about 4°C.

Solar energy

Solar energy refers to energy that is collected from sunlight. Solar energy can be applied in many ways, including to:

1. Generate electricity using photovoltaic solar cells.
2. Generate electricity using concentrated solar power.
3. Generate electricity by heating trapped air which rotates turbines in a solar updraft tower.
4. Generate electricity in geosynchronous orbit using solar power satellites.
5. Generate hydrogen using photoelectrochemical cells.
6. Heat and cool air through use of solar chimneys.
7. Heat buildings, directly, through passive solar building design.
8. Heat foodstuffs, through solar ovens.
9. Heat water or air for domestic hot water and space heating needs using solar-thermal panels.
10. Solar air conditioning.

Biofuel

Plants use photosynthesis to grow and produce biomass. Also known as biomatter, biomass can be used directly as fuel or to produce biofuels. Agriculturally produced biomass fuels, such as biodiesel, ethanol and bagasse (often a by-product of sugarcane cultivation) can be burned in internal combustion engines or boilers. Typically biofuel is burned to release its stored chemical energy. Research into more efficient methods of converting biofuels and other fuels into electricity utilising fuel cells is an area of very active work.

Liquid biofuel

Liquid biofuel is usually either a bioalcohol such as ethanol fuel or an oil such as biodiesel or straight vegetable oil. Biodiesel can be used in modern diesel vehicles with little or no modification to the engine. It can be made from waste and virgin vegetable and animal oils and fats (lipids). Virgin vegetable oils can be used in modified diesel engines. In fact the diesel engine was originally designed to run on vegetable oil rather than fossil fuel. A major benefit of biodiesel use is the reduction in net CO_2 emissions, since all the carbon emitted was recently captured during the growing phase of the biomass. The use of biodiesel also reduces emission of carbon monoxide and other pollutants by 20 to 40 per cent.

In some areas corn, cornstalks, sugarbeets, sugarcane, and switchgrasses are grown specifically to produce ethanol (also known as grain alcohol) a liquid which can be used in internal combustion engines and fuel cells. Ethanol is being phased into the current energy infrastructure. E85 is a fuel composed of 85 per cent ethanol and 15 per cent gasoline that is sold to consumers. Biobutanol is being developed as an alternative to bioethanol.

Solid biomass

Solid biomass is usually used directly as a combustible fuel, producing 10–20 MJ/kg of heat. Its forms and sources include wood fuel, the biogenic portion of municipal solid waste, or the unused portion of field crops. Field crops may or may not be grown intentionally as an energy crop, and the remaining plant by-product used as a fuel. Most types of biomass contain energy. Even cow manure still contains two-thirds of the original energy consumed by the cow. Energy harvesting via a bioreactor is a cost-effective solution to the waste disposal issues faced by the dairy farmer, and can produce enough biogas to run a farm.

With current technology, it is not ideally suited for use as a transportation fuel. Most transportation vehicles require power sources with high power density, such as that provided by internal combustion engines. These engines generally require clean burning fuels, which are generally in liquid form, and to a lesser extent, compressed gaseous phase. Liquids are more portable because they can have a high energy density, and they can be pumped, which makes handling easier.

Biogas

Biogas can easily be produced from current waste streams, such as paper production, sugar production, sewage, animal waste and so forth. These various waste streams have to be slurried together and allowed to naturally ferment, producing methane gas. This can be done by converting current sewage plants into biogas plants. When a biogas plant has extracted all the methane it can, the remains are sometimes more suitable as fertiliser than the original biomass.

Alternatively biogas can be produced via advanced waste processing systems such as mechanical biological treatment. These systems recover the recyclable elements of household waste and process the biodegradable fraction in anaerobic digesters.

Renewable natural gas is a biogas which has been upgraded to a quality similar to natural gas. By upgrading the quality to that of natural gas, it becomes possible to distribute the gas to the mass market via the existing gas grid.

Geothermal energy

Geothermal energy is energy obtained by tapping the heat of the earth itself, usually from kilometers deep into the earth's crust. It is expensive to build a power station but operating costs are low resulting in low energy costs for suitable sites. Ultimately, this energy derives from heat in the earth's core.

Three types of power plants are used to generate power from geothermal energy: dry steam, flash, and binary. Dry steam plants take steam out of fractures in the ground and use it to directly drive a turbine that spins a generator. Flash plants take hot water, usually at temperatures over 200°C, out of the ground, and allows it to boil as it rises to the surface then separates the steam phase in steam/water separators and then runs the steam through a turbine. In binary plants, the hot water flows through heat exchangers, boiling an organic fluid that spins the turbine. The condensed steam and remaining geothermal fluid from all three types of plants are injected back into the hot rock to pick up more heat.

The geothermal energy from the core of the earth is closer to the surface in some areas than in others. Where hot underground steam or water can be tapped and brought to the surface it may be used to generate electricity. Such geothermal power sources exist in certain geologically unstable parts of the world such as Chile, Iceland, New Zealand, United States, the Philippines and Italy. The two most prominent areas for this in the United States are in the Yellowstone basin and in northern California. Iceland produced 170 MW geothermal power and heated 86 per cent of all houses in the year 2005 through geothermal energy. Some 8000 MW of capacity is operational in total.

There is also the potential to generate geothermal energy from hot dry rocks. Holes at least 3 km deep are drilled into the earth. Some of these holes pump water into the earth, while other holes pump hot water out. The heat resource consists of hot underground radiogenic granite rocks, which heat up when there is enough sediment between the rock and the earths surface. Several companies in Australia are exploring this technology.

While most renewable energy sources do not produce pollution directly, the materials, industrial processes, and construction equipment used to create them may generate waste and pollution. Some renewable energy systems actually create environmental problems.

AIR CONDITIONING

The term air conditioning refers to the cooling and dehumidification of indoor air for thermal comfort. In a broader sense, the term can refer to any form of cooling, heating, ventilation or disinfection that modifies the condition of air. An air conditioner (often referred to as AC) is an appliance, system, or mechanism designed to stabilise the air temperature and humidity within an area (used for cooling as well as heating depending on the air properties at a given time), typically using a refrigeration cycle but sometimes using evaporation, most commonly for comfort cooling in buildings and motor-cars.

Air Conditioning Applications

Air conditioning engineers broadly divide air conditioning applications into comfort and process. Comfort applications aim to provide a building indoor environment that remains relatively constant in a range preferred by humans despite changes in external weather conditions or in internal heat loads. Air conditioning makes deep plan buildings feasible, for otherwise they'd have to be be built narrower or with light wells so that inner spaces receive sufficient outdoor air via natural ventilation. Air conditioning also allows buildings to be taller since wind speed increases significantly with altitude making natural ventilation impractical for very tall buildings. In addition to buildings, air conditioning can be used for many types of transportation—motor-cars and other land vehicles, trains, ships, aircraft, and spacecraft.

Process applications aim to provide a suitable environment for a process being carried out, regardless of internal heat and humidity loads and external weather conditions. Although often in the comfort range, it is the need of the process that determine conditions, not human preference. Control of the temperature, humidity and cleanliness is very important in many chemical processes, particulary in artificial-fibre and paper manufacture. Textile fibres are also important to efficient industrial organisations.

Cleanrooms for the production of integrated circuits, pharmaceuticals, and the like, in which very high levels of air cleanliness and control of temperature and humidity are required for the success of the process. In both comfort and process applications the objective may be to not only control temperature, but also humidity, air quality and air movement from space to space.

Humidity Control

Refrigeration air conditioning equipment usually reduces the humidity of the air processed by the system. The relatively cold (below the dewpoint) evaporator coil condenses water vapour from the processed air, (much like an ice-cold drink will condense water on the outside of a glass), sending the water to a drain and removing water vapour from the cooled space and lowering the relative humidity. Since humans perspire to provide natural cooling by the evaporation of perspiration from the skin, drier air (up to a point) improves the comfort provided. The comfort air conditioner is designed to create a 40 to 60 per cent relative humidity in the occupied space. In food retailing establishments large open chiller cabinets act as highly effective air dehumidifying units.

A specific type of air conditioner that is used only for dehumidifying is called a dehumidifier. A dehumidifier is different from a regular air conditioner in that both the evaporator and condenser coils are placed in the same air path, and the entire unit is placed in the environment that is intended to be conditioned (in this case dehumidified), rather than requiring the condenser coil to be outdoors. Having the condenser coil in the same air path as the evaporator coil produces warm, dehumidified air. The evaporator (cold) coil is placed first in the air path, dehumidifying the air exactly as a regular air conditioner does.

The air next passes over the condenser coil re-warming the now dehumidified air. Note that the terms 'condenser coil' and 'evaporator coil' do not refer to the behaviour of water in the air as it passes over each coil; instead they refer to the phases of the refrigeration cycle. Having the condenser coil in the main air path rather than in a separate, outdoor air path (as in a regular air conditioner) results in two consequenses—the output air is warm rather than cold, and the unit is able to be placed anywhere in the environment to be conditioned, without a need to have the condenser outdoors.

Unlike a regular air conditioner, a dehumidifier will actually heat a room just as an electric heater that draws the same amount of power (watts) as the dehumidifier. A regular air conditioner transfers energy out of the room by means of the condenser coil, which is outside the room (outdoors). This is a thermodynamic system where the room serves as the system and energy is transferred out of the system. Conversely with a dehumidifier, no energy is transferred out of the thermodynamic system (room) because the air conditioning unit (dehumidifier) is entirely inside the room. Therefore all of the power consumed by the dehumidifier is energy that is input into the thermodynamic system (the room), and remains in the room (as heat).

In addition, if the condensed water has been removed from the room, the amount of heat needed to boil that water has been added to the room. This is the inverse of adding water to the room with an evaporative cooler. Dehumidifiers are commonly used in cold, damp climates to prevent mould growth indoors, especially in basements. They are also sometimes used in hot, humid climates for comfort because they reduce the humidity which causes discomfort (just as a regular air conditioner, but without cooling the room). The engineering of physical and thermodynamic properties of gas-vapour mixtures is named Psychrometrics.

Health Implications

A poorly maintained air-conditioning system can occasionally promote the growth and spread of microorganisms, such as *Legionella pneumophila*, the infectious agent responsible for Legionnaires' disease, or thermophilic actinomycetes, but as long as the air conditioner is kept clean these health hazards can be avoided.

Conversely, air conditioning, including filtration, humidification, cooling, disinfection, etc. can be used to provide a clean, safe, hypoallergenic atmosphere in hospital operating rooms and other environments where an appropriate atmosphere is critical to patient safety and well-being. Air conditioning can have a positive effect on sufferers of allergies and asthma.

Energy Use

It should be noted that in a thermodynamically closed system, any energy input into the system that is being maintained at a set temperature (which is a standard mode of operation for modern air conditioners) requires that the energy removal rate from the air conditioner increase. This increase has the effect that for each unit of energy input into the system (say to power a light bulb in the closed system) requires the air conditioner to remove that energy. In order to do that the air conditioner must increase its consumption by the inverse of its efficiency times the input unit of energy. As an example, presume that inside the closed system a 100 watt light bulb is activated, and the air conditioner has an efficiency of 200 per cent. The air conditioner's energy consumption will increase by 50 watts to compensate for this, thus making the 100 W light bulb utilise a total of 150 W of energy. Note that it is typical for air conditioners to operate at efficiencies of significantly greater than 100 per cent.

REFRIGERATION

Refrigeration is the process of removing heat from an enclosed space, or from a substance, and moving it to a place where it is unobjectionable. The primary purpose of refrigeration is lowering the temperature of the enclosed space or substance and then maintaining that lower temperature. The term cooling refers generally to any natural or artificial process by which heat is dissipated. The process of artificially producing extreme cold temperatures is referred to as cryogenics.

Cold is the absence of heat, hence in order to decrease a temperature, one removes heat, rather than adding cold. In order to satisfy the second law of thermodynamics, some form of work must be performed to accomplish this. This work is traditionally done by mechanical work but can also be done by magnetism, laser or other means.

It is a vital factor in many chemical processes where cold or the removal of heat is necessary for optimum reaction control. Examples are the manufacture of azo dyes, the separation of an easily frozen product from liquid isomers or impurities, and the food and beverage industries. Further examples are the catalytic manufacture of ethyl chloride from liquid ethylene and anhydrous hydrogen chloride under pressure and at $-5°C$, the production of cold rubber by polymerisation at $41°F$ or lower, and the freezing of mercury at $-100°F$ into complex moulds which are coated by repeated dipping in a ceramic slurry, the mercury then being allowed the melt and run out. Refrigeration operations involve a change in phase in a body so that it will be capable of abstracting heat, exemplified by the vapourisation of liquid ammonia and the melting of ice. Mechanical refrigeration can be divided into two general types: the compression system and the absorption system. Both systems cause the refrigerant to absorb heat at the low temperature by vapourisation and give up this heat to the higher temperature by condensation. The absorption system is used mainly in household units, but it finds economical industrial application where exhaust steam is available. Table 3.2 gives properties of usual refrigerating agents.

Table 3.2. Properties of refrigerating agents.

Refrigerant	Boiling point at 760 mm, F	Critical temperature, F	Critical pressure, atm.	Latent heat at 760 mm, Btu/lb
Ammonia (NH₃)	−28	270.5	111.5	589
Carbon dioxide (CO₂)	−108.8*	88.0	73.0	126†
Sulphur dioxide (SO₂)	+14	315.0	77.7	167
Methyl chloride (CH₃Cl)	−11	289.6	65.8	184
Ethyl chloride (C₂H₅Cl)	+55	369.0	52.0	168
Freon-12 (CCl₂F₂)	−21	232.7	39.6	71.9
Propane (C₃H₈)	−44.4	206.2	42.0	159

*Sublimes. †Latent heat −20°F and 220.6 psia.

Current Applications of Refrigeration

In commerce and manufacturing, there are many uses for refrigeration. Refrigeration is used to liquify gases like oxygen, nitrogen, propane and methane for example. In compressed air purification, it is used to condense water vapour from compressed air to reduce its moisture content. In oil refineries, chemical plants, and petrochemical plants, refrigeration is used to maintain certain processes at their required low temperatures (for example, in the alkylation of butenes and butane to produce a high octane gasoline component). Metal workers use refrigeration to temper steel and cutlery. In transporting temperature sensitive foodstuffs and other materials by trucks, trains, airplanes and sea-going vessels, refrigeration is a necessity.

Dairy products are constantly in need of refrigeration, and it was only discovered in the past few decades that eggs needed to be refrigerated during shipment rather than waiting to be refrigerated after arrival at the grocery store. Meats, poultry and fish all must be kept in climate-controlled environments before being sold. Refrigeration also helps keep fruits and vegetables edible longer.

Methods of Refrigeration

Methods of refrigeration can be classified as non-cyclic, cyclic and thermoelectric.

Non-cyclic refrigeration

In these methods, refrigeration can be accomplished by melting ice or by subliming dry ice. These methods are used for small-scale refrigeration such as in laboratories and workshops, or in portable coolers.

Ice owes its effectiveness as a cooling agent to its constant melting point of 0°C (32°F). In order to melt, ice must absorb 333.55 kJ/kg (approximately 144 Btu/lb) of heat. Foodstuffs maintained at this temperature or slightly above have an increased storage life. Solid carbon dioxide, known as dry ice, is used also as a refrigerant. Having no liquid phase at normal atmospheric pressure, it sublimes directly from the solid to vapour phase at a temperature of –78.5°C (–109.3°F). Dry ice is effective for maintaining products at low temperatures during the period of sublimation.

Cyclic refrigeration

This consists of a refrigeration cycle, where heat is removed from a low-temperature space or source and rejected to a high-temperature sink with the help of external work, and its inverse, the thermodynamic power cycle. In the power cycle, heat is supplied from a high-temperature source to the engine, part of the heat being used to produce work and the rest being rejected to a low-temperature sink. This satisfies the second law of thermodynamics.

A refrigeration cycle describes the changes that take place in the refrigerant as it alternately absorbs and rejects heat as it circulates through a refrigerator. It is also applied to HVACR work, when describing the 'process' of refrigerant flow through an HVACR unit, whether it is a packaged or split system.

Heat naturally flows from hot to cold. Work is applied to cool a living space or storage volume by pumping heat from a lower temperature heat source into a higher temperature heat sink. Insulation is used to reduce the work and energy required to achieve and maintain a lower temperature in the cooled space.

The most common types of refrigeration systems use the reverse-Rankine vapour-compression refrigeration cycle although absorption heat pumps are used in a minority of applications.

Cyclic refrigeration can be classified as:
1. Vapour cycle.
2. Gas cycle.

Vapour cycle refrigeration can further be classified as:
1. Vapour compression refrigeration.
2. Vapour absorption refrigeration.

Vapour compression cycle

The vapour compression cycle is used in most household refrigerators as well as in many large commercial and industrial refrigeration systems. Figure 3.1 provides a schematic diagram of the components of a typical vapour-compression refrigeration system.

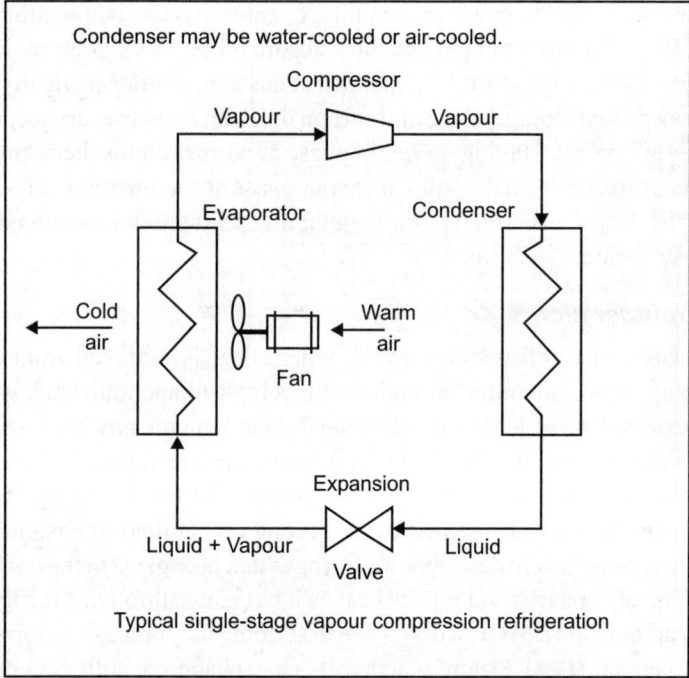

Fig. 3.1. Vapour compression refrigeration.

The thermodynamics of the cycle can be analysed on a diagram as shown in Fig. 3.2. In this cycle, a circulating refrigerant such as Freon enters the compressor as a vapour. From point 1 to point 2, the vapour is compressed at constant entropy and exits the compressor superheated. From point 2 to 3 and on to point 4, the superheated vapour travels through the condenser which first cools and removes the superheat and then condenses the vapour into a liquid by removing additional heat at constant pressure and temperature. Between points 4 and 5, the liquid refrigerant goes through the expansion valve (also called a throttle valve) where its pressure abruptly decreases, causing flash evaporation and auto-refrigeration of typically, less than half of the liquid. That results in a mixture of liquid and vapour at a lower temperature and pressure as shown at point 5.

The cold liquid-vapour mixture then travels through the evaporator coil or tubes and is completely vapourised by cooling the warm air (from the space being refrigerated) being blown by a fan across the evaporator coil or tubes. The resulting refrigerant vapour returns to the compressor inlet at point 1 to complete the thermodynamic cycle.

The above discussion is based on the ideal vapour-compression refrigeration cycle, and does not take into account real-world effects like frictional pressure drop in the system, slight thermodynamic irreversibility during the compression of the refrigerant vapour, or non-ideal gas behaviour (if any).

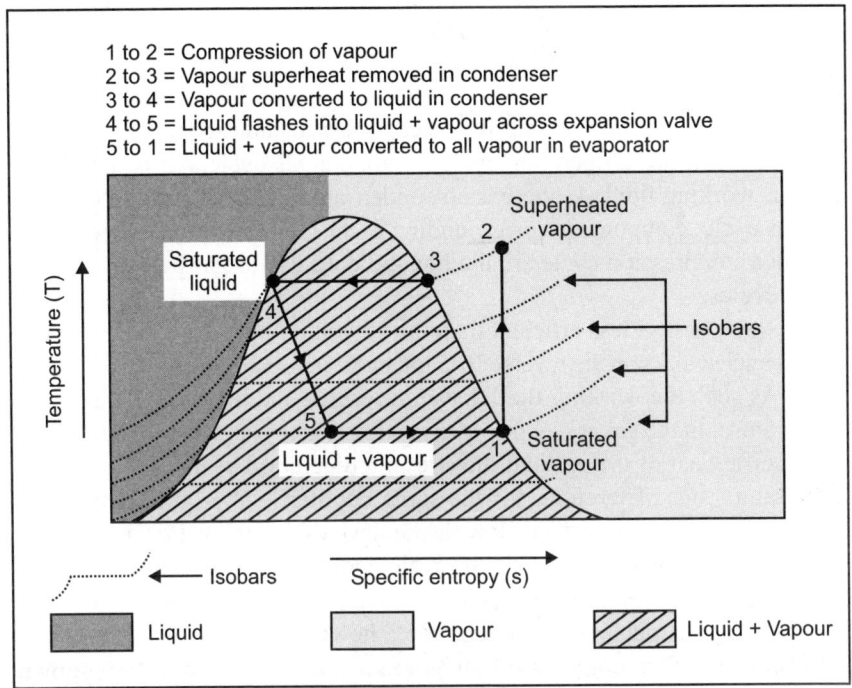

1 to 2 = Compression of vapour
2 to 3 = Vapour superheat removed in condenser
3 to 4 = Vapour converted to liquid in condenser
4 to 5 = Liquid flashes into liquid + vapour across expansion valve
5 to 1 = Liquid + vapour converted to all vapour in evaporator

Fig. 3.2. Temperature-entropy diagram.

Vapour absorption cycle

In the early years of the twentieth century, the vapour absorption cycle using water-ammonia systems was popular and widely used. After the development of the vapour compression cycle, the vapour absorption cycle lost much of its importance because of its low coefficient of performance (about one fifth of that of the vapour compression cycle). Today, the vapour absorption cycle is used mainly where fuel for heating is available but electricity is not, such as in recreational vehicles that carry LP gas. It is also used in industrial environments where plentiful waste heat overcomes its inefficiency.

The absorption cycle is similar to the compression cycle, except for the method of raising the pressure of the refrigerant vapour. In the absorption system, the compressor is replaced by an absorber which dissolves the refrigerant in a suitable liquid, a liquid pump which raises the pressure and a generator which, on heat addition, drives off the refrigerant vapour from the high-pressure liquid. Some work is required by the liquid pump but, for a given quantity of refrigerant, it is much smaller than needed by the compressor in the vapour compression cycle. In an absorption refrigerator, a suitable combination of refrigerant and absorbent is used. The most common combinations are ammonia (refrigerant) and water (absorbent), and water (refrigerant) and lithium bromide (absorbent).

Gas cycle

When the working fluid is a gas that is compressed and expanded but doesn't change phase, the refrigeration cycle is called a gas cycle. Air is most often like this working fluid. As there is no condensation and evaporation intended in a gas cycle, components corresponding to the condenser and evaporator in a vapour compression cycle are the hot and cold gas-to-gas heat exchangers in gas cycles.

The gas cycle is less efficient than the vapour compression cycle because the gas cycle works on the reverse Brayton cycle instead of the reverse Rankine cycle. As such the working fluid does not receive and reject heat at constant temperature. In the gas cycle, the refrigeration effect is equal to the product of the specific heat of the gas and the rise in temperature of the gas in the low temperature side. Therefore, for the same cooling load, a gas refrigeration cycle will require a large mass flow rate and would be bulky. Because of their lower efficiency and larger bulk, air cycle coolers are not often used now-a-days in terrestrial cooling devices.

The air cycle machine is very common, however, on gas turbine-powered jet aircraft because compressed air is readily available from the engines' compressor sections. These jet aircraft's cooling and ventilation units also serve the purpose of pressurising the aircraft.

Thermoelectric refrigeration

Thermoelectric cooling uses the Peltier effect to create a heat flux between the junction of two different types of materials. This effect is commonly used in camping and portable coolers and for cooling electronic components and small instruments.

Magnetic refrigeration

Magnetic refrigeration, or adiabatic demagnetisation, is a cooling technology based on the magnetocaloric effect, an intrinsic property of magnetic solids. The refrigerant is often a paramagnetic salt, such as cerium magnesium nitrate. The active magnetic dipoles in this case are those of the electron shells of the paramagnetic atoms.

A strong magnetic field is applied to the refrigerant, forcing its various magnetic dipoles to align and putting these degrees of freedom of the refrigerant into a state of lowered entropy. A heat sink then absorbs the heat released by the refrigerant due to its loss of entropy. Thermal contact with the heat sink is then broken so that the system is insulated, and the magnetic field is switched off. This increases the heat capacity of the refrigerant, thus decreasing its temperature below the temperature of the heat sink.

Because few materials exhibit the required properties at room temperature, applications have so far been limited to cryogenics and research.

Other methods

Other methods of refrigeration include the air cycle machine used in aircraft; the vortex tube used for spot cooling, when compressed air is available; and thermoacoustic refrigeration using sound waves in a pressurised gas to drive heat transfer and heat exchange.

Unit of Refrigeration

Domestic and commercial refrigerators may be rated in kJ/s, or Btu/hr of cooling. Commercial refrigerators in the US are mostly rated in tons of refrigeration, but elsewhere in kW. One ton of refrigeration capacity can freeze one short ton of water at 0°C (32°F) in 24 hours. Based on that:

> Latent heat of ice (i.e. heat of fusion) = 333.55 kJ/kg ≈ 144 Btu/lb
> One short ton = 2000 lb
> Heat extracted = (2000)(144)/24 hr = 288000 Btu/24 hrs = 12000 Btu/hr
> = 200 Btu/min
> 1 ton refrigeration = 200 Btu/min = 3.517 kJ/s = 3.517 kW

A much less common definition is: 1 ton of refrigeration is the rate of heat removal required to freeze a metric ton (i.e. 1000 kg) of water at 0°C in 24 hours.

Based on the heat of fusion being 333.55 kJ/kg, 1 ton of refrigeration = 13,898 kJ/hr = 3.861 kW. As can be seen, 1 ton of refrigeration is 10 per cent larger than 1 ton of refrigeration.

Most residential air conditioning units range in capacity from about 1 to 5 tons of refrigeration.

Chapter 4

Coal Chemicals

INTRODUCTION

Chemicals from coal were initially and mostly obtained by destructive distillation of coal, furnishing chiefly aromatics. In recent years substantial production of aromatics, particularly benzene, toluene, xylene, naphthalene, and methylnaphthalenes, has been obtained by processing petrochemicals. With the advancing application of chemical conversion of coal, many more chemicals can be made from coal whenever it is economical to do so. However, coal chemicals, except for metallurgical coke, are now in a very competitive field.

COAL

Coal is a fossil fuel formed in ecosystems where plant remains were preserved by water and mud from oxidisation and biodegradation, and which its chemical and physical properties have been changed as a result of geological action overtime, thus sequestering atmospheric carbon. Coal is a readily combustible black or brownish-black rock. It is a sedimentary rock, but the harder forms, such as anthracite coal, can be regarded as metamorphic rock because of later exposure to elevated temperature and pressure. It is composed primarily of carbon and hydrogen along with small quantities of other elements, notably sulphur. Coal is extracted from the ground by coal mining, either underground mining or open pit mining (surface mining). Coal is the largest source of fuel for the generation of electricity worldwide, as well as the largest worldwide source of carbon dioxide emissions. Carbon dioxide is a greenhouse gas and the major contributor to an increase in global average temperature and related climate changes. Gross carbon dioxide emissions from coal usage is slightly more than that from petroleum and about double the amount from natural gas.

Types of Coal

As geological processes apply pressure to dead biotic matter overtime, under suitable conditions it is transformed successively into:

1. Peat, considered to be a precursor of coal, has industrial importance as a fuel in some countries.

2. Lignite, also referred to as brown coal, is the lowest rank of coal and used almost exclusively as fuel for electric power generation.

3. Sub-bituminous coal, whose properties range from those of lignite to those of bituminous coal and are used primarily as fuel for steam-electric power generation. Additionally, it is an important source of light aromatic hydrocarbons for the chemical synthesis industry.

4. Bituminous coal, dense mineral, black but sometimes dark brown, often with well-defined bands of bright and dull material, used primarily as fuel in steam-electric power generation, with substantial quantities also used for heat and power applications in manufacturing and to make coke.

5. Anthracite, the highest rank; a harder, glossy, black coal used primarily for residential and commercial space heating. It may be divided further into metamorphically altered bituminous coal and petrified oil, as from the deposits in Pennsylvania.

6. Graphite, technically the highest rank, but difficult to ignite and is not so commonly used as fuel: it is mostly used in pencils and, when powdered, as a lubricant.

Applications of Coal

Coal as fuel

Coal is primarily used as a solid fuel to produce electricity and heat through combustion.

When coal is used for electricity generation, it is usually pulverised and then burned in a furnace with a boiler. The furnace heat converts boiler water to steam, which is then used to spin turbines which turn generators and create electricity. The thermodynamic efficiency of this process has been improved over time. Standard steam turbines have topped out with some of the most advanced reaching about 35 per cent thermodynamic efficiency for the entire process, which means 65 per cent of the coal energy is waste heat released into the surrounding environment.

Other efficient ways to use coal are combined cycle power plants, combined heat and power cogeneration, and an MHD topping cycle.

The total known deposits recoverable by current technologies, including highly polluting, low energy content types of coal (i.e. lignite, bituminous), might be sufficient for 300 years use at current consumption levels, although maximal production could be reached within decades.

A more energy-efficient way of using coal for electricity production would be via solid oxide fuel cells or molten-carbonate fuel cells (or any oxygen ion transport based fuel cells that do not discriminate between fuels, as long as they consume oxygen), which would be able to get 60–85 per cent combined efficiency (direct electricity + waste heat steam turbine). Currently these fuel cell technologies can only process gaseous fuels, and they are also sensitive to sulphur poisoning, issues which would first have to be worked out before large scale commercial success is possible with coal. As far as gaseous fuels go, one idea is pulverised coal in a gas carrier, such as nitrogen. Another option is coal gasification with water, which may lower fuel cell voltage by introducing oxygen to the fuel side of the electrolyte, but may also greatly simplify carbon sequestration. However, this technology has been criticised as being inefficient, slow, risky and costly, while doing nothing about total emissions from mining, processing and combustion. Another efficient and clean way of coal combustion in a form of coal-water slurry (CWS) fuel which significantly reduces emissions saving the heating value of coal.

Coking and use of coke

Coke is a solid carbonaceous residue derived from low-ash, low-sulphur bituminous coal from which the volatile constituents are driven off by baking in an oven without oxygen at temperatures as high as 1000°C (1832°F) so that the fixed carbon and residual ash are fused together. Metallurgic coke is used as a fuel and as a reducing agent in smelting iron ore in a blast furnace. Coke from coal is grey, hard, and porous and has a heating value of 24.8 million Btu/ton (29.6 MJ/kg). Some coke-making processes produce valuable by-products that include coal tar, ammonia, light oils and 'coal gas'. Petroleum coke is the solid residue obtained in oil refining, which resembles coke but contains too many impurities to be useful in metallurgical applications.

Gasification

Coal gasification can be used to produce syngas, a mixture of carbon monoxide (CO) and hydrogen (H_2) gas. This syngas can then be converted into transportation fuels like gasoline and diesel through the Fischer-Tropsch process. Currently, this technology is being used by the Sasol chemical company of South Africa to make gasoline from coal and natural gas. Alternatively, the hydrogen obtained from gasification can be used for various purposes such as powering a hydrogen economy, making ammonia, or upgrading fossil fuels.

During gasification, the coal is mixed with oxygen and steam (water vapour) while also being heated and pressurised. During the reaction, oxygen and water molecules oxidise the coal into carbon monoxide (CO) while also releasing hydrogen (H_2) gas. This process has been conducted in both underground coal mines and in coal refineries.

$$(Coal) + O_2 + H_2O \rightarrow H_2 + CO$$

If the refiner wants to produce gasoline, the syngas is collected at this state and routed into a Fischer-Tropsch reaction. If hydrogen is the desired end-product, however, the syngas is fed into the water gas shift reaction where more hydrogen is liberated.

$$CO + H_2O \rightarrow CO_2 + H_2$$

High prices of oil and natural gas are leading to increased interest in 'Btu conversion' technologies such as gasification, methanation and liquefaction.

Liquefaction Coal-to-Liquids (CTL)

Coals can also be converted into liquid fuels like gasoline or diesel by several different processes. In the direct liquefaction processes, the coal is either hydrogenated or carbonised. Alternatively, coal can be converted into a gas first, and then into a liquid, by using the Fischer-Tropsch process.

The NUS Corporation developed another hydrogenation process which was patented by Wilburn C. Schroeder. The process involved dried, pulverised coal mixed with roughly 1 per cent by weight of molybdenum catalysts. Hydrogenation occurred by use of high temperature and pressure synthesis gas produced in a separate gasifier. The process ultimately yielded a synthetic crude product, naphtha, a limited amount of C_3/C_4 gas, light-medium weight liquids (C_5–C_{10}) suitable for use as fuels, small amounts of NH_3 and significant amounts of CO_2.

The process of low temperature carbonisation (LTC) can also convert coal into a liquid fuel. Coal is coked at temperatures between 450° and 700°C compared to 800° to 1000°C for metallurgical coke. These temperatures optimise the production of coal tars richer in lighter hydrocarbons than normal coal tar. The coal tar is then further processed into fuels.

All of these liquid fuel production methods release carbon dioxide (CO_2) in the conversion process, far more than is released in the extraction and refinement of liquid fuel production from petroleum. If these methods were adopted to replace declining petroleum supplies, carbon dioxide emissions would be greatly increased on a global scale. For future liquefaction projects, carbon dioxide sequestration is proposed to avoid releasing it into the atmosphere, though no pilot projects have confirmed the feasibility of this approach on a wide scale. As CO_2 is one of the process streams, sequestration is easier than from flue gases produced in combustion of coal with air, where CO_2 is diluted by nitrogen and other gases. Sequestration will, however, add to the cost. The reaction of coal and water using high temperature heat from a nuclear reactor offers promise of liquid transport fuels that could prove carbon-neutral compared to petroleum use. The development of a reliable nuclear reactor that could provide 900° to 1000°C process heat, such as the pebble bed reactor, would be necessary.

Refined Coal

Refined coal is the product of a coal upgrading technology that removes moisture and certain pollutants from lower-rank coals such as sub-bituminous and lignite (brown) coals. It is one form of several precombustion treatments and processes for coal that alter coal's characteristics before it is burned. The goals of precombustion coal technologies are to increase efficiency and reduce emissions when the coal is burned. Depending on the situation, pre-combustion technology can be used in place of or as a supplement to post-combustion technologies to control emissions from coal-fueled boilers.

Economic aspects

Coal liquefaction is one of the backstop technologies that could potentially limit escalation of oil prices and mitigate the effects of transportation energy shortage that some authors have suggested could occur under peak oil. This is contingent on liquefaction production capacity becoming large enough to satiate the very large and growing demand for petroleum.

Energy Density

The energy density of coal can also be expressed in kilowatt-hours for some unit of mass, the units that electricity is most commonly sold in, to estimate how much coal is required to power electrical appliances. One kilowatt-hour is 3.6 MJ, so the energy density of coal is 6.67 kW·h/kg. The typical thermodynamic efficiency of coal power plants is about 30 per cent, so of the 6.67 kW·h of energy per kilogram of coal, 30 per cent of that—2.0 kW·h/kg—can successfully be turned into electricity; the rest is waste heat. So coal power plants obtain approximately 2.0 kW·h per kilogram of burned coal.

BIOCHAR

Biochar is charcoal created by pyrolysis of biomass. The resulting charcoal-like material can be used as a soil improver to create terra-preta, and is a form of carbon capture and storage. Charcoal is a stable solid and rich in carbon content, and thus, can be used to lock carbon in the soil. Biochar is of increasing interest because of concerns about climate change caused by emissions of carbon dioxide and other greenhouse gases. Biochar is a way for carbon to be drawn from the atmosphere and is a solution to reducing the global impact of farming (and in reducing the impact from all agricultural waste). However, it is not a longterm solution, possessing carbon capture capabilities for a mere few hundred years. The burning and natural decomposition of trees and agricultural matter contributes a large amount of CO_2 released to the atmosphere. Biochar can store this carbon in the ground, potentially making a significant reduction in atmospheric green house gases (GHG) levels; at the same time its presence in the earth can improve water quality, increase soil fertility, raise agricultural productivity and reduce pressure on old growth forests.

Pyrolysis of Biomass as a Carbon Sink

Biochar can be used to sequester carbon on centurial or even millennial time scales. Plant matter absorbs CO_2 from the atmosphere while growing. In the natural carbon cycle, plant matter decomposes rapidly after the plant dies, which emits CO_2; the overall natural cycle is carbon neutral. Instead of allowing the plant matter to decompose, pyrolysis can be used to sequester the carbon in a much more stable form. Biochar thus removes circulating CO_2 from the atmosphere and stores it in virtually permanent soil carbon pools, making it a carbon-negative process. In places like the rocky mountains, where beetles have been killing off vast swathes of pine trees, the utilisation of pyrolysis to char the trees instead of letting them decompose into the atmosphere would offset substantial amounts of CO_2 emissions. Although some organic matter is necessary for agricultural soil to maintain its productivity, much of the agricultural waste can be turned directly into biochar, bio-oil, and syngas. The use of pyrolysis also provides an opportunity for the processing of municipal waste into useful clean energy rather than increased problems with land space for storage.

Biochar is believed to have long mean residence times in the soil. While the methods by which biochar mineralises (turns into CO_2) are not completely known. The amount of time the biochar will remain in the soil depends on the feedstock material, how charred the material is, the surface:volume ratio of the particles, and the conditions of the soil the biochar is placed in. Estimates for the residence time range from 100 to 10,000 years, with 5000 being a common estimate.

Laboratory experiments confirm a decrease in carbon mineralisation with increasing temperature, so carefully controlled charring of plant matter can increase the soil residence time of the biochar carbon. Under some circumstances, the addition of biochar to the soil has been found to accelerate the mineralisation of the existing soil organic matter, but this would only reduce and not suppress the net benefit gained by sequestering carbon in the soil by this method. Furthermore, the suggested soil conditions for the integration of biochar are in heavily degraded tropical soils used for agriculture, not organic matter-rich boreal forest soils.

Cobenefits of pyrolysis

Biochar can be used as a soil amendment to increase plant growth yield, improve water quality, reduce soil emissions of GHGs, reduce leaching of nutrients, reduce soil acidity, and reduce irrigation and fertiliser requirements.

These properties are very dependent on the properties of the biochar, and may depend on regional conditions including soil type, condition (depleted or healthy), temperature, and humidity. Modest additions of biochar to soil were found to reduce N_2O emissions by up to 80 per cent and completely suppress methane emissions.

Switching to slash-and-char can sequester up to 50 per cent of the carbon in a highly stable form. Adding the biochar back into the soil rather than removing it all for energy production is necessary to avoid heavy increases in the cost and emissions from more required nitrogen fertilisers. Additionally, by improving the soil tilth, fertility, and productivity, the biochar enhanced soils can sustain agricultural production, whereas non-amended soils quickly become depleted of nutrients, and the fields are abandoned, leading to a continuous slash-and-burn cycle and the continued loss of tropical rainforest. Using pyrolysis to produce bio-energy also has the added benefit of not requiring infrastructure changes the way processing biomass for cellulosic ethanol does.

Additionally, the biochar produced can be applied by the currently used tillage machinery or equipment used to apply fertilisers.

Pyrolysis for the production of energy

Bio-oil can be used as a replacement for numerous applications where fuel oil is used, including fuelling space heaters, furnaces, and boilers. Additionally, these biofuels can be used to fuel some combustion turbines and reciprocating engines, and as a source to create several chemicals. If bio-oil is used without modification, care must be taken to prevent emissions of black carbon and other particulates. Syngas and bio-oil can also be 'upgraded' to transportation fuels like biodiesel and gasoline substitutes. If biochar is used for the production of energy rather than as a soil amendment, it can be directly substituted for any application that uses coal. Pyrolysis also may be the most cost-effective way of producing electrical energy from biomaterial. Syngas can be burned directly, used as a fuel for gas engines and gas turbines, converted to clean diesel fuel through Fischer Tropsch or potentially used in the production of methanol and hydrogen.

Bio-oil has a much higher energy density than the raw biomass material. Mobile pyrolysis units can be used to lower the costs of transportation of the biomass itself if the biochar is returned to the soil and the syngas stream is used to power the process. Bio-oil contains organic acids which are corrosive to steel containers, has a high water vapour content which is detrimental to ignition, and, unless carefully cleaned, contains some biochar particles which can block injectors. The greatest potential for bio-oil seems to be its use in a biorefinery, where compounds that are valuable chemicals, pesticides, pharmaceuticals or food additives are first extracted, and the remainder is either upgraded to fuel or reformed to syngas.

Coke

Cokes are the solid carbonaceous material derived from destructive distillation of low-ash, low-sulphur bituminous coal. Cokes from coal are grey, hard, and porous.

Coking

Volatile constituents of the coal—including water, coal-gas, and coal-tar—are driven off by baking in an airless furnace or oven at temperatures as high as 2000 degrees Celsius. This fuses together the fixed carbon and residual ash. Most cokes in modern facilities are produced in by-product coking ovens and the resultant cokes are used as the main fuel in iron-making blast furnaces. Today, the hydrocarbons are considered to be by-products of modern coke-making facilities (though they are usually captured and used to produce valuable products. Non by-product coking furnaces or cokes furnaces (ovens) burn the hydrocarbon off gases produced by the coke-making process to drive the carbonisation process.

Properties and usage

The most important properties of coke are ash and sulphur content, which are linearly dependent on the coal used for production. Coke with less ash and sulphur content is highly priced on the market. Other important characteristics of coke include M10, M25, and M40 test crush indexes as they convey the strength of coke during transportation into the blast furnaces (BF); depending on BF size, there are certain requirements for coke size before entering the blast furnace, finely crushed coke pieces must not be allowed into the BF because gas dynamics would be impeded inside. Coke strength after reaction or CSR index is another important characteristics of coke as it represents coke's ability to withstand the violent conditions inside the blast furnace before turning into fine particles.

The volatility of coke reaches minimum levels at the end of the coking process. Volatility is an important property index for bituminous coal used in coke production, the greater volatile matter inside coal the more by-product could be produced, but there are some limitations on acceptable volatile content depending on the technical specifications of the coke batteries. Coal is being blended in proportions among different types of coal to reach acceptable levels of volatility before the coking process begins. Too low or too high levels of volatile matter in the coal blend results in inferior coke produced in respect to coke quality properties, it is generally considered that levels of 26–29 per cent of volatile matter in the coal blend is good for coking purposes.

The water content in coke is practically zero at the end of the coking process, but coke is often water quenched to reduce its temperature so that it can be transported inside the BF. The porous structure of coke absorbs some water, usually to 3–6 per cent of its mass. In modern coke plants an advanced method of coke cooling is by air quenching. Since smoke producing constituents are driven off during the coking of coal, coke forms a desirable fuel for stoves and furnaces in which conditions are not suitable for the complete burning of bituminous coal itself. Coke may be burned with little or no smoke under combustion conditions which would result in a large amount of smoke if bituminous coal were the fuel. Bituminous coal must meet a set of criteria for use as coking coal, determined by particular coal assay techniques. These include moisture content, ash content, sulphur content, volatile content, tar, and plasticity. Coke is used as a fuel and as a reducing agent in smelting iron ore in a blast furnace.

Other processes

The solid residue remaining from refinement of petroleum by the cracking process is also a form of coke. Petroleum coke has many uses besides being a fuel, such as the manufacture of dry cells, electrodes, etc. Fluid coking is a process which converts heavy residual crude into lighter products such as naphtha, kerosene, heating oil, and hydrocarbon gases. The 'fluid' term refers to the fact that coke particles are in a continuous system versus older batch-coking technology.

GASES PRODUCED

Coke may be used to make fuel gases such as:
1. Water gas: a mixture of carbon monoxide and hydrogen, made by passing steam over red-hot coke (or any carbon based char).
2. Producer gas, wood gas, generator gas, synthetic gas, suction gas: a mixture of carbon monoxide, hydrogen and nitrogen, made by passing air over red-hot coke (or any carbon based char).

These are useful gases but require careful handling because of the risk of carbon monoxide poisoning.

Coal Gas

Coal gas is a flammable gaseous fuel made from coal and supplied to the user via a piped distribution system. Town gas is a more general term referring to manufactured gaseous fuels produced for sale to consumers and municipalities. It is also known as manufactured gas, syngas (SNG), hygas, and producer gas in some countries.

Originally a by-product of the coking process, coal gas was extensively exploited in the 19th and early 20th centuries for lighting, cooking and heating.

The development of manufactured gas paralleled that of the industrial revolution and urbanisation; and the by-products, coal tars and ammonia, were at some times an important chemical feedstock for the dye and chemical industry. The whole rainbow of artificial dye colours is made from coal gas and coal tar.

Depending on the processes used for its creation, coal gas is a mixture of the calorific gases: hydrogen, carbon monoxide, methane and volatile hydrocarbons, with small amounts of noncalorific gases—carbon dioxide and nitrogen—as impurities. Coal gas plants, especially those that operated in the past, are commonly referred to, by environmental professionals and within the utility industry, as manufactured gas plants (MGPs).

Manufacturing processes

Manufactured gas can be made by two processes: carbonisation or gasification. Carbonisation refers to the devolatilisation of an organic feedstock to yield gas and char. Gasification is the process of subjecting a feedstock to chemical reactions that produce gas.

The first process used was the carbonisation and partial pyrolysis of coal. The off gases liberated in the high-temperature carbonisation (coking) of coal in coke ovens were collected, scrubbed and used as fuel. Depending on the goal of the plant, the desired product was either a high quality coke for metallurgical use, with the gas being a side product or the production of a high quality gas with coke being the side product. Coke plants are typically associated with metallurgical facilities such as smelters, and blast furnaces, while gas works typically served urban areas.

A facility used to manufacture coal gas, carburetted water gas (CWG), and oil gas is today generally referred to as a manufactured gas plant (MGP).

Gas for industrial use

Fuel gas for industrial use was made using producer gas technology. Producer gas is made by blowing air through an incandescent fuel bed (commonly coke or coal) in a gas producer. The reaction of fuel with insufficient air for total combustion produces carbon monoxide (CO); this reaction is exothermic and self sustaining. It was discovered that adding steam to the input air of a gas producer would increase the CV of the fuel gas by enriching it with CO and hydrogen (H_2) produced by water gas reactions. Producer gas has a very low CV of 3.7 to 5.6 MJ/m^3 [100–150 Btu/ft^3 (std)]; because the calorific gases CO/H_2 are diluted with lots of inert nitrogen (from air) and carbon dioxide (CO_2) (from combustion).

The problem of nitrogen dilution was overcome by the blue water gas (BWG) process. The incandescent fuel bed would be alternately blasted with air followed by steam. The air reactions during the blow cycle are exothermic, heating up the bed, while the steam reactions during the make cycle, are endothermic and cool down the bed. The products from the air cycle contain noncalorific nitrogen and are exhausted out the stack while the products of the steam cycle are kept as blue water gas. This gas is composed almost entirely of CO and H_2, and burns with a pale blue flame similar to natural gas. BWG has a CV of 11 MJ/m^3 (300 Btu/ft^3).

During a make run, steam would be passed through the generator to make blue water gas. From the generator the hot water gas would pass into the top of the carburettor where light petroleum oils would be injected into the gas stream. The light oils would be thermocracked as they came in contact with the hot checkerwork fire bricks inside the carburettor. The hot enriched gas would then flow into the superheater, where the gas would be further cracked by more hot fire bricks.

Composition

The composition of coal gas varied according to the type of coal and the temperature of carbonisation. Typical figures were:
1. Hydrogen 50 per cent.
2. Methane 35 per cent.
3. Carbon monoxide 10 per cent.
4. Ethylene 5 per cent.

In a plain burner, only the ethylene produced a luminous flame but the light output could be greatly increased by using a gas mantle.

By-products

The by-products of coal gas manufacture included coke, coal tar, sulphur and ammonia and these were all useful products. Dyes, medicines such as sulpha drugs, saccharine and sugar free soda drink, and dozens of organic compounds are made from coal gas. Sulphur is used in the manufacture of sulphuric acid and ammonia is used in manufacture of fertilisers. Coke is used as a smokeless fuel and for the manufacture of water gas and producer gas.

Coal Tar

Coal tar is a brown or black liquid of high viscosity, which smells of naphthalene and aromatic hydrocarbons. Coal tar is among the by-products when coal is carbonised to make coke or gasified to make coal gas. Coal tars are complex and variable mixtures of phenols, polycyclic aromatic hydrocarbons (PAHs), and heterocyclic compounds. Coal tar was subjected to fractional distillation to recover various products, including:
1. Tar, for roads.

2. Benzole, a motor fuel.
3. Creosote, a wood preservative.
4. Phenol, used in the manufacture of plastics.
5. Cresols, disinfectants.

Applications

Being flammable, coal tar is sometimes used for heating or to fire boilers. Like most heavy oils, it must be heated before it will flow easily.

It can be used in medicated shampoo, soap and ointment, as a treatment for dandruff and psoriasis, as well as being used to kill and repel head lice. Today, petroleum derived binders and sealers are more commonly used. These sealers are used to extend the life and lower maintenance cost associated with asphalt pavements, primarily in asphalt road paving, parking lots and walkways.

CLEAN COAL TECHNOLOGY

Clean coal technology is an umbrella term used to describe technologies being developed that aim to reduce the environmental impact of coal energy generation. These include chemically washing minerals and impurities from the coal, gasification, treating the flue gases with steam to remove sulphur dioxide, carbon capture and storage technologies to capture the carbon dioxide from the flue gas and dewatering lower rank coals (brown coals) to improve the calorific quality, and thus the efficiency of the conversion into electricity.

Clean coal technology usually addresses atmospheric problems resulting from burning coal. Historically, the primary focus was on sulphur dioxide and particulates, due to the fact that it is the most important gas which leads to acid rain. More recent focus has been on carbon dioxide (due to its likely impact on global warming) as well as other pollutants. Concerns exist regarding the economic viability of these technologies and the timeframe of delivery, potentially high hidden economic costs in terms of social and environmental damage, and the costs and viability of disposing of removed carbon and other toxic matter.

Coal, which is primarily used for the generation of electricity, is the second largest domestic contributor to carbon dioxide emissions in the USA, Europe and other developing countries. The public has become more concerned about global warming which has led to new legislation. The coal industry has responded by running advertising touting clean coal in an effort to counter negative perceptions, and deployment of clean coal technologies, including carbon capture and storage. The expenditure has been unsuccessful to date in that there is not a single commercial scale coal fired power station in the US or any other country of the world that captures and stores more than token amounts of CO_2.

Changing meanings of the term clean coal and questions about motives have provoked skepticism from environmentalists. The term 'clean coal' is often stated in quotation marks by its critics due to claims that it is a misnomer and a public relations term.

Clean Coal and the Environment

According to the United nations intergovernmental panel on climate change, the burning of coal, a fossil fuel, is blamed for climate change and global warming. As 25.5 per cent of the world's electrical generation in 2004 was from coal-fired generation, reaching the carbon dioxide reduction targets of the Kyoto protocol will require modifications to how coal is utilised.

The latest in clean coal technologies, carbon capture and sequestration, is a means to capture carbon dioxide emissions from coal-fired plants and permanently bury them underground. Currently, there are more than 80 carbon capture and sequestration projects underway in the United States. Sequestration technology has yet to be tested on a large scale and may not be safe or successful. Sequestered CO_2 may eventually leak up through the ground, may lead to unexpected geological instability or may cause contamination of aquifers used for drinking water supplies. There are also concerns that plans to pump some of the sequestered CO_2 into certain oil and gas reserves, to help make the fuels easier to pump out of the ground, will lead to increased concentrations of CO_2 in potential fuel supplies. This would have to be removed, or released during the refining process. Supporters of clean coal use the great plains synfuels plant to support the technical feasibility of carbon dioxide sequestration. Carbon dioxide from the coal gasification is shipped to Canada where it is injected into the ground to aid in oil recovery. Supporters acknowledge that economics can be problematic for carbon sequestration.

By-products

The by-products of coal combustion are considerably hazardous to the environment if not properly contained. This is clean coal's largest challenge, both from the practical and public relations perspectives.

While it is possible to remove most of the sulphur dioxide (SO_2), nitrogen oxides (NO_x) and particulate matter (PM) emissions from the coal-burning process, carbon dioxide (CO_2) emissions and radionuclides will be more difficult to address. Technologies exist to capture and store CO_2, but they have not yet been utilised on a large-scale commercial basis due to the high economic costs.

Coal burning industries have previously succeeded in significantly reducing pollutants. Current coal fired electric generating plants emit 70 per cent fewer regulated emissions (total mass per energy produced). This factoid includes values for NO_x, SO_x, volatile organic compounds, particulate matter, and carbon monoxide emissions only. SO_x formed the greatest proportion of these emissions, where significant gains had been made in order to combat acid rain.

Coal burning power plants produce large amounts of solid and liquid waste products, mostly in the form of fly-ash and bottom ash. These waste products are stored in landfills and large ponds. Increasing emission controls at the plants results in increased waste products. While the hazard content of ash by percentage is very low, the concentration of millions of tons at plant sites creates the danger of significant pollution in the event of containment failures. A small fraction of coal ash is beneficially used in the manufacture of concrete and other construction materials. The use of ash in construction materials sequesters the hazardous ingredients and prevents their release in quantities large enough to be hazardous. Unfortunately the economics of beneficial use are such that some subsidy would be required from the power plant for widespread use. As it stands now, most electric utilities prefer to store the ash rather than subsidise beneficial use and in most cases charge for sale of the ash. As a result of these policies, most power plant waste continues to be stored in ponds and landfills.

Refined Coal

Refined coal is the product of a coal upgrading technology that removes moisture and certain pollutants from lower-rank coals such as sub-bituminous and lignite (brown) coals. It is one form of several precombustion treatments and processes for coal that alter coal's characteristics before it is burned. The goals of precombustion coal technologies are to increase efficiency and reduce emissions when the coal is burned. Depending on the situation, pre-combustion technology can be used in place of or as a supplement to post-combustion technologies to control emissions from coal-fuelled boilers.

Chapter 5

Fuel Gases

INTRODUCTION

Fuel gas can refer to any of several gases (e.g. natural gas, twon gas, syngas, propane, butane, producer gas, and water gas) burned to produce thermal energy. Natural gas (methane) is the most common fuel gas.

During the 1950s a deep and far-reaching change took place in the fuel-gas industries, involving the domination of these widespread markets by natural gas. This was made economical through the countrywide installation of gas pipelines, reaching from large gas fields to most homes and factories in the United States. Local peak demands in winter are satisfied by the use of natural gas stored in neighbouring underground depleted production wells, by the use of liquefied natural gas (LNG), by the use of liquid petroleum gas (LPG), or by peak production. LPG also meets needs in areas pipelines cannot reach. The convenience, cleanliness, and reasonable price of natural gas have been a boon to Americans. This change has competitively restricted coke-oven gas to those coproduct areas where coke is made from coal for the steel and foundry industries. Also, this competition has reduced water gas mostly to use in the making of peak gas and synthesis gas. Producer gas and retort coal gas have almost disappeared in the United States.

NATURAL GAS

Natural gas is a gas consisting primarily of methane. It is found associated with fossil fuels, in coal beds, as methane clathrates, and is created by methanogenic organisms in marshes, bogs, and landfills. It is an important fuel source, a major feedstock for fertilisers, and a potent greenhouse gas.

Natural gas is often informally referred to as simply gas, especially when compared to other energy sources such as electricity. Before natural gas can be used as a fuel, it must undergo extensive processing to remove almost all materials other than methane. The by-products of that processing include ethane, propane, butanes, pentanes and higher molecular weight hydrocarbons, elemental sulphur, and sometimes helium and nitrogen.

Fossil Natural Gas

Fossil natural gas can be associated (found in oil fields) or non-associated (isolated in natural gas fields), and is also found in coal beds (as coalbed methane). It sometimes contains significant quantities of ethane, propane, butane, and pentane—heavier hydrocarbons removed prior to use as a consumer fuel—as well as carbon dioxide, nitrogen, helium and hydrogen sulphide. Natural gas is commercially produced from oil fields and natural gas fields. Gas produced from oil wells is called casinghead gas or associated gas. The natural gas industry is producing gas from increasingly more challenging resource types: sour gas, tight gas, shale gas and coalbed methane.

Because natural gas is not a pure product, when non-associated gas is extracted from a field under supercritical (pressure/temperature) conditions, it may partially condense upon isothermic depressurising—an effect called retrograde condensation.

The liquids thus formed may get trapped by depositing in the pores of the gas reservoir. One method to deal with this problem is to reinject dried gas free of condensate to maintain the underground pressure and to allow re-evaporation and extraction of condensates.

Biogas

When methane-rich gases are produced by the anaerobic decay of non-fossil organic matter (biomass), these are referred to as biogas (or natural biogas). Sources of biogas include swamps, marshes, and landfills, as well as sewage sludge and manure by way of anaerobic digesters, in addition to enteric fermentation particularly in cattle.

Methanogenic archaea are responsible for all biological sources of methane, some in symbiotic relationships with other life forms, including termites, ruminants, and cultivated crops. Methane released directly into the atmosphere would be considered a pollutant, however, methane in the atmosphere is oxidised, producing carbon dioxide and water. Methane in the atmosphere has a half-life of seven years, meaning that every seven years, half of the methane present is converted to carbon dioxide and water.

Future sources of methane, the principal component of natural gas, include landfill gas, biogas and methane hydrate. Biogas, and especially landfill gas, are already used in some areas, but their use could be greatly expanded. Landfill gas is a type of biogas, but biogas usually refers to gas produced from organic material that has not been mixed with other waste.

Landfill gas is created from the decomposition of waste in landfills. If the gas is not removed, the pressure may get so high that it works its way to the surface, causing damage to the landfill structure, unpleasant odour, vegetation die-off and an explosion hazard.

The gas can be vented to the atmosphere, flared or burned to produce electricity or heat.

Once water vapour is removed, about half of landfill gas is methane. Almost all of the rest is carbon dioxide, but there are also small amounts of nitrogen, oxygen and hydrogen. There are usually trace amounts of hydrogen sulphide and siloxanes, but their concentration varies widely. Landfill gas cannot be distributed through natural gas pipelines unless it is cleaned up to the same quality. It is usually more economical to combust the gas on site or within a short distance of the landfill using a dedicated pipeline. Water vapour is often removed, even if the gas is combusted on site. If low temperatures condense water out of the gas, siloxanes can be lowered as well because they tend to condense out with the water vapour. Other non-methane components may also be removed in order to meet emission standards, to prevent fouling of the equipment or for environmental considerations. Cofiring landfill gas with natural gas improves combustion, which lowers emissions.

Biogas is usually produced using agricultural waste materials, such as otherwise unusable parts of plants and manure. Biogas can also be produced by separating organic materials from waste that otherwise goes to landfills. This is more efficient than just capturing the landfill gas it produces. Using materials that would otherwise generate no income, or even cost money to get rid of, improves the profitability and energy balance of biogas production.

Anaerobic lagoons produce biogas from manure, while biogas reactors can be used for manure or plant parts. Like landfill gas, biogas is mostly methane and carbon dioxide, with small amounts of nitrogen, oxygen and hydrogen. However, with the exception of pesticides, there are usually lower levels of contaminants.

Natural Gas Processing

Flow diagram of a typical natural gas processing plant (Fig. 5.1) shows the various unit processes used to convert raw natural gas into sales gas pipelined to the end user markets. The flow diagram also shows how processing of the raw natural gas yields by-product sulphur, by-product ethane, and natural gas liquids (NGL) propane, butanes and natural gasoline (denoted as pentanes).

Natural gas processing plants, or fractionators, are used to purify the raw natural gas extracted from underground gas fields and brought up to the surface by gas wells. The processed natural gas, used as fuel by residential, commercial and industrial consumers, is almost pure methane and is very much different from the raw natural gas.

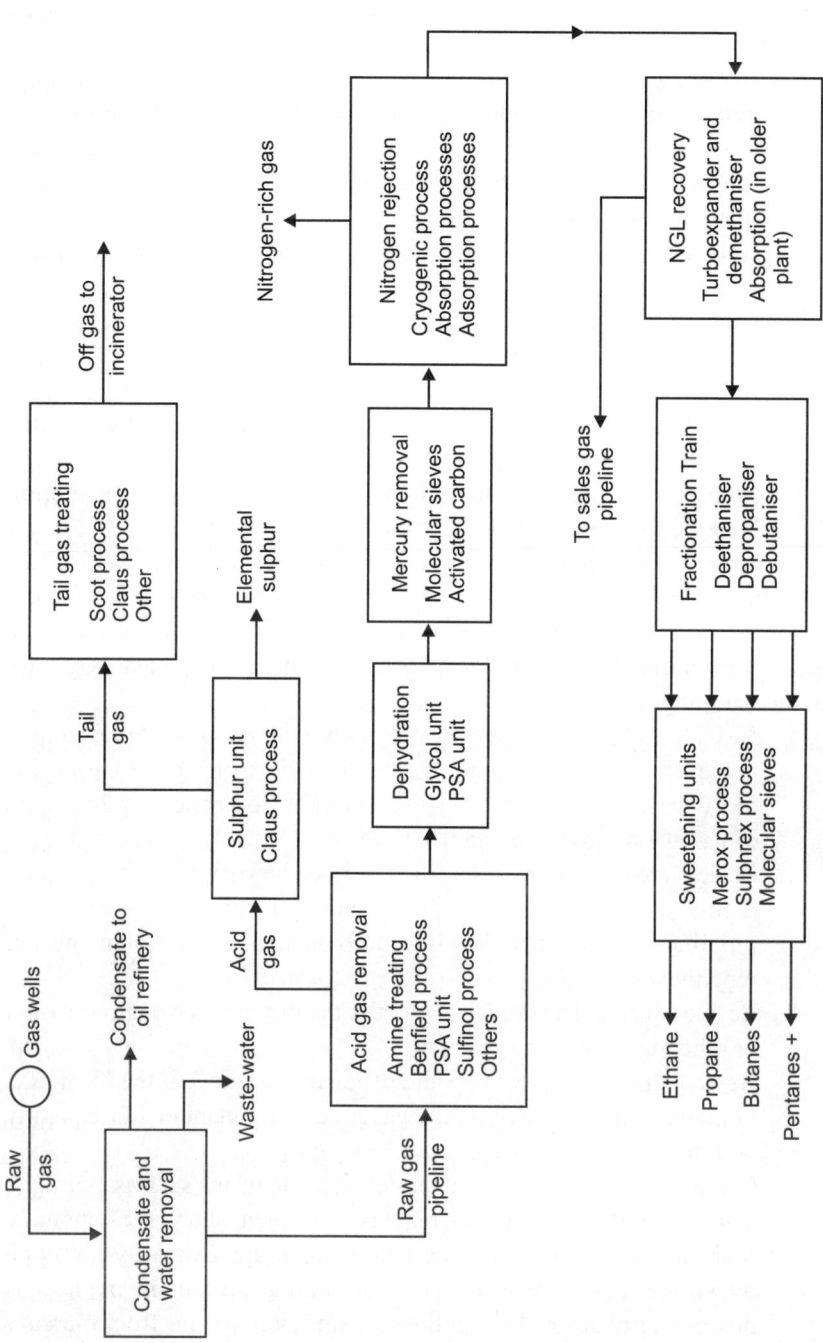

Fig. 5.1. Flow diagram of a typical natural gas processing plant.

Raw natural gas typically consists primarily of methane (CH_4), the shortest and lightest hydrocarbon molecule. It also contains varying amounts of:

1. Heavier gaseous hydrocarbons: Ethane (C_2H_6), propane (C_3H_8), normal butane (n-C_4H_{10}), isobutane (i-C_4H_{10}), pentanes and even higher molecular weight hydrocarbons. When processed and purified into finished by-products, all of these are collectively referred to NGL (natural gas liquids).
2. Acid gases: Carbon dioxide (CO_2), hydrogen sulphide (H_2S) and mercaptans such as methanethiol (CH_3SH) and ethanethiol (C_2H_5SH).
3. Other gases: Nitrogen (N_2) and helium (He).
4. Water: Water vapour and liquid water.
5. Liquid hydrocarbons: Perhaps some natural gas condensate (also referred to as casinghead gasoline or natural gasoline) and/or crude oil.
6. Mercury: Very small amounts of mercury primarily in elemental form, but chlorides and other species are possibly present.

The raw natural gas must be purified to meet the quality standards specified by the major pipeline transmission and distribution companies. Those quality standards vary from pipeline to pipeline and are usually a function of a pipeline system's design and the markets that it serves. In general, the standards specify that the natural gas:

1. Be within a specific range of heating value (caloric value). For example, in the United States, it should be about 1035 ± 5 per cent Btu per cubic foot of gas at 1 atmosphere and 60°F ($41 \text{ MJ} \pm 5$ per cent per cubic metre of gas at 1 atmosphere and 0°C).
2. Be delivered at or above a specified hydrocarbon dew point temperature (below which some of the hydrocarbons in the gas might condense at pipeline pressure forming liquid slugs which could damage the pipeline).
3. Be free of particulate solids and liquid water to prevent erosion, corrosion or other damage to the pipeline.
4. Be dehydrated of water vapour sufficiently to prevent the formation of methane hydrates within the gas processing plant or subsequently within the sales gas transmission pipeline.
5. Contain no more than trace amounts of components such as hydrogen sulphide, carbon dioxide, mercaptans, nitrogen, and water vapour.
6. Maintain mercury at less than detectable limits (approximately 0.001 ppb by volume) primarily to avoid damaging equipment in the gas processing plant or the pipeline transmission system from mercury amalgamation and embrittlement of aluminum and other metals.

Types of Raw Natural Gas Wells

Raw natural gas comes primarily from any one of three types of wells crude oil wells, gas wells, and condensate wells. Natural gas that comes from crude oil wells is typically termed associated gas. This gas can exist separate from the crude oil in the underground formation, or dissolved in the crude oil.

Natural gas from gas wells and from condensate wells, in which there is little or no crude oil, is termed non-associated gas. Gas wells typically produce only raw natural gas, while condensate wells produce raw natural gas along with a very low density liquid hydrocarbon called natural gas condensate (sometimes also called natural gasoline or simply condensate).

Raw natural gas can also come from methane deposits in the pores of coal seams. Such gas is referred to as coalbed gas and it is also called sweet gas because it is relatively free of hydrogen sulphide.

Uses of natural gas

Power generation

Natural gas is a major source of electricity generation through the use of gas turbines and steam turbines. Particularly high efficiencies can be achieved through combining gas turbines with a steam turbine in combined cycle mode. Natural gas burns cleaner than other fossil fuels, such as oil and coal, and produces less carbon dioxide per unit energy released. For an equivalent amount of heat, burning natural gas produces about 30 per cent less carbon dioxide than burning petroleum and about 45 per cent less than burning coal. Combined cycle power generation using natural gas is thus the cleanest source of power available using fossil fuels, and this technology is widely used wherever gas can be obtained at a reasonable cost. Fuel cell technology may eventually provide cleaner options for converting natural gas into electricity, but as yet it is not price-competitive.

Residential domestic use

Natural gas is supplied to homes, where it is used for such purposes as cooking in natural gas-powered ranges and/or ovens, natural gas-heated clothes dryers, heating/cooling and central heating. Home or other building heating may include boilers, furnaces, and water heaters.

Natural gas vehicles

Compressed natural gas (methane) is a cleaner alternative to other automobile fuels such as gasoline (petrol) and diesel. As of 2005, the countries with the largest number of natural gas vehicles were Argentina, Brazil, India, Pakistan, Italy, Iran, and the United States. The energy efficiency is generally equal to that of gasoline engines, but lower compared with modern diesel engines.

Gasoline/petrol vehicles converted to run on natural gas suffer because of the low compression ratio of their engines, resulting in a cropping of delivered power while running on natural gas (10–15 per cent). CNG-specific engines, however, use a higher compression ratio due to this fuel's higher octane number of 120–130.

Fertiliser

Natural gas is a major feedstock for the production of ammonia, via the Haber process, for use in fertiliser production.

Hydrogen

Natural gas can be used to produce hydrogen, with one common method being the hydrogen reformer. Hydrogen has various applications: it is a primary feedstock for the chemical industry, a hydrogenating agent, an important commodity for oil refineries, and a fuel source in hydrogen vehicles.

Other uses

Natural gas is also used in the manufacture of fabrics, glass, steel, plastics, paint, and other products.

Storage and transport

The major difficulty in the use of natural gas is transportation and storage because of its low density. Natural gas pipelines are economical, but are impractical across oceans. LNG carriers can be used to transport liquefied natural gas (LNG) across oceans, while tank trucks can carry liquefied or compressed natural gas (CNG) over shorter distances. They may transport natural gas directly to end-users, or to distribution points such as pipelines for further transport. These may have a higher cost, requiring additional facilities for liquefaction or compression at the production point, and then gasification or decompression at enduse facilities or into a pipeline.

Environmental effects

Global climate change

Natural gas is often described as the cleanest fossil fuel, producing less carbon dioxide per joule delivered than either coal or oil. However, in absolute terms it does contribute substantially to global emissions, and this contribution is projected to grow. In addition, natural gas itself is a greenhouse gas far more potent than carbon dioxide when released into the atmosphere but is not of large concern due to the small amounts in which this occurs.

When drilled in the US, the CO_2 pumped out with the natural gas is released directly into the atmosphere. This amount of CO_2 is not counted with the release of the CO_2 when natural gas is burned.

Safety

In any form, a minute amount of odourant such as *t*-butyl mercaptan, with a rotting-cabbage-like smell, is added to the otherwise colourless and almost odourless gas, so that leaks can be detected before a fire or explosion occurs. Sometimes a related compound, thiophane is used, with a rotten-egg smell.

Explosions caused by natural gas leaks occur a few times each year. Individual homes, small businesses and boats are most frequently affected when an internal leak builds up gas inside the structure. Frequently, the blast will be enough to significantly damage a building but leave it standing. In these cases, the people inside tend to have minor to moderate injuries. Occasionally, the gas can collect in high enough quantities to cause a deadly explosion, disintegrating one or more buildings in the process. The gas usually dissipates readily outdoors, but can sometimes collect in dangerous quantities if weather conditions are right. However, considering the tens of millions of structures that use the fuel, the individual risk of using natural gas is very low. Some gas fields yield sour gas containing hydrogen sulphide (H_2S). This untreated gas is toxic. Amine gas treating, an industrial scale process which removes acidic gaseous components, is often used to remove hydrogen sulphide from natural gas. Extraction of natural gas (or oil) leads to decrease in pressure in the reservoir. This in turn may lead to subsidence at ground level. Subsidence may affect ecosystems, waterways, sewer and water supply systems, foundations, etc.

Natural gas heating systems are the leading cause of carbon monoxide deaths in the United States, according to the US Consumer Product Safety Commission. When a natural gas heating system malfunctions, it produces odourless carbon monoxide. With no fumes or smoke to give warning, poisoning victims are easily asphyxiated by the carbon monoxide. Detectors are available that warn of carbon monoxide and/or explosive gas (methane, propane, etc.)

PRODUCER GAS

A gas obtained by burning coal or coke with a restricted supply of air, or by passing air and steam through a bed of incandescent fuel under such conditions that the carbon dioxide formed is converted into carbon monoxide. The water vapour reacts to form carbon monoxide and hydrogen. Producer gas is cheap but has low Btu and is used where transportation is not required. Highly flammable and toxic. Explosive range 20 to 73 per cent of air.

WATER GAS (BLUE GAS)

A mixture of gases made from coke, air, and steam. The steam is decomposed by passing it over a bed of incandescent coke, at a temperature of about 1000°C

(1832°F), or by high-temperature reaction with natural gas or similar hydrocarbons.

LIQUEFIED PETROLEUM GAS

Liquefied petroleum gas (also called LPG, GPL, LP Gas, or autogas) is a mixture of hydrocarbon gases used as a fuel in heating appliances and vehicles, and increasingly replacing chlorofluorocarbons as an aerosol propellant and a refrigerant to reduce damage to the ozone layer.

Varieties of LPG bought and sold include mixes that are primarily propane, mixes that are primarily butane, and the more common, mixes including both propane (60 per cent) and butane (40 per cent), depending on the season — in winter more propane, in summer more butane. Propylene and butylenes are usually also present in small concentration. A powerful odourant, ethanethiol, is added so that leaks can be detected easily. The international standard is EN 589.

LPG is usually derived from fossil fuel sources, being manufactured during the refining of crude oil, or extracted from oil or gas streams as they emerge from the ground. At normal temperatures and pressures, LPG will evaporate. Because of this, LPG is supplied in pressurised steel bottles. In order to allow for thermal expansion of the contained liquid, these bottles are not filled completely; typically, they are filled to between 80 and 85 per cent of their capacity. The ratio between the volumes of the vapourised gas and the liquefied gas varies depending on composition, pressure and temperature, but is typically around 250:1. The pressure at which LPG becomes liquid, called its vapour pressure, likewise varies depending on composition and temperature; for example, it is approximately 220 kilopascals (2.2 bar) for pure butane at 20°C (68°F), and approximately 2.2 megapascals (22 bar) for pure propane at 55°C (131°F). LPG is heavier than air, and thus will flow along floors and tend to settle in low spots, such as basements.

This can cause ignition or suffocation hazards if not dealt with. LPG is a low carbon emitting hydrocarbon fuel available in rural areas, emitting 19 per cent less CO_2 per kWh than oil, 30 per cent less than coal and more than 50 per cent less than coal-generated electricity distributed via the grid. Being a mix of propane and butane, LPG emits more carbon per joule than propane and LPG emits less carbon per joule than butane.

LPG burns cleanly with no soot and very few sulphur emissions, posing no ground or water pollution hazards. Large amounts of LPG can be stored in bulk tanks and can be buried underground if required. Alternatively, gas cylinders can be used. LPG has a typical specific calorific value of 46.1 MJ/kg compared to 42.5 MJ/kg for diesel and 43.5 MJ/kg for premium grade petrol (gasoline).

Usage in Vehicles

When LPG is used to fuel internal combustion engines, it is often referred to as autogas. In some countries, it has been used since the 1940s as an alternative fuel for spark ignition engines. More recently, it has also been used in diesel engines. Its advantage is that it is non-toxic, non-corrosive and free of tetra-ethyl lead or any additives, and has a high octane rating (108 RON). It burns more cleanly than petrol or diesel and is especially free of the particulates from the latter.

LPG as a vehicle fuel has two main disadvantages.

1. It has a lower energy density than either petrol or diesel, so the equivalent fuel consumption is higher, but since many governments impose less tax, it is still usually more cost effective.
2. Some designs of internal combustion engine required the lubrication of petrol or diesel with lead or lead substitute, and LPG's lack thereof can damage valves or shorten their life. Engines designed for unleaded fuel, equipped with stellite valve seats, are usually suitable for use with LPG without added upper cylinder lubrication.

Refrigerant

In highly purified form, various blends of the LPG constituents propane and isobutane are used to make hydrocarbon refrigerants. Hydrocarbons are more energy efficient, run at the same or lower pressure and are generally cheaper than HFC.

LPG Cooking Fuel

LPG is the most common cooking fuel in developing countries. LPG is the most common cooking fuel in Brazilian urban areas, being used in virtually all households.

LPG and SNG

LPG has a higher calorific value (94 MJ/m^3 equivalent to 26.1 kWh/m^3) than natural gas (methane) (38 MJ/m^3 equivalent to 10.6 kWh/m^3), which means that LPG cannot simply be substituted for natural gas. In order to allow the use of the same burner controls and to provide for similar combustion characteristics, LPG can be mixed with air to produce a synthetic natural gas (SNG) that can be easily substituted. LPG/air mixing ratios average 60/40, though this is widely variable based on the gases making up the LPG. The method for determining the mixing ratios is by calculating the Wobbe index of the mix. Gases having the same Wobbe index are held to be interchangeable.

LPG-based SNG is used in emergency back-up systems for many public, industrial, and military installations, and many utilities use LPG peak shaving plants in times of high demand to make up shortages in natural gas supplied to their distributions systems. LPG-SNG installations are also used during initial gas system introductions, when the distribution infrastructure is in place before gas supplies can be connected. Developing markets in India and China (among others) use LPG-SNG systems to build up customer bases prior to expanding existing natural gas systems.

LPG and renewable energy

As the name LPG suggests it is almost exclusively derived from fossil fuels and as such is not in itself renewable. LPG can, however, be used in conjunction with renewable technologies, releasing less carbon dioxide than purely fossil fuel alternatives.

LPG and solar thermal water heating

In cold weather there is usually not enough energy from sunlight for solar cylinders to heat a home's full water supply. LPG can be used as a supplementary fuel to heat the water needed, with the solar panel reducing carbon emissions by 16 per cent. The LPG supply can then be used to fuel the central heating, cooker and fire.

LPG and ground source heat pumps

LPG can be used as a back up to ground source heat pumps, which use the earth's natural energy to provide space and water heating. LPG ensures there is enough heat output for both the central heating and hot water supply and can also be used for that real flame cooker and fire.

LPG and air source heat pumps

Air source heat pumps extract heat from the surrounding air to concentrate heat. Again, LPG can be used as a supplementary heat source.

Fire risk and mitigation

LPG containers that are subjected to fire of sufficient duration and intensity can undergo a boiling liquid expanding vapour explosion (BLEVE). Due to the destructive nature of LPG explosions, the substance is classified as a dangerous good. This is typically a concern for large refineries and petro-chemical plants that maintain very large containers. The remedy is to equip such containers with a measure to provide a fire-resistance rating. If the containers are cylindrical and horizontal, they are referred to as cigars or bullets, whereas circular ones are spheres. Large, spherical LPG containers may have up to a 15 cm steel wall thickness. Ordinarily, they are equipped with an approved pressure relief valve on the top, in the centre. One of the main dangers is that accidental spills of hydrocarbons may ignite and heat an LPG container, which increases its temperature and pressure, following the basic gas laws. The relief valve on the top is designed to vent off excess pressure in order to prevent the rupture of the tank itself. Given a fire of sufficient duration and intensity, the pressure being generated by the boiling and expanding gas can exceed the ability of the valve to vent the excess. When that occurs, an overexposed tank may rupture violently, launching pieces at high velocity, while the released products can ignite as well, potentially causing catastrophic damage to anything nearby, including other tanks. In the case of 'cigars', a midway rupture may send two 'rockets' going off each way, with plenty of fuel in each to propel each segment at high speed until the fuel is spent.

Mitigation measures include separating LPG tanks from potential sources of fire. In the case of rail transport, for instance, LPG tanks can be staggered, so that other goods are put in between them. This is not always done, but it does represent a low-cost remedy to the problem. LPG rail cars are easy to spot from the relief valves on top, typically with railings all around.

In the case of new LPG containers, one may simply bury them, only leaving valves and armatures exposed, for easy maintenance. Great care must be taken there though, as mechanical damage can occur to the primers, which can result in hazardous corrosion of the containers. For the buried container, only the exposed parts need to be treated with approved fireproofing materials, such as intumescent and/or endothermic coatings, or even fireproofing plasters. The rest are amply protected by soil. Speciality removable covers exist for easy access to the dials and components that must be accessed for proper maintenance and operation of the equipment.

SYNTHETIC NATURAL GAS

Synthetic natural gas (SNG), also referred to as substitute natural gas, is a manufactured form of natural gas. It is created through converting or reforming hydrocarbons and can be used in almost every way that natural gas can be used.

Distribution

It is advantageous to distribute SNG and bio-SNG together with natural gas in a gas grid. In this way, the production of renewable gas can be phased in at the same rate as the production capacity is increased. The gas market and infrastructure the natural gas has contributed with is a condition for large scale introduction of renewable biomethane produced through anaerobic digestion (biogas) or gasification and methanation bio-SNG. Any manufactured fuel gas of approximately the same composition and Btu value as that obtained naturally from oil fields (about 85 per cent methane and 15 per cent ethane). There are two major methods of synthesis involving catalysts, high temperature and high pressure: (i) direct hydrogenation of coal, and (ii) methanation of synthesis gas, which is obtained by hydrogenolysis of coal or steam-reforming of naphtha or similar petroleum distillate. Other methods involve pyrolysis of solid wastes and extraction from oil shale and manures.

Although the above processes have features that are different, they accomplish the same objective as indicated in the very simplified coal gasification flowsheet in Fig. 5.2.

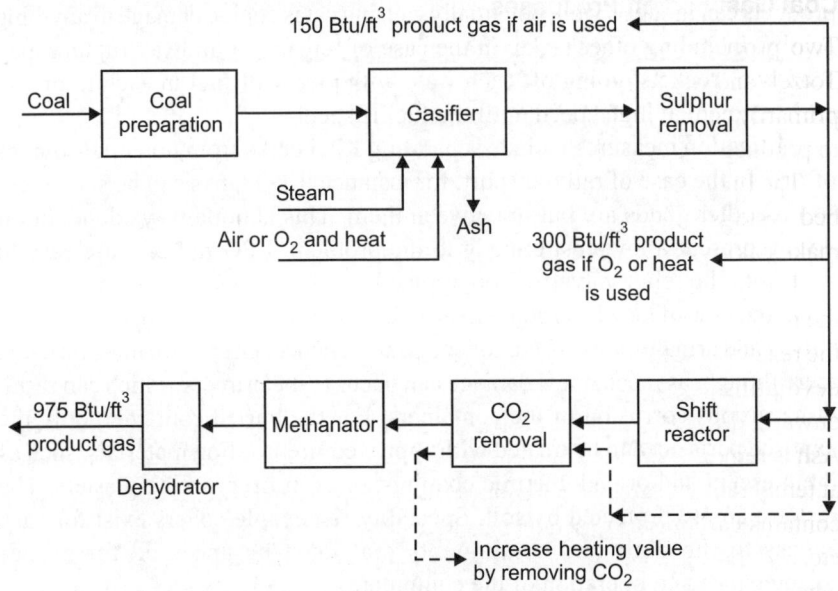

Fig. 5.2. Simplified coal gasification flow sheet.

Gas with about 150 Btu/ft^3 can be produced by using coal, air, and steam in the gasification reaction. Some of the coal is burned with air to H_2O and CO_2 to provide the heat required by the gasification step. To make the gas product a clean-burning fuel, similar to producer gas, the sulphur (H_2S) is removed. Low-Btu gas of about 300 Btu/ft^3 is made in the same manner, except that oxygen is used rather than air, or some other method is used for supplying the heat of reaction. For example, an inert solid can be heated in a furnace, using coal and air, and the hot inert material then transferred to the gasifier. The principal difference between the 150– and 300 Btu/ft^3 gas is that the 150 Btu/ft^3.

To make a gas with 900 to 1000 Btu/ft^3, a low-Btu gas is first made by gasifying coal with steam and oxygen or heat and removing the sulphur. This gas is processed further in a water-gas shift reactor to change the ratio of hydrogen to carbon monoxide to 3:1 by the following reaction:

$$CO + H_2O \longrightarrow H_2 + CO_2$$

The carbon dioxide formed in this step is then removed, and the hydrogen and carbon monoxide are converted to methane using a catalyst. The operating pressure in the gasifier, depending on the process, can be atmospheric to over 1000 lb, and temperatures can vary from about 1500 to about 3000°F. The higher pressure and lower temperature result in a larger amount of methane being formed in the gasifier.

Coal Gasification Processes

Two processes available today for making gas from coal are the Koppers-Totzek and the Lurgi process. The gas produced in this slagging gasifier is primarily carbon monoxide and hydrogen which can be shifted and methanated to produce pipeline quality gas.

The Lurgi pressurised (up to 400 psi) coal gasification process has a moving-bed system that cannot use strongly caking coals, although modifications to make caking coals acceptable.

The Lurgi process requires that the coal be properly sized and then fed into the gasifier through a pressurised lock hopper. Steam and oxygen are fed into the reactor under the ash grate. As the coal gravitates downward and is heated, devolatilisation commences and, from a temperature of 1150° to 1400°F onward, devolatilisation is accompanied by gasification of the resulting char. Ash is removed through the ash lock hopper. The crude gas leaves the gasifier at temperatures between 700° and 1100°F, depending upon the type of coal. It contains carbonisation products such as tar, oil, naphtha, phenols, ammonia, etc. and traces of coal and ash dust. This crude gas is passed through a scrubber where it is washed by circulating gas liquor and then cooled to a temperature at which the gas is saturated with steam. As higher-boiling tar fractions are condensed, the wash water contains tar to which the coal and ash dust is bonded. The steam-saturated gas is passed to a waste heat boiler in which waste heat at a temperature of 320° to 360°F is recovered. The gas liquor condensed in this boiler is pumped to the scrubber, and surplus gas liquor is routed to a tar-gas-liquor separator. The mixture of tar and dust is returned to the gasifier for cracking and gasification. The gas leaving the gasifier is mainly CO_2, CO, CH_4, H_2 and H_2O. The resulting H_2/CO ratio of the gas is not suitable for subsequent methane synthesis, so a shift conversion reaction ($CO + H_2O \longrightarrow CO_2 + H_2$) is necessary.

Impurities are removed from the gas by the Rectisol process. It employs a physical gas absorption system using organic solvents, preferably methanol, at low temperatures between approximately +30° and –80°F. The Rectisol unit has three sections: one for the removal of gas naphtha, unsaturated hydrocarbons, and other high-boiling gas impurities, one for the removal of CO_2, H_2S, and COS, and one to remove the last portion of CO_2 and to dehydrate the gas following the methane synthesis. The final product gas can have a heating value of 970 Btu/ft^3 and is almost all methane.

Oil Gasification

As already indicated, there is now a need for SNG. The oil gasification, especially when naphtha is used, is that such a plant requires a smaller investment and a shorter time to construct than a coal-to-gas plant. Generally, the degree of difficulty in gasifying oil increases as the weight of the oil used increases. When making gas from oil most of the cost of the product gas is due to the cost of oil.

Naphtha gasification process is the catalytic rich gas (CRG) double methanation process. In this process route the CRG reactor is followed by two separate stages of methanation. The purified naphtha feedstock, after admixture with steam to give a steam/naphtha ratio of 2 lb/1 lb, is preheated to 450°C (840°F) and gasified in a CRG reactor. The rich gas leaving this reactor at 505°C (940°F) is cooled in a waste heat boiler to 300°C (570°F) before entering the first methanator in which part of the hydrogen reacts with the carbon oxides to cause a temperature rise of 74°C (133°F). The gas is then not only cooled in a second waste heat boiler, but part of the undecomposed steam is also rejected to allow further methane formation to take place in the second methanator; the temperature rise in this reactor is 40°C (72°F).

The catalyst used for the methanation stages is identical with the CRG catalyst. It has a very low threshold temperature and, since it is protected from sulphur poisoning, has a very long life. The final stages of the process then comprise carbon dioxide removal and drying. There is a progressive reduction in the hydrogen and an increase in the methane content. After the removal of carbon dioxide the gas contains 98 per cent methane with a calorific value of just under 1000 Btu/ft^3.

SAFE HANDLING OF MODERN INDUSTRIAL FUELS

Modern fuels such as liquefied petroleum gas (LPG), compressed natural gas (CNG) and liquefied natural gas (LNG) have high fuel efficiency, are cost effective and environment friendly. At the same time, all three are flammable and carry the risk of large fires and explosions. Fortunately, such accidents are eminently preventable by safety in design and layout during production and storage, safe operating procedures in operation and maintenance, and by efficiently following emergency procedures.

Another important aspect of safety is proper education and training of operating and other personnel involved in maintenance, inspection, etc. Any operating person will run the plant correctly under normal operation, but efficiency and proper judgment becomes absolutely important during an emergency, such as gas leak or fire. On the safety aspects of the three fuels and pointed out that LNG is difficult to handle since its liquefaction and storage temperature is −160°C. LNG is also a potential health hazard, as it damages tissue on contact with skin, due to its very low storage temperature. CNG, on the other hand, involves only a simple gas compression system. Compared to gasoline and other alternate fuels, CNG poses least concern regarding toxicity, accidental ignition and propagation of flame.

A few physical and chemical properties that make the fuel inherently safer are its low density, high ignition temperature and non-toxic nature. In the following section, salient features of industrial fuels (LPG, LNG, CNG, etc.) are presented.

Safety in Refrigerated Storage of LPG

LPG is a mixture of commercial butane and commercial propane having both saturated and unsaturated hydrocarbons. LPG marketed in India is governed by Indian Standard Code IS-4576 and the test methods by IS-1448. LPG is inherently dangerous on account of fire, explosion and other hazards (Table 5.1).

Table 5.1. Safety considerations in design of LPG storage.

Minimum flanges to reduce possibility of leaks

Adequate instrumentation—level, temperature, pressure indicators

Safety valves, minimum 2 nos. each having 100 per cent capacity

Fire-proofing should be adequate to protect material from overheating

Fire protection and detection system

Electrical equipment as per hazardous area classification—follow and maintain rigidly

Weight of water. Weight should be taken for design and not LPG contents

Double block system is recommended

Design and layout in the most important part of LPG facilities—follow appropriate standards—IS 2825, OISD-144, also ASME etc.

Safe procedures, practices and maintenance: Establish safe procedures and practices in process operation, maintenance and inspection. For safe maintenance follow rigidly work permit system for cold and hot work. In general this very important point is not given proper attention. Conduct periodical safety audit to ensure established procedures are followed. Another most important aspect is safety education and thorough training of all concerned personnel. Emergency procedures should be established, an appropriate emergency manual prepared and each one who has a part informed, trained and periodical mock drills conducted and deficiencies noticed corrected

Presently, we have two types of LPG:

1. Domestic LPG: It has propane fraction below 50 per cent, along with butane and some unsaturates (C5 hydrocarbons and heavier maximum 2.5 per cent).

2. Auto LPG: It has propane fractions varying between 30–50 per cent, along with butane and unsaturates (C5 hydrocarbons and heavier maximum 2 per cent).

Refrigerated storage of LPG

LPG products, namely propane and butane, have their critical temperatures above ambient. They can be stored either at ambient temperature under pressure, or at atmospheric pressure with refrigeration. In order to minimise storage space, it is most advantageous to store as liquefied gases. For fully refrigerated storage there are various options available: single containment tank, double containment tank and full containment tank.

The full containment type of storage tanks are the most commonly used design. These tanks consist of two separate liquid containers. The inner tank is a metallic open-top vertical cylindrical container set within an outer container. It has a top deck suspended from the roof. All the nozzles are mounted on the top.

There are several ways of recovering the boil-off-gases (BOG); usually it is either integrated into the main process plant refrigeration unit or kept as a stand alone system local to the storage tanks or a combination of both.

Refrigerated LPG storage is normally used in association with bulk shipping for long haul international transportation.

Chapter 6

Industrial Gases

INTRODUCTION

Industrial gases have performed varied and essential functions in our economy. Some are raw materials for the manufacture of other chemicals. This is particularly true of oxygen, nitrogen, and hydrogen. Nitrogen preserves the flavour of packaged foods by reducing chemical action leading to rancidity of canned fats. Some gases are essential medicaments, like oxygen and helium. However, many of these gases, their liquids, and their solids have a common application in creating cold largely by absorbing heat, upon evaporation, by performing work, or by melting. In past decades, the outstanding examples of this have been liquid carbon dioxide and dry ice. On the other hand, with the modern expansion of industry, a new division of engineering arisen called cryogenics.

CARBON DIOXIDE

Carbon dioxide is a chemical compound composed of two oxygen atoms covalently bonded to a single carbon atom. It is a gas at standard temperature and pressure and exists in earth's atmosphere in this state.

Carbon dioxide is used by plants during photosynthesis to make sugars which may either be consumed again in respiration or used as the raw material to produce polysaccharides such as starch and cellulose, proteins and the wide variety of other organic compounds required for plant growth and development. It is produced during respiration by plants, and by all animals, fungi and micro-organisms that depend on living and decaying plants for food, either directly or indirectly. It is, therefore, a major component of the carbon cycle. Carbon dioxide is generated as a by-product of the combustion of fossil fuels or the burning of vegetable matter, among other chemical processes. Large amounts of carbon dioxide are emitted from volcanoes and other geothermal processes such as hot springs and geysers and by the dissolution of carbonates in crustal rocks.

121

Carbon dioxide has no liquid state at pressures below 5.1 atm. At 1 atm it is a solid at temperatures below –78°C. In its solid state, carbon dioxide is commonly called dry ice.

CO_2 is an acidic oxide as an aqueous solution turns litmus from blue to pink.

CO_2 is toxic in higher concentrations: 1 per cent (10,000 ppm) will make some people feel drowsy. Concentrations of 7 to 10 per cent cause dizziness, headache, visual and hearing dysfunction, and unconsciousness within a few minutes to an hour.

Carbon dioxide is a colourless, odourless gas. When inhaled at concentrations much higher than usual atmospheric levels, it can produce a sour taste in the mouth and a stinging sensation in the nose and throat. These effects result from the gas dissolving in the mucous membranes and saliva, forming a weak solution of carbonic acid. Amounts above 5000 ppm are considered very unhealthy, and those above about 50,000 ppm (equal to 5 per cent by volume) are considered dangerous to animal life.

At standard temperature and pressure, the density of carbon dioxide is around 1.98 kg/m^3, about 1.5 times that of air. The carbon dioxide molecule (O=C=O) contains two double bonds and has a linear shape. It has no electrical dipole, and as it is fully oxidised, it is moderately reactive and is non-flammable, but will support the combustion of metals such as magnesium.

At –78.51°C or –109.3°F, carbon dioxide changes directly from a solid phase to a gaseous phase through sublimation or from gaseous to solid through deposition. Solid carbon dioxide is normally called dry ice, a generic trademark. Dry ice is commonly used as a cooling agent, and it is relatively inexpensive. A convenient property for this purpose is that solid carbon dioxide sublimes directly into the gas phase leaving no liquid. Liquid carbon dioxide forms only at pressures above 5.1 atm; the triple point of carbon dioxide is about 518 kPa at –56.6°C. The critical point is 7.38 MPa at 31.1°C.

An alternative form of solid carbon dioxide, an amorphous glass-like form, is possible, although not at atmospheric pressure. This form of glass, called carbonia, was produced by supercooling heated CO_2 at extreme pressure (40–48 GPa or about 4,00,000 atmospheres) in a diamond anvil. This discovery confirmed the theory that carbon dioxide could exist in a glass state similar to other members of its elemental family, like silicon (silica glass) and germanium. Unlike silica and germania glasses, however, carbonia glass is not stable at normal pressures and reverts back to gas when pressure is released.

Isolation and Production

Carbon dioxide may be obtained from air distillation. However, this yields only very small quantities of CO_2. A large variety of chemical reactions yield carbon dioxide, such as the reaction between most acids and most metal carbonates. For example, the reaction between hydrochloric acid and calcium carbonate (limestone or chalk) is depicted below:

$$2HCl + CaCO_3 \rightarrow CaCl_2 + H_2CO_3$$

The H_2CO_3 then decomposes to water and CO_2. Such reactions are accompanied by foaming or bubbling, or both. In industry such reactions are widespread because they can be used to neutralise waste acid streams.

The production of quicklime (CaO), a chemical that has widespread use, from limestone by heating at about 850°C also produces CO_2:

$$CaCO_3 \rightarrow CaO + CO_2$$

The combustion of all carbon containing fuels, such as methane (natural gas), petroleum distillates (gasoline, diesel, kerosene, propane), but also of coal and wood, will yield carbon dioxide and, in most cases, water. As an example the chemical reaction between methane and oxygen is given below.

$$CH_4 + 2O_2 \rightarrow CO_2 + 2H_2O$$

Iron is reduced from its oxides with coke in a blast furnace, producing pig iron and carbon dioxide:

$$2Fe_2O_3 + 3C \rightarrow 4Fe + 3CO_2$$

Yeast metabolises sugar to produce carbon dioxide and ethanol, also known as alcohol, in the production of wines, beers and other spirits, but also in the production of bioethanol.

All aerobic organisms produce CO_2 when they oxidise carbohydrates, fatty acids, and proteins in the mitochondria of cells.

Although there are many methods of producing CO_2, but the most important is from flue gas which results from burning carbonaceous material (fuel oil, gas, and coke) and contain 10 to 18 per cent of CO_2.

Carbon dioxide is soluble in water, in which it spontaneously interconverts between CO_2 and H_2CO_3 (carbonic acid). The relative concentrations of CO_2, H_2CO_3, and the deprotonated forms HCO_3^- (bicarbonate) and CO_3^{2-} (carbonate) depend on the pH. In neutral or slightly alkaline water (pH > 6.5), the bicarbonate form predominates (>50 per cent) becoming the most prevalent (>95 per cent) at the pH of seawater, while in very alkaline water (pH > 10.4) the predominant (>50 per cent) form is carbonate. The bicarbonate and carbonate forms are very soluble, such that air-equilibrated ocean water (mildly alkaline with typical pH = 8.2 – 8.5) contains about 120 mg of bicarbonate per litre. It is used in drinks, foods, pneumatic systems, fire extinguisher, welding, caffeine removal, pharmaceutical and other chemical processing, polymers and plastics, oil recovery, as refrigerants, coal-bed methane recovery, and wine making.

Typical Uses of Carbon Dioxide

Carbon dioxide, which two decades back was not considered as a pollutant and was well accepted as the end-product of carbon combustion, is of great concern today as one of the most significant contributors towards the greenhouse effect.

Due to growing proportions of this gas in our atmosphere, very serious environmental consequences like global warming are being associated. Carbon dioxide concentration is increasing due to growing industrialisation, cutting of plants and growing population on earth.

The natural balance that maintains these atmospheric gases within their accepted limits is disturbed and we are witnessing the adverse consequences of such changes like rise in temperature and erratic climatic changes.

Controlling CO_2 production is always a topic of debate, but dependence of our present lifestyles on technology mostly run by carbon combustion (in which CO_2 is the ultimate end-product) makes it impossible for policy makers to make a decision to abandon it.

The other option, i.e. safe disposal of CO_2 from the atmosphere has been mooted by planners and scientists and engineers are working hard to develop technologies making it possible, but with no great success so far.

Solar energy is well known and so are the industrial uses of carbon monoxide (CO). Carbon monoxide is a valuable commodity chemical that is widely used to make plastics and other products.

It is also a key ingredient in a process for making synthetic fuels, including syngas (a mixture largely of carbon monoxide and hydrogen), methanol and gasoline. While CO easily gets converted into CO_2, conversion of CO_2 to CO or into carbon is not practically so easy due to the endothermic nature of chemical reaction.

Plants do change CO_2 with the help of sunlight and plant chemistry, directly to carbon and oxygen. Heat has to be supplied along with maintaining some other favourable conditions for the reaction to proceed.

Researchers at the University of California, San Diego (UCSD), USA, recently demonstrated that light absorbed and converted into electricity by a silicon electrode can help drive a reaction that converts carbon dioxide into carbon monoxide and oxygen.

Chemists have shown that it is possible to use solar energy, paired with the right catalyst, to convert carbon dioxide into a raw material for making a wide range of products, including plastics and gasoline.

Scientists agree that such a process will not make a significant impact on reducing greenhouse gases in the atmosphere because that would take quite large-scale operations, but any chemical process that uses CO_2 as a feedstock, rather than having it be an end-product, is probably worth doing.

It may be possible that in future chemical manufacturers can make millions of pounds of plastics from greenhouse gases rather than making tons of greenhouse gases in the process.

The system may also be part of a solution to a continuing problem with solar energy. For solar panels to be useful when the sun is not shining, the electricity they produce has to be stored. A potentially practical way of doing that is by converting the electrical energy into chemical energy. One popular approach is to use solar cells to produce hydrogen, which could then be used in fuel cells. But hydrogen gas is much more difficult to transport and store than are liquid fuels, such as gasoline, which contain far more energy by volume than hydrogen does.

The UCSD system shows that it is possible to use solar energy to make carbon monoxide that then, together with hydrogen, can be converted into gasoline. Currently, carbon monoxide is made from natural gas and coal. But carbon dioxide is a more attractive raw material in part because it is very cheap and every one wants to get rid of its extra amount.

In the prototype device, sunlight passes through carbon dioxide dissolved in a solution before being absorbed by a semiconductor cathode, which converts photons into electrons. Aided by a catalyst, the electrons react with carbon dioxide to form carbon monoxide at the electrode.

At the anode — a catalyst made of platinum — water is converted into oxygen. In the first prototype, only about half of the energy needed for the reactions was supplied by the sun, with the rest coming from outside electricity.

That is because the researchers decided to prove the concept using silicon as the semiconductor. They are now working with a gallium-phosphide semiconductor, which has exactly the right electronic properties to drive the necessary reactions using sunlight alone.

To make a fuel, the carbon monoxide so produced can be combined with hydrogen to create syngas in a well-known technology called the Fischer-Tropsch process, which has been widely used to make gasoline from coal.

At this early stage, scientists involved say that commercial systems for converting CO_2 into CO making use of sunlight in presence of some suitable catalysts could be 10 years away. Modes of operation, under what conditions CO_2 from the atmosphere (directly or indirectly) can be used and the efficiency and economics of making fuels this way are not known.

Scientists are hopeful that for large-scale applications use catalyst-coated nanoparticles may be helpful in speeding up reactions of CO_2 conversion into CO directly from the atmosphere so that this atmospheric CO_2 menace may be put under control without sacrificing the advantages of carbon lifestyle.

CARBON MONOXIDE

Carbon monoxide, with the chemical formula CO, is a colourless and odourless, tasteless, yet highly toxic gas. Its molecules consist of one carbon atom covalently bonded to one oxygen atom. There are two covalent bonds and a coordinate covalent bond between the oxygen and carbon atoms.

Manufacture

Carbon monoxide is so fundamentally important that many methods have been developed for its production. Producer gas is formed by combustion of carbon in oxygen at high temperatures when there is an excess of carbon. In an oven, air is passed through a bed of coke. The initially produced CO_2 equilibrates with the remaining hot carbon to give CO.

Carbon monoxide is also produced from the partial oxidation of carbon-containing compounds, notably in internal-combustion engines. Carbon monoxide forms in preference to the more usual carbon dioxide when there is a reduced availability of oxygen present during the combustion process. Carbon monoxide has significant fuel value, burning in air with a characteristic blue flame, producing carbon dioxide. Despite its serious toxicity, CO plays a highly useful role in modern technology, being a precursor to myriad products.

Principal Chemical Reactions

Industrial uses

Carbon monoxide (CO) is a major industrial gas that has many applications in bulk chemicals manufacturing. High volume aldehydes are produced by the hydroformylation reaction of alkenes, CO, and H_2. In one of many applications of this technology, hydroformylation is coupled to the shell higher olefin process to give precursors to detergents.

Methanol is produced by the hydrogenation of CO. In a related reaction, the hydrogenation of CO is coupled to C—C bond formation, as in the Fischer-Tropsch process where CO is hydrogenated to liquid hydrocarbon fuels. This technology allows coal to be converted to petrol.

In the Monsanto process, carbon monoxide and methanol react in the presence of a homogeneous rhodium catalyst and HI to give acetic acid. This process is responsible for most of the industrial production of acetic acid.

Carbon monoxide is a principal component of syngas, which is often used for industrial power. Carbon monoxide is also used in industrial scale operations for purifying nickel (Mond process).

Toxicity

Carbon monoxide is a significantly toxic gas and has no odour or colour. It is the most common type of fatal poisoning in many countries. Exposures can lead to significant toxicity of the central nervous system and heart.

HYDROGEN

Hydrogen is the chemical element with atomic number 1. It is represented by the symbol H. At standard temperature and pressure, hydrogen is a colourless, odourless, nonmetallic, tasteless, highly flammable diatomic gas with the molecular formula H_2. With an atomic weight of 1.00794, hydrogen is the lightest element. Hydrogen is the most abundant chemical element, constituting roughly 75 per cent of the universe's elemental mass. Stars in the main sequence are mainly composed of hydrogen in its plasma state. Elemental hydrogen is relatively rare on earth. Industrial production is from hydrocarbons such as methane with most being used captively at the production site. The two largest uses are in fossil fuel processing (e.g. hydrocracking) and ammonia production mostly for the fertiliser market. Hydrogen may be produced from water by electrolysis at substantially greater cost than production from natural gas.

Hydrogen is important in metallurgy as it can embrittle many metals, complicating the design of pipelines and storage tanks. Hydrogen is highly soluble in many rare earth and transition metals and is soluble in both crystalline and amorphous metals.

Combustion

Hydrogen gas (dihydrogen) is highly flammable and will burn in air at a very wide range of concentrations between 4 and 75 per cent by volume. The enthalpy of combustion for hydrogen is 286 kJ/mol:

$$2H_2(g) + O_2(g) \rightarrow 2H_2O(l) + 572 \text{ kJ (286 kJ/mol)}$$

Hydrogen/oxygen mixtures are explosive across a wide range of proportions. Its autoignition temperature, the temperature at which it ignites spontaneously in air, is 560°C (1040°F). Pure hydrogen-oxygen flames emit ultraviolet light and are nearly invisible to the naked eye as illustrated by the faint plume of the space shuttle main engine compared to the highly visible plume of a space shuttle Solid Rocket Booster.

The detection of a burning hydrogen leak may require a flame detector; such leaks can be very dangerous. H_2 reacts with every oxidising element. Hydrogen can react spontaneously and violently at room temperature with chlorine and fluorine to form the corresponding halides: hydrogen chloride and hydrogen fluoride.

Manufacture of Hydrogen

Hydrogen can be prepared in several different ways, but economically the most important processes involve removal of hydrogen from hydrocarbons. Commercial bulk hydrogen is usually produced by the steam reforming of natural gas. At high temperatures (700°–1100°C; 1300°–2000°F), steam (water vapour) reacts with methane to yield carbon monoxide and H_2 (Fig. 6.1).

$$CH_4 + H_2O \rightarrow CO + 3H_2$$

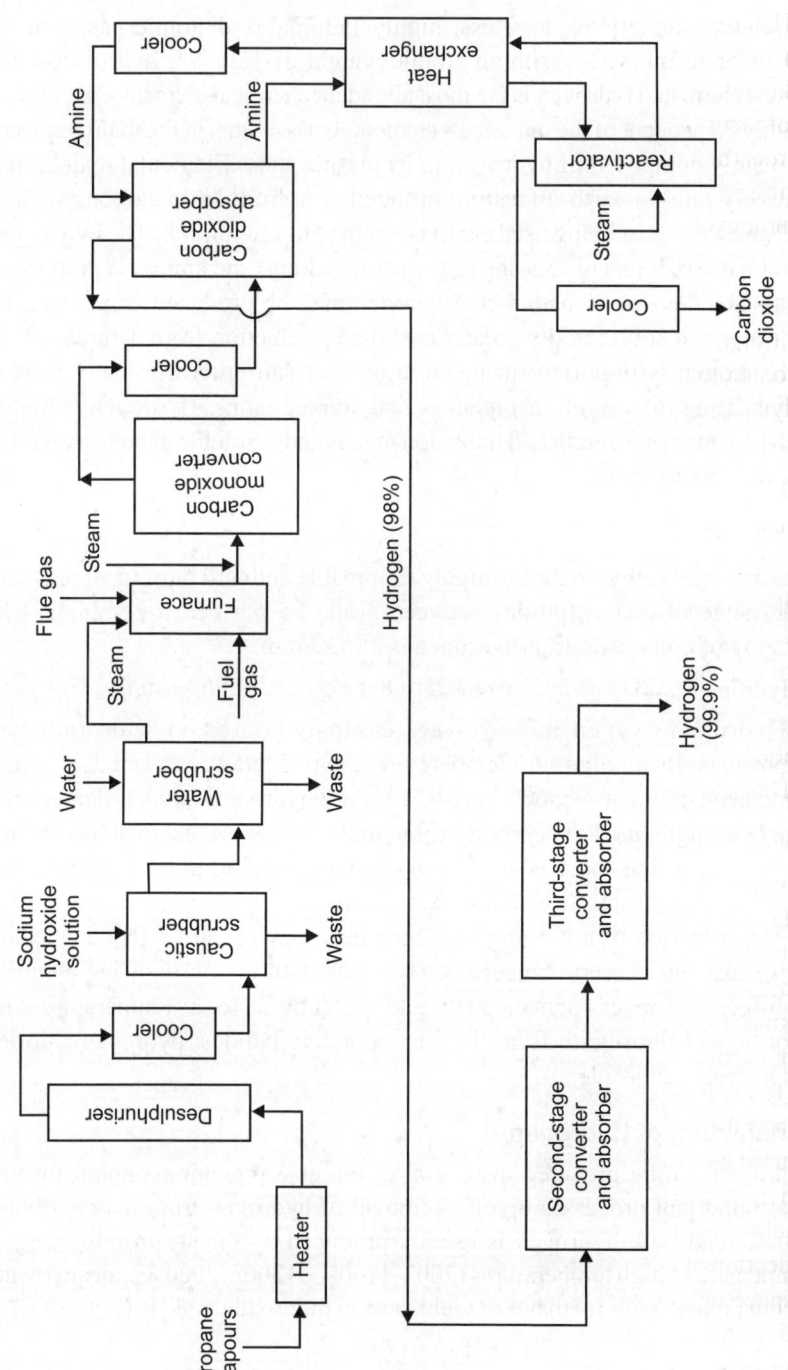

Fig. 6.1. Flow diagram for manufacture of hydrogen from hydrocarbons and steam (steam reformer process).

This reaction is favoured at low pressures but is nonetheless conducted at high pressures (20 atm; 600 in Hg) since high pressure H_2 is the most marketable product. The product mixture is known as 'synthesis gas' because it is often used directly for the production of methanol and related compounds. Hydrocarbons other than methane can be used to produce synthesis gas with varying product ratios. One of the many complications to this highly optimised technology is the formation of coke or carbon.

$$CH_4 \rightarrow C + 2H_2$$

Consequently, steam reforming typically employs an excess of H_2O. Additional hydrogen can be recovered from the steam by use of carbon monoxide through the water gas shift reaction, especially with an iron oxide catalyst. This reaction is also a common industrial source of carbon dioxide:

$$CO + H_2O \rightarrow CO_2 + H_2$$

Other important methods for H_2 production include partial oxidation of hydrocarbons:

$$2CH_4 + O_2 \rightarrow 2CO + 4H_2$$

and the coal reaction, which can serve as a prelude to the shift reaction above:

$$C + H_2O \rightarrow CO + H_2$$

Hydrogen is sometimes produced and consumed in the same industrial process, without being separated. In the Haber process for the production of ammonia, hydrogen is generated from natural gas. Electrolysis of brine to yield chlorine also produces hydrogen as a coproduct.

Applications

Large quantities of H_2 are needed in the petroleum and chemical industries. The largest application of H_2 is for the processing ('upgrading') of fossil fuels, and in the production of ammonia. The key consumers of H_2 in the petrochemical plant include hydrodealkylation, hydrodesulphurisation, and hydrocracking. H_2 has several other important uses. H_2 is used as a hydrogenating agent, particularly in increasing the level of saturation of unsaturated fats and oils (found in items such as margarine), and in the production of methanol. It is similarly the source of hydrogen in the manufacture of hydrochloric acid. H_2 is also used as a reducing agent of metallic ores.

In more recent applications, hydrogen is used pure or mixed with nitrogen (sometimes called forming gas) as a tracer gas for minute leak detection. Applications can be found in the automotive, chemical, power generation, aerospace, and telecommunications industries. Hydrogen is an authorised food additive that allows food package leak testing among other anti-oxidising properties.

Safety and Precautions

Inhalation of air with high concentration of hydrogen displaces oxygen and may cause the above symptoms as an asphyxant. Hydrogen poses a number of hazards to human safety, from potential detonations and fires when mixed with air to being an asphyxant in its pure, oxygen-free form. In addition, liquid hydrogen is a cryogen and presents dangers (such as frostbite) associated with very cold liquids. Hydrogen dissolves in some metals, and, in addition to leaking out, may have adverse effects on them, such as hydrogen embrittlement.

NATURAL GAS

Natural gas is often informally referred to as simply gas, especially when compared to other energy sources such as electricity. Before natural gas can be used as a fuel, it must undergo extensive processing to remove almost all materials other than methane.

The by-products of that processing include ethane, propane, butanes, pentanes and higher molecular weight hydrocarbons, elemental sulphur, and sometimes helium and nitrogen. Natural gas is already discussed in chapter 5. Here we are dicussing the various application of natural gas.

Biogas

When methane-rich gases are produced by the anaerobic decay of non-fossil organic matter (biomass), these are referred to as biogas (or natural biogas). Sources of biogas include swamps, marshes, and landfills, as well as sewage sludge and manure by way of anaerobic digesters, in addition to enteric fermentation particularly in cattle.

Methanogenic archaea are responsible for all biological sources of methane, some in symbiotic relationships with other life forms, including termites, ruminants, and cultivated crops. Methane released directly into the atmosphere would be considered a pollutant, however, methane in the atmosphere is oxidised, producing carbon dioxide and water. Biogas is already discuss in detail in chapter 5.

Methane in the atmosphere has a half-life of seven years, meaning that every seven years, half of the methane present is converted to carbon dioxide and water. Future sources of methane, the principal component of natural gas, include landfill gas, biogas and methane hydrate. Biogas, and especially landfill gas, are already used in some areas, but their use could be greatly expanded. Landfill gas is a type of biogas, but biogas usually refers to gas produced from organic material that has not been mixed with other waste.

Landfill gas is created from the decomposition of waste in landfills. If the gas is not removed, the pressure may get so high that it works its way to the surface, causing damage to the landfill structure, unpleasant odour, vegetation die-off and an explosion hazard. The gas can be vented to the atmosphere, flared or burned to produce electricity or heat.

Uses of Natural Gas

Power generation

Natural gas is a major source of electricity generation through the use of gas turbines and steam turbines. Particularly high efficiencies can be achieved through combining gas turbines with a steam turbine in combined cycle mode. Natural gas burns cleaner than other fossil fuels, such as oil and coal, and produces less carbon dioxide per unit energy released. For an equivalent amount of heat, burning natural gas produces about 30 per cent less carbon dioxide than burning petroleum and about 45 per cent less than burning coal. Combined cycle power generation using natural gas is thus the cleanest source of power available using fossil fuels, and this technology is widely used wherever gas can be obtained at a reasonable cost. Fuel cell technology may eventually provide cleaner options for converting natural gas into electricity, but as yet it is not price-competitive.

Residential domestic use

Natural gas is supplied to homes, where it is used for such purposes as cooking in natural gas-powered ranges and/or ovens, natural gas-heated clothes dryers, heating/cooling and central heating. Home or other building heating may include boilers, furnaces, and water heaters. CNG is used in rural homes without connections to piped-in public utility services, or with portable grills. However, due to CNG being less economical than LPG, LPG (propane) is the dominant source of rural gas.

Other uses

Natural gas vehicles, fertiliser, aviation, hydrogen, fabrics, glass, steel, plastics, paint, and other products.

Environmental Effects

Global climate change

Natural gas is often described as the cleanest fossil fuel, producing less carbon dioxide per joule delivered than either coal or oil. However, in absolute terms it does contribute substantially to global emissions, and this contribution is projected to grow. In addition, natural gas itself is a greenhouse gas far more potent than carbon dioxide when released into the atmosphere but is not of large concern due to the small amounts in which this occurs.

HYDROGEN

Hydrogen is manufactured from hydrocarbons and the processes is known as steam reformer process.

The largest quantities of hydrogen are manufactured by catalytically reacting hydrocarbons and steam, to yield hydrogen and carbon oxides, followed by the water-gas shift reaction. The most commonly used raw material is natural gas, although other natural and refinery hydrocarbons, including naphtha cuts may be used. The following description is based on propane as a raw material.

Reaction

$$CH_3CH_2CH_3 + 3H_2O \rightarrow 3CO + 7H_2$$

$$3CO + 3H_2O \rightarrow 3CO_2 + 3H_2$$

$$CH_3CH_2CH_3 + 6H_2O \rightarrow 3CO_2 + 10H_2$$

Commercial propane, obtained from either natural-gasoline plants or oil refineries, contains small amounts of organic sulphur compounds which are removed before processing. The propane from storage passes in the form of vapours through a heater at a temperature of about 370°C. The hot gases then pass over a bauxite or metallic oxide catalyst which converts the sulphur compounds (mercaptans, organic sulphides, and carbonyl sulphide) to hydrogen sulphide. After cooling, the gases are scrubbed with aqueous sodium hydroxide and water to remove the soluble sulphides. When natural gas is used as the process feed, sulphur is removed by passing the gas through drums containing activated carbon.

The sulphur-free propane vapours are mixed with steam and passed into the top of a reforming furnace. One type of furnace consists of a number of vertical nickel–chromium–iron alloy tubes, 7.5 to 20 cm in diameter and about 7.5 m long, mounted in a refractory furnace. Heat for the endothermic reaction is supplied by multiple horizontal burners located at various levels, with the flue gases passing upward counter-current to the process gas. At temperatures of 760 to 980°C, the propane gas passes down the tubes over a supported nickel catalyst at a space velocity of about 600 volumes/hour per volume of catalyst. The propane is converted to hydrogen, carbon monoxide, and carbon dioxide, with a trace of a methane remaining in the mixture.

Operating pressure in the reformer may be as high as 600 psi (4.2 MPa). Although the steam-hydrocarbon reaction is favoured by low pressure, there are compensating economic advantages in operating at high pressure. Accordingly, new plants usually are designed for high-pressure operation.

In any event, the reformed gases are cooled to about 370°C by mixing with steam, and then passed into the first-stage carbon monoxide converter containing an iron oxide catalyst promoted with chromium oxide. Here the exothermic conversion reaction (water-gas shift) takes place at a temperature of about 425°C and a space velocity of 100 (or greater) volumes of gas/volume of catalyst per hour. Both this catalyst and the nickel reforming catalyst are rugged and have a normal life of 1 year or more.

From the converter, the gases containing a small amount of carbon monoxide are cooled to about 38°C and passed into a packed (or bubble-tray) tower. Here aqueous 15 to 20 per cent monoethanolamine is circulated down through the counter-currently blowing gas (Girbotol process). The amine solution absorbs the carbon dioxide and, after passing through heat exchangers, is run to the top of a reactivating tower. Here the carbon dioxide is desorbed by steam generated by heating the solution in a reboiler at the bottom of the tower. The carbon dioxide removed amounts to about 30 volumes/100 volumes of hydrogen and, since it is recovered at a purity of 99.8 per cent, it is available as a useful by-product. The regenerated amine solution is then returned to the system. At atmospheric pressure and 38°C hydrogen gas containing 20 per cent carbon dioxide may be purified to 0.1 per cent carbon dioxide by scrubbing, with the monoethanolamine absorbing 15 to 30 m^3 carbon monoxide/m^3 of solution circulated. Approximately 0.12 kg steam/litre of solution is required for regeneration. The carbon dioxide-free hydrogen coming from the absorber still contains about 1 per cent carbon monoxide. This is removed by passing through two more stages of carbon monoxide conversion, followed by carbon dioxide removal.

OXYGEN

It is non-metallic gaseous element of atomic number 8. Atmospheric oxygen is the result of photosynthesis. It is colourless, odourless, tasteless diatomic gas. It constituents about 20 per cent by volume of air at sea level.

It is soluble in density 1.429 g/l at 0°C. It is soluble in water and alcohol. It is noncombustible, but actively supports combustion.

Manufacture

Two major methods are employed to produce O_2 extracted from air for industrial uses annually. The most common method is to fractionally-distil liquefied air into its various components, with nitrogen N_2 distilling as a vapour while oxygen O_2 is left as a liquid (Fig. 6.2). The other major method of producing O_2 gas involves passing a stream of clean, dry air through one bed of a pair of identical zeolite molecular sieves, which absorbs the nitrogen and delivers a gas stream that is 90 to 93 per cent O_2. Simultaneously, nitrogen gas is released from the other nitrogen-saturated zeolite bed, by reducing the chamber operating pressure and diverting part of the oxygen gas from the producer bed through it, in the reverse direction of flow.

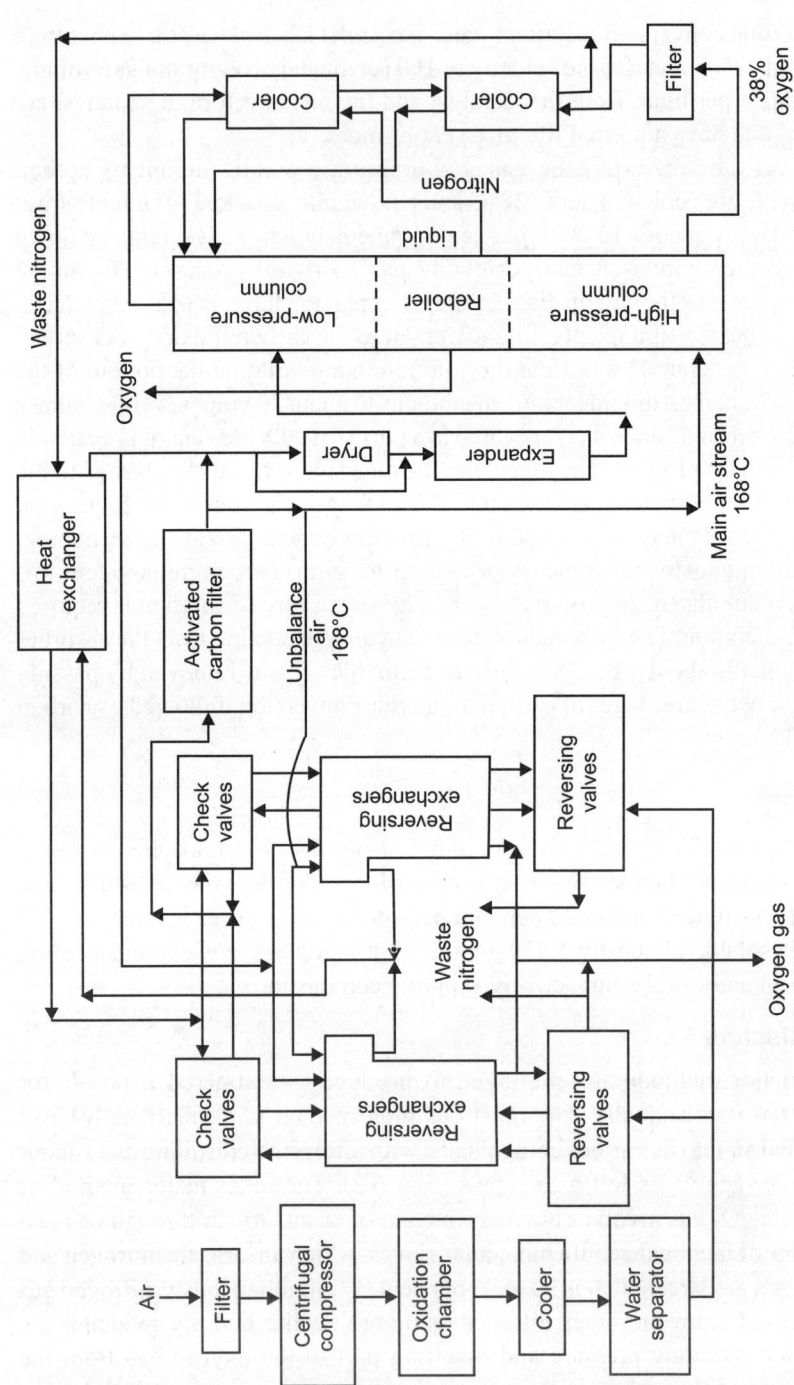

Fig. 6.2. Flow diagram for manufacture of oxygen from air by liquefaction (modified Linde-Frankl cycle).

After a set cycle time the operation of the two beds is interchanged, thereby allowing for a continuous supply of gaseous oxygen to be pumped through a pipeline.

This is known as pressure swing adsorption. Oxygen gas is increasingly obtained by these non-cryogenic technologies.

Oxygen gas can also be produced through electrolysis of water into molecular oxygen and hydrogen. A similar method is the electrocatalytic O_2 evolution from oxides and oxoacids. Chemical catalysts can be used as well, such as in chemical oxygen generators or oxygen candles that are used as part of the life-support equipment on submarines, and are still part of standard equipment on commercial airliners in case of depressurisation emergencies. Another air separation technology involves forcing air to dissolve through ceramic membranes based on zirconium dioxide by either high pressure or an electric current, to produce nearly pure O_2 gas.

For reasons of economy, oxygen is often transported in bulk as a liquid in specially-insulated tankers, since one litre of liquefied oxygen is equivalent to 840 litres of gaseous oxygen at atmospheric pressure and 20°C. Such tankers are used to refill bulk liquid oxygen storage containers, which stand outside hospitals and other institutions with a need for large volumes of pure oxygen gas. Liquid oxygen is passed through heat exchangers, which convert the cryogenic liquid into gas before it enters the building. Oxygen is also stored and shipped in smaller cylinders containing the compressed gas; a form that is useful in certain portable medical applications and oxy-fuel welding and cutting.

Uses

Medical

Uptake of O_2 from the air is the essential purpose of respiration, so oxygen supplementation is used in medicine. Oxygen therapy is used to treat emphysema, pneumonia, some heart disorders, and any disease that impairs the body's ability to take up and use gaseous oxygen. Treatments are flexible enough to be used in hospitals, the patient's home, or increasingly by portable devices. Oxygen is also used medically for patients who require mechanical ventilation, often at concentrations above 21 per cent found in ambient air.

Life support and recreational use

A notable application of O_2 as a low-pressure breathing gas is in modern space suits, which surround their occupant's body with pressurised air. These devices use nearly pure oxygen at about one-third normal pressure, resulting in a normal blood partial pressure of O_2. This trade-off of higher oxygen concentration for lower pressure is needed to maintain flexible spacesuits. People who climb mountains or fly in non-pressurised fixed-wing aircraft sometimes have supplemental O_2 supplies.

Industrial

Smelting of iron ore into steel consumes 55 per cent of commercially-produced oxygen. In this process, O_2 is injected through a high-pressure lance into molten iron, which removes sulphur impurities and excess carbon as the respective oxides, SO_2 and CO_2. The reactions are exothermic, so the temperature increases to 1700°C. Another 25 per cent of commercially-produced oxygen is used by the chemical industry. Ethylene is reacted with O_2 to create ethylene oxide, which, in turn, is converted into ethylene glycol; the primary feeder material used to manufacture a host of products, including antifreeze and polyester polymers (the precursors of many plastics and fabrics).

Most of the remaining 20 per cent of commercially-produced oxygen is used in medical applications, metal cutting and welding, as an oxidiser in rocket fuel, and in water treatment. Oxygen is used in oxyacetylene welding burning acetylene with O_2 to produce a very hot flame. In this process, metal up to 60 cm thick is first heated with a small oxy-acetylene flame and then quickly cut by a large stream of O_2. Rocket propulsion requires a fuel and an oxidiser. Larger rockets use liquid oxygen as their oxidiser, which is mixed and ignited with the fuel for propulsion.

NITROGEN

Nitrogen is a chemical element that has the symbol N and atomic number 7 and atomic mass 14.00674 µ. Elemental nitrogen is a colourless, odourless, tasteless and mostly inert diatomic gas at standard conditions, constituting 78 per cent by volume of earth's atmosphere.

Many industrially important compounds, such as ammonia, nitric acid, organic nitrates (propellants and explosives), and cyanides, contain nitrogen. The extremely strong bond in elemental nitrogen dominates nitrogen chemistry, causing difficulty for both organisms and industry in converting the N_2 into useful compounds, and releasing large amounts of energy when these compounds burn or decay back into nitrogen gas.

It is a constituent element of amino acids and thus of proteins, and of nucleic acids (DNA and RNA). It resides in the chemical structure of almost all neurotransmitters, and is a defining component of alkaloids, biological molecules produced by many organisms.

Properties

Nitrogen is a nonmetal, with an electronegativity of 3.04. It has five electrons in its outer shell and is therefore trivalent in most compounds. The triple bond in molecular nitrogen (N_2) is the strongest in nature. The resulting difficulty of converting (N_2) into other compounds, and the ease (and associated high energy release) of converting nitrogen compounds into elemental N_2, have dominated the role of nitrogen in both nature and human economic activities.

At atmospheric pressure molecular nitrogen condenses (liquifies) at 77 K (−195.8°C) and freezes at 63 K (−210.0°C) into the β-hexagonal close-packed crystal allotropic form.

Uses

Nitrogen gas is an industrial gas produced by the fractional distillation of liquid air, or by mechanical means using gaseous air (i.e. pressurised reverse osmosis membrane or pressure swing adsorption). Commercial nitrogen is often a by-product of air-processing for industrial concentration of oxygen for steelmaking and other purposes. When supplied compressed in cylinders it is often referred to as OFN (oxygen-free nitrogen).

Nitrogen gas has a wide variety of applications, including serving as an inert replacement for air where oxidation is undesirable:

1. To preserve the freshness of packaged or bulk foods (by delaying rancidity and other forms of oxidative damage)
2. In ordinary incandescent light bulbs as an inexpensive alternative to argon.
3. On top of liquid explosives for safety measures.
4. The production of electronic parts such as transistors, diodes, and integrated circuits.
5. Dried and pressurised, as a dielectric gas for high voltage equipment.
6. The manufacturing of stainless steel.
7. Use in military aircraft fuel systems to reduce fire hazard.
8. Filling automotive and aircraft tyres due to its inertness and lack of moisture or oxidative qualities, as opposed to air, though this is not necessary for consumer automobiles.

Nitrogen molecules are less likely to escape from the inside of a tyre compared with the traditional air mixture used. Air consists mostly of nitrogen and oxygen. Nitrogen molecules have a larger effective diameter than oxygen molecules and therefore diffuse through porous substances more slowly.

Molecular nitrogen, a diatomic gas, is apt to dimerise into a linear four nitrogen long polymer. This is an important phenomenon for understanding high-voltage nitrogen dielectric switches because the process of polymerisation can continue to lengthen the molecule to still longer lengths in the presence of an intense electric field. A nitrogen polymer fog is thereby created. The second virial coefficient of nitrogen also shows this effect as the compressibility of nitrogen gas is changed by the dimerisation process at moderate and low temperatures. Nitrogen tanks are also replacing carbon dioxide as the main power source for paintball guns. The downside is that nitrogen must be kept at higher pressure than CO_2, making N_2 tanks heavier and more expensive.

RARE GASES OF ATMOSPHERE

Some of the rare earth gases are discussed below.

Neon

Neon is the chemical element that has the symbol Ne and atomic number 10. Although not a very common element in the universe, it is rare on earth. A colourless, inert noble gas under standard conditions, neon gives a distinct reddish-orange glow when used in discharge tubes and neon lamps. It is commercially extracted from air, in which it is found in trace amounts.

Neon is the second-lightest noble gas. According to recent studies, neon is the least reactive noble gas and thus the least reactive of all elements. Also, neon has the narrowest liquid range of any element. It liquefies at $-245.92°C$, f.p. $-248.6°C$. It has over 40 times the refrigerating capacity of liquid helium and three times that of liquid hydrogen (on a per unit volume basis). In most applications it is a less expensive refrigerant than helium.

Two quite different kinds of neon lights are in common use. Glow-discharge lamps are typically tiny, and often designed to operate at 120 volts; they are widely used as power-on indicators and in circuit-testing equipment. Neon signs and other arc-discharge devices operate instead at high voltages, often 3–15 kilovolts (3000–15,000 volts); they can be made into (often bent) tubes a few metres long.

Neon is actually abundant on a universal scale: the fifth most abundant chemical element in the universe by mass, after hydrogen, helium, oxygen, and carbon. Its relative rarity on earth, like that of helium, is due to its relative lightness and chemical inertness, both properties keeping it from being trapped in the condensing gas and dust clouds of the formation of smaller and warmer solid planets like earth.

Neon is a monatomic gas at standard conditions, making it lighter than the molecules of diatomic nitrogen and oxygen which form the bulk of earth's atmosphere; a balloon filled with neon will rise up into the air, albeit more slowly than a helium balloon.

Neon is rare on earth, found in the earth's atmosphere at 1 part in 65,000 (by volume) or 1 part in 83,000 by mass. It is industrially produced by cryogenic fractional distillation of liquefied air.

Neon is often used in signs and produces an unmistakable bright reddish-orange light. Although still referred to as 'neon', all other colours are generated with the other noble gases or by many colours of fluorescent lighting.

Neon is used in vacuum tubes, high-voltage indicators, lightning arrestors, wave metre tubes, television tubes, and helium-neon lasers. Liquefied neon is commercially used as a cryogenic refrigerant in applications not requiring the lower temperature range attainable with more extreme liquid helium refrigeration. Liquid neon is actually quite expensive, and nearly impossible to obtain in small quantities for laboratory tests.

Krypton

Krypton is a chemical element with the symbol Kr and atomic number 36. It is a member of group 18 of 'Periodic Table'. A colourless, odourless, tasteless noble gas, krypton occurs in trace amounts in the atmosphere, is isolated by fractionally distilling liquified air, and is often used with other rare gases in fluorescent lamps. Krypton is inert for most practical purposes. Krypton can also form clathrates with water when atoms of it are trapped in a lattice of the water molecules.

Krypton, like the other noble gases, can be used in lighting and photography. Krypton light has a large number of spectral lines, and krypton's high light output in plasmas allows it to play an important role in many high-powered gas lasers, which pick out one of the many spectral lines to amplify. There is also a specific krypton fluoride laser. The high power and relative ease of operation of krypton discharge tubes caused, the official metre (metric distance) to be defined in terms of one orange-red spectral line of krypton-86.

Krypton is characterised by a brilliant green and orange spectral signature. It is one of the products of uranium fission. Solidified krypton is white and crystalline with a face-centered cubic crystal structure, which is a common property of all noble gases.

The earth has retained all of the noble gases that were present at its formation except for helium. Helium atoms are very light, and move fast enough to escape the earth's gravity readily. Krypton's concentration in the atmosphere is about 1 ppm. It can be extracted from liquid air by fractional distillation. The amount of krypton in space is uncertain, as the amount is derived from the meteoritic activity and that from solar winds. The first measurements suggest an overabundance of krypton in space. Like the other noble gases, krypton is chemically unreactive.

Krypton's multiple emission lines make ionised krypton gas discharges appear whitish, which in turn makes krypton-based bulbs useful in photography as a brilliant white light source. Krypton is thus used in some types of photographic flashes used in high speed photography. Krypton gas is also combined with other gases to make luminous signs that glow with a bright greenish-yellow light.

Krypton is mixed with argon as the fill gas of energy saving fluorescent lamps. This reduces their operating voltage and power consumption. Unfortunately it also reduces their light output and raises their cost. Krypton costs 100 times as much as argon. Krypton (along with xenon) is also used to fill incandescent lamps to reduce filament evaporation and allow higher operating temperatures to be used for the filament. A brighter light results which contains more blue than conventional lamps.

Krypton's white discharge is often used to good effect in coloured gas discharge tubes, which are then simply painted or stained in other ways to allow the desired colour (for example, 'neon' type advertising signs where the letters appear in differing colours, are often entirely krypton-based). Krypton is also capable of much higher light power density than neon in the red spectral line region, and for this reason, red lasers for high power laser light shows are often krypton lasers with mirrors which select out the red spectral line for laser amplification and emission, rather than the more familiar helium-neon variety, which could never practically achieve the multi-watt red laser light outputs needed for this application.

Krypton has an important role in production and usage of the krypton fluoride laser. The laser has been important in the nuclear fusion energy research community in confinement experiments. The laser has high beam uniformity, short wavelength, and the ability to modify the spot size to track an imploding pellet. In experimental particle physics, liquid krypton is used to construct quasi-homogeneous electromagnetic calorimeters.

Krypton as a gas may also cause asphyxiate if large amounts of the gas has been inhaled, this is due to it's equivalent inert properties that it shares with other noble gases such as xenon and argon that allows krypton gas to displace oxygen.

Xenon

Xenon is a chemical element represented by the symbol Xe. Its atomic number is 54. A colourless, heavy, odourless noble gas, xenon occurs in the Earth's atmosphere in trace amounts. Although generally unreactive, xenon can undergo a few chemical reactions such as the formation of xenon hexafluoroplatinate, the first noble gas compound to be synthesised.

Naturally occurring xenon consists of nine stable isotopes. There are also over 40 unstable isotopes that undergo radioactive decay. The isotope ratios of xenon are an important tool for studying the early history of the solar system. Xenon-135 is produced as a result of nuclear fission and acts as a neutron absorber in nuclear reactors. Xenon is used in flash lamps and arc lamps, and as a general anesthetic. The first excimer laser design used a xenon dimer molecule (Xe_2) as its lasing medium, and the earliest laser designs used xenon flash lamps as pumps. Xenon is also being used to search for hypothetical weakly interacting massive particles and as the propellant for ion thrusters in spacecraft.

Xenon is a trace gas in earth's atmosphere, occurring at 0.087 ± 0.001 parts per million (μl/l), or approximately 1 part per 11.5 million, and is also found in gases emitted from some mineral springs. Some radioactive species of xenon, for example, ^{133}Xe and ^{135}Xe, are produced by neutron irradiation of fissionable material within nuclear reactors.

Xenon is obtained commercially as a by-product of the separation of air into oxygen and nitrogen. After this separation, generally performed by fractional distillation in a double-column plant, the liquid oxygen produced will contain small quantities of krypton and xenon. By additional fractional distillation steps, the liquid oxygen may be enriched to contain 0.1–0.2 per cent of a krypton/xenon mixture, which is extracted either via adsorption onto silica gel or by distillation. Finally, the krypton/xenon mixture may be separated into krypton and xenon via distillation. Extraction of a litre of xenon from the atmosphere requires 220 watt-hours of energy.

Xenon is relatively rare in the sun's atmosphere, on earth, and in asteroids and comets. The atmosphere of Mars shows a xenon abundance similar to that of earth: 0.08 parts per million, however, Mars shows a higher proportion of ^{129}Xe than the earth or the sun. As this isotope is generated by radioactive decay, the result may indicate that Mars lost most of its primordial atmosphere, possibly within the first 100 million years after the planet was formed. By contrast, the planet Jupiter has an unusually high abundance of xenon in its atmosphere; about 2.6 times as much as the Sun. This high abundance remains unexplained and may have been caused by an early and rapid build-up of planetesimals—small, subplanetary bodies—before the presolar disk began to heat up. (Otherwise, xenon would not have been trapped in the planetesimal ices.) Within the solar system, the nucleon fraction for all isotopes of xenon is 1.56×10^{-8}, or one part in 64 millions of the total mass. The problem of the low terrestrial xenon may potentially be explained by covalent bonding of xenon to oxygen within quartz, hence reducing the outgassing of xenon into the atmosphere.

Unlike the lower mass noble gases, the normal stellar nucleo-synthesis process inside a star does not form xenon. Elements more massive than iron-56 have a net energy cost to produce through fusion, so there is no energy gain for a star to create xenon. Instead, many isotopes of xenon are formed during supernova explosions.

An atom of xenon is defined as having a nucleus with 54 protons. At standard temperature and pressure, pure xenon gas has a density of 5.761 kg/m^3, about

4.5 times the surface density of the earth's atmosphere, 1.217 kg/m^3. As a liquid, xenon has a density of up to 3.100 g/ml, with the density maximum occurring at the triple point. Under the same conditions, the density of solid xenon, 3.640 g/cm^3, is larger than the average density of granite, 2.75 g/cm^3. Using gigapascals of pressure, xenon has been forced into a metallic phase.

Xenon is a member of the zero-valence elements that are called noble or inert gases. It is inert to most common chemical reactions (such as combustion, for example) because the outer valence shell contains eight electrons. This produces a stable, minimum energy configuration in which the outer electrons are tightly bound. However, xenon can be oxidised by powerful oxidising agents, and many xenon compounds have been synthesised.

In a gas-filled tube, xenon emits a blue or lavenderish glow when the gas is excited by electrical discharge. Xenon emits a band of emission lines that span the visual spectrum, but the most intense lines occur in the region of blue light, which produces the colouration. Although xenon is rare and relatively expensive to extract from the earth's atmosphere, it still has a number of applications. Xenon is used in light-emitting devices called xenon flash lamps, which are used in photographic flashes and stroboscopic lamps. Xenon has been used as a general anaesthetic, although it is expensive.

Gamma emission from the radioisotope ^{133}Xe of xenon can be used to image the heart, lungs, and brain, for example, by means of single photon emission computed tomography. ^{133}Xe has also been used to measure blood flow. In nuclear energy applications, xenon is used in bubble chambers, probes, and in other areas where a high molecular weight and inert nature is desirable.

Xenon gas can be safely kept in normal sealed glass or metal containers at standard temperature and pressure. However, it readily dissolves in most plastics and rubber, and will gradually escape from a container sealed with such materials.

Argon

Argon is a chemical element designated by the symbol Ar. Argon has atomic number 18 and is the third element in group 18 of the periodic table (noble gases). Argon is present in the Earth's atmosphere at 0.94 per cent. Terrestrially, it is the most abundant and most frequently used of the noble gases. Argon's full outer shell makes it stable and resistant to bonding with other elements. Its triple point temperature of 83.8058 K is a defining fixed point in the International Temperature Scale of 1990.

Argon has approximately the same solubility in water as oxygen gas and is 2.5 times more soluble in water than nitrogen gas. Argon is colourless, odourless, tasteless and nontoxic in both its liquid and gaseous forms. Argon is inert under most conditions and forms no confirmed stable compounds at room temperature. Although argon is a noble gas, it has been found to have the capability of forming some compounds. For example, the creation of argon fluorohydride (HArF), a metastable compound of argon with fluorine and hydrogen. Argon constitutes 0.934 per cent by volume and 1.29 per cent by mass of the Earth's atmosphere, and air is the primary raw material used by industry to produce purified argon products. Argon is isolated from air by fractionation, most commonly by cryogenic fractional distillation, a process that also produces purified nitrogen, oxygen, neon, krypton and xenon.

Argon is produced industrially by the partial distillation of liquid air, a process that separates liquid nitrogen, which boils at 77.3 K, from argon, which boils at 87.3 K and oxygen, which boils at 90.2 K.

Argon-40, the most abundant isotope of argon, is produced by the decay of potassium-40 with a half-life of 1.26×10^9 years by electron capture or positron emission. Because of this, it is used in potassium-argon dating to determine the age of rocks. Argon is used in some high-temperature industrial processes, where ordinarily unreactive substances become reactive. For example, an argon atmosphere is used in graphite electric furnaces to prevent the graphite from burning. For some of these processes, the presence of nitrogen or oxygen gases might cause defects within the material. Argon is used in various types of metal inert gas welding such as tungsten inert gas welding, as well as in the processing of titanium and other reactive elements. An argon atmosphere is also used for growing crystals of silicon and germanium.

Argon is an asphyxiant in the poultry industry, either for mass culling following disease outbreaks, or as a means of slaughter more humane than the electric bath. Argon's relatively high density causes it to remain close to the ground during gassing.

Its non-reactive nature makes it suitable in a food product, and since it replaces oxygen within the dead bird, argon also enhances shelf-life.

Argon is also available in aerosol-type cans, which may be used to preserve compounds such as varnish, polyurethane, paint, etc. for storage after opening.

Helium

Helium (He) is a colourless, odourless, tasteless, non-toxic, inert monatomic chemical element that heads the noble gas group in the periodic table and whose atomic number is 2. Its boiling and melting points are the lowest among the elements and it exists only as a gas except in extreme conditions.

Helium is the second lightest element and is the second most abundant in the observable universe. Most helium was formed during the big bang, but new helium is being created as a result of the nuclear fusion of hydrogen in stars. On earth, helium is relatively rare and is created by the natural radioactive decay of some elements, as alpha particles that are emitted consist of helium nuclei. This radiogenic helium is trapped with natural gas in concentrations up to seven per cent by volume, from which it is extracted commercially by a low-temperature separation process called fractional distillation. Helium is the least reactive noble gas after neon and thus the second least reactive of all elements; it is inert and monoatomic in all standard conditions. Due to helium's relatively low molar (atomic) mass, in the gas phase its thermal conductivity, specific heat, and sound speed are all greater than any other gas except hydrogen. For similar reasons, and also due to the small size of helium atoms, helium's diffusion rate through solids is three times that of air and around 65 per cent that of hydrogen.

Helium is less water soluble than any other gas known, and helium's index of refraction is closer to unity than that of any other gas. Helium has a negative Joule-Thomson coefficient at normal ambient temperatures, meaning it heats up when allowed to freely expand. Only below its Joule-Thomson inversion temperature (of about 32 to 50 K at 1 atmosphere) does it cool upon free expansion. Once pre-cooled below this temperature, helium can be liquefied through expansion cooling.

Nearly all helium on earth is a result of radioactive decay. The decay product is primarily found in minerals of uranium and thorium, including cleveites, pitchblende, carnotite and monazite, because they emit alpha particles, which consist of helium nuclei (He^{2+}) to which electrons readily combine. There are also small amounts in mineral springs, volcanic gas, and meteoric iron. Because helium is trapped in a similar way by non-permeable layer of rock like natural gas the greatest concentrations on the planet are found in natural gas, from which most commercial helium is derived.

For large-scale use, helium is extracted by fractional distillation from natural gas, which contains up to 7 per cent helium. Since helium has a lower boiling point than any other element, low temperature and high pressure are used to liquefy nearly all the other gases (mostly nitrogen and methane). The resulting crude helium gas is purified by successive exposures to lowering temperatures, in which almost all of the remaining nitrogen and other gases are precipitated out of the gaseous mixture. Activated charcoal is used as a final purification step, usually resulting in 99.995 per cent pure Grade-A helium. The principal impurity in Grade-A helium is neon. In a final production step, most of the helium that is produced is liquefied via a cryogenic process. This is necessary for applications requiring liquid helium and also allows helium suppliers to reduce the cost of long distance transportation, as the largest liquid helium containers have more than five times the capacity of the largest gaseous helium tube trailers.

In the United States, most helium is extracted from natural gas of the Hugoton and nearby gas fields in Kansas, Oklahoma, and Texas. Diffusion of crude natural gas through special semipermeable membranes and other barriers is another method to recover and purify helium. Helium can be synthesised by bombardment of lithium or boron with high-velocity protons, but this is not an economically viable method of production. Helium is used for many purposes that require some of its unique properties, such as its low boiling point, low density, low solubility, high thermal conductivity, or inertness. Helium is commercially available in either liquid or gaseous form.

Because of its low density and incombustibility, helium is the gas of choice to fill airships such as the Goodyear blimp.

Because it is lighter than air, airships and balloons are inflated with helium for lift. While hydrogen gas is approximately 7 per cent more buoyant, helium has the advantage of being non-flammable (in addition to being fire retardant). In rocketry, helium is used as an ullage medium to displace fuel and oxidisers in storage tanks and to condense hydrogen and oxygen to make rocket fuel. It is also used to purge fuel and oxidiser from ground support equipment prior to launch and to pre-cool liquid hydrogen in space vehicles.

Helium is used as a protective gas in growing silicon and germanium crystals, in titanium and zirconium production, and in gas chromatography, because it is inert. Because of its inertness, thermally and calorically perfect nature, high speed of sound, and high value of the heat capacity ratio, it is also useful in supersonic wind tunnels and impulse facilities.

Because it diffuses through solids at three times the rate of air, helium is used as a tracer gas to detect leaks in high-vacuum equipment and high-pressure containers. Helium, mixed with a heavier gas such as xenon, is useful for thermoacoustic refrigeration due to the resulting high heat capacity ratio and low Prandtl number. The inertness of helium has environmental advantages over conventional refrigeration systems which contribute to ozone depletion or global warming. Liquid helium is used to cool the superconducting magnets in modern MRI scanners.

The use of helium reduces the distorting effects of temperature variations in the space between lenses in some telescopes, due to its extremely low index of refraction. This method is especially used in solar telescopes where a vacuum tight telescope tube would be too heavy.

The age of rocks and minerals that contain uranium and thorium can be estimated by measuring the level of helium with a process known as helium dating. Liquid helium is used to cool certain metals to the extremely low temperatures required for superconductivity, such as in superconducting magnets for magnetic resonance imaging.

Neutral helium at standard conditions is non-toxic, plays no biological role and is found in trace amounts in human blood. If enough helium is inhaled that oxygen needed for normal respiration is replaced asphyxia is possible.

The safety issues for cryogenic helium are similar to those of liquid nitrogen; its extremely low temperatures can result in cold burns and the liquid to gas expansion ratio can cause explosions if no pressure-relief devices are installed.

ACETYLENE

Acetylene, C_2H_2, is the simplest hydrocarbon alkyne chemical compound. As an alkyne, acetylene is unsaturated because its two carbon atoms are bonded together in a triple bond.

Preparation

The principal raw materials for acetylene manufacture are calcium carbonate (limestone) and coal. The calcium carbonate is first converted into calcium oxide and the coal into coke, then the two are reacted with each other to form calcium carbide and carbon monoxide:

$$CaO + 3C \rightarrow CaC_2 + CO$$

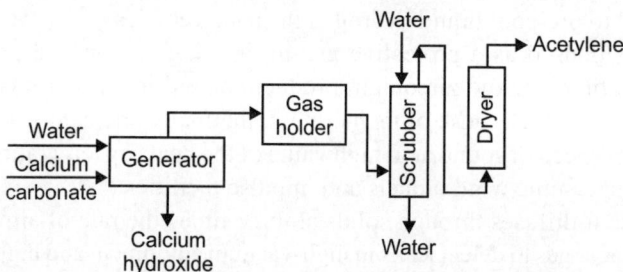

Fig. 6.3. Flow diagram for manufacture of acetylene from calcium carbide.

Calcium carbide (or calcium acetylide) and water are then reacted by any of several methods to produce acetylene and calcium hydroxide (Fig. 6.3).

$$CaC_2 + 2H_2O \rightarrow Ca(OH)_2 + C_2H_2$$

Calcium carbide synthesis requires an extremely high temperature, 2000°C, so the reaction is performed in an electric arc furnace. This reaction was an important part of the late-1800s revolution in chemistry enabled by the massive hydroelectric power project at Niagra Falls. Acetylene can also be manufactured by the partial combustion of methane with oxygen, or by the cracking of hydrocarbons. Berthelot was able to prepare acetylene from methyl alcohol, ethyl alcohol, ethylene, or ether, when he passed any one of these as a gas or vapour through a red-hot tube. Berthelot also found that acetylene was formed by sparking electricity through mixed cyanogen and hydrogen gases. He was also able to form acetylene directly by combining pure hydrogen with carbon using electrical discharge of a carbon arc.

Reactions

1. Above 400°C (673 K) the pyrolysis of acetylene will start, which is relatively low for a hydro-carbon. The main products are the dimer vinylacetylene (C_4H_4) and benzene. At temperatures above 900°C (1173 K), the main product will be soot.

2. Using acetylene, Berthelot was the first to show that an aliphatic compound could form an aromatic compound when he heated acetylene in a glass tube to produce benzene with some toluene. Berthelot oxidised acetylene to yield acetic acid and oxalic acid. He found acetylene could be reduced to form ethylene and ethane.

3. Polymerisation of acetylene with Ziegler-Natta catalysts produces polyacetylene films. Polyacetylene, a chain of carbon molecules with alternating single and double bonds, was the first organic semiconductor to be discovered; reaction with iodine produces an extremely conductive material.

4. Formation of acetylides with many metal ions, when bubbled through the solutions of their salts. Several, e.g. silver acetylide and copper acetylide, are powerful explosives. Copper acetylide is also formed by reacting acetylene with metallic copper or its alloys; these materials are therefore unsuitable for installations for handling acetylene.

Uses

Approximately 80 per cent of the acetylene produced annually is used in chemical synthesis. The remaining 20 per cent is used primarily for oxyacetylene gas welding and cutting due to the high temperature of the flame; combustion of acetylene with oxygen produces a flame of over 3300°C (6000°F), releasing 11.8 kJ/g. Oxyacetylene is the hottest burning common fuel gas. In modern times acetylene is sometimes used for carburisation (that is, hardening) of steel when the object is too large to fit into a furnace.

Acetylene is the third hottest chemical flame behind Cyanogen at 4525°C (8180°F) and dicyanoacetylene's 5260 K (4990°C, 9010°F). Acetylene has been proposed as a carbon feedstock for molecular manufacturing using nanotechnology. Acetylene is used to volatilise carbon in radiocarbon dating. The carbonaceous material in an archeological sample is reacted with lithium metal in a small specialised research furnace to form lithium carbide (also known as lithium acetylide).

The carbide can then be reacted with water, as usual, to form acetylene gas to be fed into mass spectrometer to sort out the isotopic ratio of carbon 14 to carbon 12. The use of acetylene is expected to continue a gradual increase in the future as new applications are developed. One new application is the conversion of acetylene to ethylene for use in making a variety of polyethylene plastics. In the past, a small amount of acetylene had been generated and wasted as part of the steam cracking process used to make ethylene.

Safety and Handling

Due to the carbon-to-carbon triple bond, acetylene gas is fundamentally unstable, and will decompose in an exothermic reaction if compressed to any great extent. Acetylene can explode with extreme violence if the pressure of the gas exceeds about 200 kPa (39 psi) as a gas or when in liquid or solid form, so it is shipped and stored dissolved in acetone or dimethylformamide (DMF), contained in a metal cylinder with porous filling (Agamassan), which renders it safe to transport and use.

At pressures above 15 psi the gas becomes extremely unstable, and can be ignited by shock.

For use in welding and cutting, the working pressures must be controlled by a regulator or the gas will spontaneously combust.

SULPHUR DIOXIDE

Sulphur dioxide is the chemical compound with the formula SO_2. It is produced by volcanoes and in various industrial processes. Since coal and petroleum often contain sulphur compounds, their combustion generates sulphur dioxide. Further oxidation of SO_2, usually in the presence of a catalyst such as NO_2, forms H_2SO_4, and thus acid rain.

This is one of the causes for concern over the environmental impact of the use of these fuels as power sources.

Preparation

Sulphur dioxide can be prepared by burning sulphur:

$$S_8 + 8O_2 \rightarrow 8SO_2$$

The combustion of hydrogen sulphide and organosulphur compounds proceeds similarly.

$$2H_2S \text{ (g)} + 3O_2\text{(g)} \rightarrow 2H_2O \text{ (g)} + 2SO_2 \text{ (g)}$$

The roasting of sulphide ores such as pyrites, sphalerite (zinc blende), and cinnabar (mercury sulphide) also releases SO_2:

$$4FeS_2 \text{ (s)} + 11O_2 \text{ (g)} \rightarrow 2Fe_2O_3 \text{ (s)} + 8SO_2 \text{ (g)}$$
$$2ZnS \text{ (s)} + 3O_2 \text{ (g)} \rightarrow 2ZnO \text{ (s)} + 2SO_2 \text{ (g)}$$
$$HgS \text{ (s)} + O_2 \text{ (g)} \rightarrow Hg \text{ (g)} + SO_2 \text{ (g)}$$

Sulphur dioxide is a by-product in the manufacture of calcium silicate cement: $CaSO_4$ is heated with coke and sand in this process:

$$2CaSO_4 \text{ (s)} + 2SiO_2 \text{ (s)} + C \text{ (s)} \rightarrow 2CaSiO_3 \text{ (s)} + 2SO_2 \text{ (g)} + CO_2 \text{ (g)}$$

Action of hot sulphuric acid on copper turnings produces sulphur dioxide.

$$Cu \text{ (s)} + 2H_2SO_4 \text{ (aq)} \rightarrow CuSO_4 \text{ (aq)} + SO_2 \text{ (g)} + 2H_2O \text{ (l)}$$

Uses

Sulphur dioxide is an intermediate in the production of sulphuric acid, being converted to sulphur trioxide, and then to oleum, which is made into sulphuric acid. Sulphur dioxide for this purpose is made when sulphur combines with oxygen. The method of converting sulphur dioxide to sulphuric acid is called the contact process. Sulphur dioxide is sometimes used as a preservative for dried apricots and other dried fruits owing to its antimicrobial properties. Sulphur dioxide is an important compound in winemaking.

NITROUS OXIDE

Nitrous oxide, commonly known as laughing gas, is a chemical compound with the chemical formula N_2O. At room temperature, it is a colourless non-flammable gas, with a pleasant, slightly sweet odour and taste. It is used in surgery and dentistry for its anesthetic and analgesic effects. It is known as 'laughing gas' due to the euphoric effects of inhaling it, a property that has led to its recreational use as an inhalant drug. It is also used as an oxidiser in rocketry and in motor racing to increase the power output of engines. Nitrous oxide reacts with ozone in the stratosphere. Nitrous oxide is the main naturally occurring regulator of stratospheric ozone. Nitrous oxide is a major greenhouse gas. Considered over a 100 year period, it has 310 times more impact per unit weight than carbon dioxide. Thus, despite its low concentration, nitrous oxide is the fourth largest contributor to the greenhouse effect. It ranks behind carbon dioxide, methane, and water vapour, the latter of which comprises greater than 95 per cent of all greenhouse gases by weight (but not by effect). Control of nitrous oxide is part of efforts to curb greenhouse gas emissions.

Manufacture

Nitrous oxide is most commonly prepared by careful heating of ammonium nitrate, which decomposes into nitrous oxide and water vapour. The addition of various phosphates favours formation of a purer gas at slightly lower temperatures.

$$NH_4NO_3(s) \rightarrow 2H_2O(g) + N_2O(g)$$

This reaction occurs between 170°–240°C, temperatures where ammonium nitrate is a moderately sensitive explosive and a very powerful oxidiser. Above 240°C the exothermic reaction may accelerate to the point of detonation, so the mixture must be cooled to avoid such a disaster. Superheated steam is used to reach reaction temperature in some turnkey production plants.

Downstream, the hot, corrosive mixture of gases must be cooled to condense the steam, and filtered to remove higher oxides of nitrogen. Ammonium nitrate smoke, as an extremely persistent colloid, will also have to be removed. The cleanup is often done in a train of three gas washes; namely base, acid and base again. Any significant amounts of nitric oxide (NO) may not necessarily be absorbed directly by the base (sodium hydroxide) washes. The nitric oxide impurity is sometimes chelated out with ferrous sulphate, reduced with iron metal, or oxidised and absorbed in base as a higher oxide. The first base wash may (or may not) react out much of the ammonium nitrate smoke, however, this reaction generates ammonia gas, which may have to be absorbed in the acid wash.

Other routes

The direct oxidation of ammonia may someday rival the ammonium nitrate pyrolysis synthesis of nitrous oxide mentioned above. This capital-intensive process, which originates in Japan, manganese dioxide-bismuth oxide catalyst:

$$2NH_3 + 2O_2 \rightarrow N_2O + 3H_2O$$

Higher oxides of nitrogen are formed as impurities. In comparison, uncatalysed ammonia oxidation (i.e. combustion or explosion) goes primarily to N_2 and H_2O. Nitrous oxide can be made by heating a solution of sulphamic acid and nitric acid. Many gases are made this way in Bulgaria.

$$HNO_3 + NH_2SO_3H \rightarrow N_2O + H_2SO_4 + H_2O$$

There is no explosive hazard in this reaction if the mixing rate is controlled. However, as usual, toxic higher oxides of nitrogen form.

Nitrous oxide is produced in large volumes as a by-product in the synthesis of adipic acid; one of the two reactants used in nylon manufacture. This might become a major commercial source, but will require the removal of higher oxides of nitrogen and organic impurities. Currently much of the gas is decomposed before release for environmental protection. Greener processes may prevail that substitute hydrogen peroxide for nitric acid oxidation; hence no generation of oxide of nitrogen by-products. Hydroxylammonium chloride can react with sodium nitrite to produce N_2O as well:

$$NH_3OH^+Cl^- + NaNO_2 \rightarrow N_2O + NaCl + H_2O$$

If the nitrite is added to the hydroxylamine solution, the only remaining by-product is salt water. However, if the hydroxylamine solution is added to the nitrite solution (nitrite is in excess), then toxic higher oxides of nitrogen are also formed.

Uses

Rocket motors, internal combustion engine, aerosol propellant, and in medicine.

Safety

The major safety hazards of nitrous oxide come from the fact that it is a compressed liquefied gas, an asphyxiation risk, and a dissociative anaesthetic. Exposure to nitrous oxide causes short-term decreases in mental performance, audiovisual ability, and manual dexterity.

Chapter 7

Industrial Carbon

INTRODUCTION

Carbon is essential to all known living systems, and without it life as we know it could not exist. Carbon is a chemical element with symbol C and atomic number 6. As a member of group 14 on the periodic table, it is nonmetallic and tetravalent—making four electrons available to form covalent chemical bonds. There are three naturally occurring isotopes, with ^{12}C and ^{13}C being stable, while ^{14}C is radioactive, decaying with a half-life of about 5730 years.

The uses of carbon and its compounds are extremely varied. It can form alloys with iron, of which the most common is carbon steel. Graphite is combined with clays to form the 'lead' used in pencils used for writing and drawing. It is also used as a lubricant and a pigment, as a moulding material in glass manufacture, in electrodes for dry batteries and in electroplating and electroforming, in brushes for electric motors and as a neutron moderator in nuclear reactors.

Carbon black is used as the black pigment in printing ink, artist's oil paint and water colours, carbon paper, automotive finishes, India ink and laser printer toner. Carbon black is also used as a filler in rubber products such as tyres and in plastic compounds. Activated charcoal is used as an absorbent and adsorbent in filter material in applications as diverse as gas masks, water purification and kitchen extractor hoods and in medicine to absorb toxins, poisons, or gases from the digestive system. Carbon is used in chemical reduction at high temperatures. Coke is used to reduce iron ore into iron. Case hardening of steel is achieved by heating finished steel components in carbon powder. Carbides of silicon, tungsten, boron and titanium, are among the hardest known materials, and are used as abrasives in cutting and grinding tools. Carbon compounds make up most of the materials used in clothing, such as natural and synthetic textiles and leather, and almost all of the interior surfaces in the built environment other than glass, stone and metal.

There are several allotropes of carbon of which the best known are graphite, diamond, and amorphous carbon. The physical properties of carbon vary widely with the allotropic form. All the allotropic forms are solids under normal conditions but graphite is the most thermodynamically stable.

All forms of carbon are highly stable, requiring high temperature to react even with oxygen. The most common oxidation state of carbon in inorganic compounds is +4, while +2 is found in carbon monoxide and other transition metal carbonyl complexes. The largest sources of inorganic carbon are limestones, dolomites and carbon dioxide, but significant quantities occur in organic deposits of coal, peat, oil and methane clathrates. Carbon forms more compounds than any other element, with almost ten million pure organic compounds described to date, which in turn are a tiny fraction of such compounds that are theoretically possible under standard conditions. Atomic carbon is a very short-lived species and therefore, carbon is stabilised in various multi-atomic structures with different molecular configurations called allotropes. Diamond and graphite are two allotropes of carbon: pure forms of the same element that differ in structure.

Carbon compounds form the basis of all life on earth and the carbon-nitrogen cycle provides some of the energy produced by the sun and other stars. Although it forms an extraordinary variety of compounds, most forms of carbon are comparatively unreactive under normal conditions. At standard temperature and pressure, it resists all but the strongest oxidisers. It does not react with sulphuric acid, hydrochloric acid, chlorine or any alkalies. At elevated temperatures carbon reacts with oxygen to form carbon oxides, and will reduce such metal oxides as iron oxide to the metal. This exothermic reaction is used in the iron and steel industry to control the carbon content of steel:

$$Fe_3O_4 + 4C_{(s)} \rightarrow 3Fe_{(s)} + 4CO_{(g)}$$

with sulphur to form carbon disulphide and with steam in the coal-gas reaction

$$C_{(s)} + H_2O_{(g)} \rightarrow CO_{(g)} + H_{2(g)}$$

Carbon combines with some metals at high temperatures to form metallic carbides, such as the iron carbide, cementite in steel, and tungsten carbide, widely used as an abrasive and for making hard tips for cutting tools. Graphene, which occurs naturally in graphite, is the strongest substance known to man. However, the process of separating it from graphite will require some technological development before it is economical enough to be used in industrial processes. The system of carbon allotropes spans a range of extremes which are given in Table 7.1.

Once considered exotic, fullerenes are now-a-days commonly synthesised and used in research; they include buckyballs, carbon nanotubes, carbon nanobuds and nanofibres. Several other exotic allotropes have also been discovered, such as lonsdaleite, glassy carbon,carbon nanofoam and linear acetylenic carbon.

Table 7.1. System of carbon allotropes spans a range of extremes.

Synthetic diamond nanorods are the hardest materials known	Graphite is one of the softest materials known
Diamond is the ultimate abrasive	Graphite is a very good lubricant
Diamond is an excellent electrical insulator	Graphite is a conductor of electricity
Diamond is the best known naturally occurring thermal conductor	Some forms of graphite are used for thermal insulation (i.e. firebreaks and heatshields)
Diamond is highly transparent	Graphite is opaque
Diamond crystallises in the cubic system	Graphite crystallises in the hexagonal system
Amorphous carbon is completely isotropic	Carbon nanotubes are among the most anisotropic materials ever produced

Carbon nanotubes are structurally similar to buckyballs, except that each atom is bonded trigonally in a curved sheet that forms a hollow cylinder. Nanobuds were first published in 2007 and are hybrid bucky tube/buckyball materials (buckyballs are covalently bonded to the outer wall of a nanotube) that combine the properties of both in a single structure.

Of the other discovered allotropes, carbon nanofoam is a ferromagnetic allotrope discovered in 1997. It consists of a low-density cluster-assembly of carbon atoms strung together in a loose three-dimensional web, in which the atoms are bonded trigonally in six- and seven-membered rings.

Coal is a significant commercial source of mineral carbon; anthracite containing 92–98 per ent carbon and the largest source of carbon in a form suitable for use as fuel.

Pure carbon has extremely low toxicity and can be handled and even ingested safely in the form of graphite or charcoal. It is resistant to dissolution or chemical attack, even in the acidic contents of the digestive tract, for example. Consequently if it gets into body tissues it is likely to remain there indefinitely. Similarly, diamond dust used as an abrasive can do harm if ingested or inhaled. Microparticles of carbon are produced in diesel engine exhaust fumes, and may accumulate in the lungs. In these examples, the harmful effects may result from contamination of the carbon particles, with organic chemicals or heavy metals for example, rather than from the carbon itself.

The great variety of carbon compounds include such lethal poisons as tetrodotoxin, the lectin ricin from seeds of the castor oil plant *Ricinus communis*, cyanide (CN^-) and carbon monoxide; and such essentials to life as glucose and protein.

LAMPBLACK

Lampblack or soot is a general term that refers to the black, impure carbon particles resulting from the incomplete combustion of a hydrocarbon. It is more properly restricted to the product of the gas-phase combustion process but is commonly extended to include the residual pyrolysed fuel particles such as cenospheres, charred wood, petroleum coke, etc. that may become airborne during pyrolysis and which are more properly identified as cokes or chars. The gas-phase soots contain polycyclic aromatic hydrocarbons (PAHs). Some soots are produced commercially to be used as pigments, such as lampblack and carbon black. These products have been used for many years as common pigments used in paints and inks, and remain in use today in toners for xerography and laser printers. The black colour of rubber tyres is due to the use of lampblack or carbon black as an ingredient in their vulcanisation; this use accounts for around 85 per cent of the market use of these products. Bone black, another black pigment and decolourising agent, is the product of charring bones and is not a soot.

Soot, as an airborne contaminant in the environment has many different sources but they are all the result of some form of pyrolysis. They include soot from internal combustion engines, diesel engines, power plant boilers, hog-fuel boilers, ship boilers, central steam heat boilers, waste incineration, local field burning, house fires, forest fires, fireplaces, furnaces, etc. These exterior sources also contribute to the indoor environment sources such as smoking of plant matter, cooking, oil lamps, candles, quartz/halogen bulbs with settled dust, fireplaces, defective furnaces, etc. Soot in very low concentrations is capable of darkening surfaces or making particle agglomerates, such as those from ventilation systems, appear black. Soot is the primary cause of 'ghosting', the discolouration of walls and ceilings or walls and flooring where they meet. It is generally responsible for the discolouration of the walls above baseboard electric heating units.

The production of soot in a flame is a complex process consisting of several chemical reactions taking place in series. In the fuel-pyrolysis zone of the flame, typically clear or blue, the fuel molecules are broken down into various fragments, including carbon-ring structures, acetylene (C_2H_2), the radical C_3H_3 (and higher order), as well as monatomic and diatomic hydrogen. As the combustion process continues the radicals quickly combine into new structures, giving off heat. These precursors polymerise into larger 'pre-soot' chains then gather into formations of hydrogen-rich spheres in the soot-inception zone. In the soot-growth zone these spheres give up their hydrogen gas through diffusion, resulting in solids consisting of several of the formerly liquid spheres stuck together into larger chains. It is this portion of the flame that has the bright yellow colour. Hydrogen-rich examples then further oxidise, releasing more heat. In perfect combustion the soot would break down into almost pure CO_2 and H_2O; it is only in incomplete combustion that the soot is able to form and escape the flame. Soot normally forms at about 1400°C, forming an excellent blackbody radiator of colours in the yellow to red spectrum. The typical yellow colour of a candle flame or wood fire is produced primarily by the hot soot forming inside.

The energy being radiated from the soot is an important contributor to the ongoing combustion process, cooling the flame above the soot-growth zone and feeding energy back into the fuel-pyrolysis zone. In 'pool fires' of open liquid fuel this process can feed as much as 50 per cent of the flame's energy back into the liquid fuel below, which vapourises it and keeps the reaction going; it would otherwise burn much more slowly. The same release of energy is responsible for quickly cooling the flame above the soot-growth region, limiting its further combustion into lighter molecules, and explaining why these fires release so much soot.

The separation of flame into zones of different chemical reactions is due to convection forcing the hot reactants upward. In microgravity or zero gravity convection no longer occurs, and such flames tend to become more blue and more efficient, producing much less soot.

Lampblack has been used since prehistoric times as the source for carbon black, collected by holding a cold surface over a cool flame. Candles or lamps using animal fats or waxes generate considerable amounts of soot that can be collected and then mixed with a lubricant to produce ink. This process can be easily duplicated today by passing some noncombustible surface, such as a tin-can lid or glass, closely through a candle flame. Lampblack produced in this way is among the darkest and least reflective substances known. Lampblack is also used to coat aluminium foil that has been previously attached to a recording drum for use in a recording barograph or other instrument. The surface is scratched clear by a pointed stylus. In this case, the sooty smoke is produced by burning a small amount of camphor. After recording the image is fixed by spraying the surface with a clear lacquer. Similar coatings were used in direct recording pendulum seismometers. While not a sensitive instrument, these were capable of directly recording the direction of significant horizontal shocks upon a smoked glass plate.

Hazards: Soot is in the general category of airborne particulate matter, and as such is considered hazardous to the lungs and general health when the particles are less than five micrometres in diameter, as such particles are not filtered out by the upper respiratory tract. Smoke from diesel engines, while composed mostly of carbon soot, is considered especially dangerous owing to both its particulate size and the many other chemical compounds present.

ACTIVATED CARBON

Activated carbon, also called activated charcoal or activated coal, is a form of carbon that has been processed to make it extremely porous and thus to have a very large surface area available for adsorption or chemical reactions. Due to its high degree of microporosity, just one gram of activated carbon has a surface area of approximately 500 m^2, as determined typically by nitrogen gas adsorption. Sufficient activation for useful applications may come solely from the high surface area, though further chemical treatment often enhances the adsorbing properties of the material. Activated carbon is usually derived from charcoal. Activated carbon is produced from carbonaceous source materials like nutshells, wood and coal. It can be produced by one of the following processes.

Physical Reactivation

The precursor is developed into activated carbons using gases. This is generally done by using one or a combination of the following processes.

Carbonisation

Material with carbon content is pyrolysed at temperatures in the range 600°–900°C, in absence of air (usually in inert atmosphere with gases like argon or nitrogen)

Activation/oxidation

Raw material or carbonised material is exposed to oxidising atmospheres (carbon dioxide, oxygen, or steam) at temperatures above 250°C, usually in the temperature range of 600°–1200°C.

Chemical Activation

Impregnation with chemicals such as acids like phosphoric acid or bases like potassium hydroxide, sodium hydroxide or salts like zinc chloride, followed by carbonisation at temperatures in the range of 450°–900°C. It is believed that the carbonisation/activation step proceeds simultaneously with the chemical activation. This technique can be problematic in some cases, because, for example, zinc trace residues may remain in the end product. However, chemical activation is preferred over physical activation owing to the lower temperatures and shorter time needed for activating material.

Manufacture

Many carbonaceous materials, such as petroleum coke, sawdust, lignite, coal, peat, wood, charcoal, nutshells, and fruit pits, may be used for the manufacture of activated carbon, but the properties of the finished material are governed not only by the raw material but by the method of activation used. Decolourising activated carbons are usually employed as powders. Thus, the raw materials for this type are either structureless or have a weak structure. Sawdust and lignite yield carbons of this kind. Vapour adsorbent carbons are used in the form of hard granules and are generally produced from coconut shells, fruit pits, and briquetted coal and charcoal.

Activation is a physical change wherein the surface of the carbon is tremendously increased by the removal of hydrocarbons. Several methods are available for this activation. The most widely employed are treatment of the carbonaceous material with oxidising gases such as air, steam, or carbon dioxide, and the carbonisation of the raw material in the presence of chemical agents such as zinc chloride or phosphoric acid. Gaseous-oxidation activation employs material that has been carbonised at a temperature high enough to remove most of the volatile constituents but not high enough to crack the evolved gases. The carbonised material is subjected to the action of the oxidising gas, usually steam or carbon dioxide, in a furnace or retort at 1475° to 1800°F.

Conditions are controlled to permit removal of substantially all the adsorbed hydrocarbons and some of the carbon, so as to increase the surface area. The use of chemical impregnating agents causes the carbonisation to proceed under conditions that prevent the deposition of hydrocarbons on the carbon surface. The raw material, sawdust or peat, is mixed with the chemical agent, dried, and calcined at temperatures up to 1560°F. When the carbonisation has been completed, the residual impregnating agent is removed by leaching with water.

Revivification: After activated carbon has become saturated with a vapour or an adsorbed colour, either the vapour can be steamed out, condensed, and recovered as shown in Fig. 7.1, or the colouration can be destroyed and the carbon made ready for reuse. The oldest example of this process uses the decolourising carbon long known as bone char, or bone black. This consists of about 10 per cent carbon deposited on a skeleton of tricalcium phosphate and is made by the carbonisation in closed retorts at 1380° to 1740°F of fat-free bones.

Classifications

Activated carbons are complex products which are difficult to classify on the basis of their behaviour, surface characteristics and preparation methods. However, some broad classification is made for general purpose based on their physical characteristics.

Powdered activated carbon (PAC)

Traditionally, active carbons are made in particular form as powders or fine granules less than 1.0 mm in size with an average diameter between 0.15 and

0.25 mm. Thus they present a large surface to volume ratio with a small diffusion distance. PAC is made up of the crushed or ground carbon particles, 95–100 per cent of which will pass through a designated mesh sieve or sieve. Granular activated carbon is defined as the activated carbon being retained on a 50-mesh sieve (0.297 mm) and PAC material as finer material, while ASTM classifies particle sizes corresponding to an 80-mesh sieve (0.177 mm) and smaller as PAC. PAC is not commonly used in a dedicated vessel, owing to the high headloss that would occur. PAC is generally added directly to other process units, such as raw water intakes, rapid mix basins, clarifiers, and gravity filters.

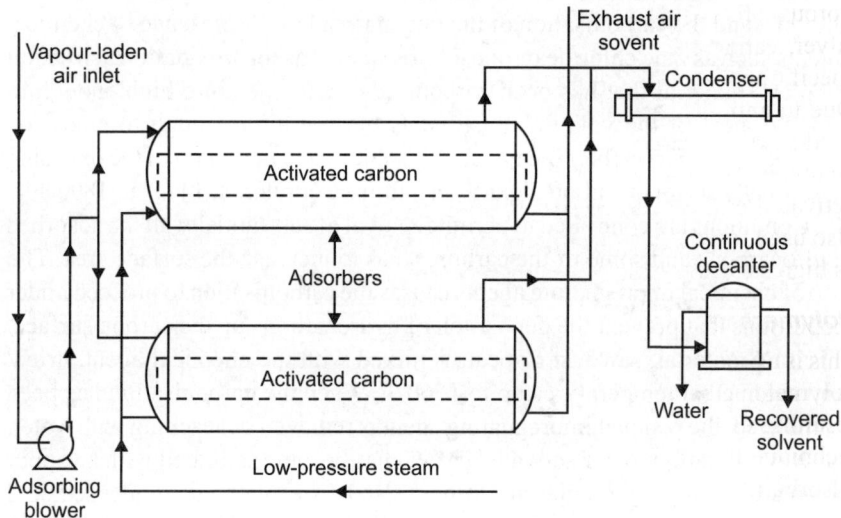

Fig. 7.1. Flow diagram of one type of solvent recovery plant employing activated carbon.

Granular activated carbon (GAC)

Granular activated carbon has a relatively larger particle size compared to powdered activated carbon and consequently, presents a smaller external surface. Diffusion of the adsorbate is thus an important factor. These carbons are therefore preferred for all adsorption of gases and vapours as their rate of diffusion are faster. Granulated carbons are used for water treatment, deodourisation and separation of components of flow system. GAC can be either in the granular form or extruded. GAC is designated by sizes such as 8×20, 20×40, or 8×30 for liquid phase applications and 4×6, 4×8 or 4×10 for vapour phase applications. A 20×40 carbon is made of particles that will pass through a US standard mesh size no. 20 sieve (0.84 mm) (generally specified as 85 per cent passing) but be retained on a US standard mesh size no. 40 sieve (0.42 mm) (generally specified as 95 per cent retained). The most popular aqueous phase carbons are the 12 × 40 and 8 × 30 sizes because they have a good balance of size, surface area, and headloss characteristics.

Extruded activated carbon (EAC)

Consists of extruded and cylindrical shaped activated carbon with diameters from 0.8 to 45 mm. These are mainly used for gas phase applications because of their low pressure drop, high mechanical strength and low dust content.

Impregnated carbon

Porous carbons containing several types of inorganic impregnant such as iodine, silver, cation such as Al, Mn, Zn, Fe, Li, Ca have also been prepared for specific application in air pollution control especially in museums and galleries. Due to antimicrobial/antiseptic properties, silver loaded activated carbon is used as an adsorbent for purifications of domestic water. Drinking water can be obtained from natural water by treating the natural water with a mixture of activated carbon and flocculating agent $Al(OH)_3$. Impregnated carbons are also used for the adsorption of H_2S and mercaptans. Adsorption rates for H_2S as high as 50 per cent by weight have been reported.

Polymers coated carbon

This is a process by which a porous carbon can be coated with a biocompatible polymer to give a smooth and permeable coat without blocking the pores. The resulting carbon is useful for hemoperfusion. Hemoperfusion is a treatment technique in which large volumes of the patient's blood are passed over an adsorbent substance in order to remove toxic substances from the blood.

Other

Activated carbon is also available in special forms such as cloths and fibres. The carbon cloth for instance is used in personnel protection for the military.

Properties of Activated Carbon

A gram of activated carbon can have a surface area in excess of 500 m^2, with 1500 m^2 being readily achievable. Carbon aerogels, while more expensive, have even higher surface areas, and are used in special applications.

Under an electron microscope, the high surface-area structures of activated carbon are revealed. Individual particles are intensely convoluted and display various kinds of porosity; there may be many areas where flat surfaces of graphite-like material run parallel to each other, separated by only a few nanometers or so. These micropores provide superb conditions for adsorption to occur, since adsorbing material can interact with many surfaces simultaneously.

Physically, activated carbon binds materials by van der Waals force or London dispersion force.

Activated carbon does not bind well to certain chemicals, including alcohols, glycols, ammonia, strong acids and bases, metals and most inorganics, such as lithium, sodium, iron, lead, arsenic, fluorine, and boric acid. Activated carbon does adsorb iodine very well and in fact the iodine number, mg/g, (ASTM D28 standard method test) is used as an indication of total surface area.

Activated carbon can be used as a substrate for the application of various chemicals to improve the adsorptive capacity for some inorganic (and problematic organic) compounds such as hydrogen sulphide (H_2S), ammonia (NH_3), formaldehyde (HCOH), radioisotopes iodine-131 (^{131}I) and mercury (Hg). This property is known as chemisorption.

Iodine number

Many carbons preferentially adsorb small molecules. Iodine number is the most fundamental parameter used to characterise activated carbon performance. It is a measure of activity level (higher number indicates higher degree of activation), often reported in mg/g (typical range 500–1200 mg/g). It is a measure of the micropore content of the activated carbon (0 to 20 Å, or up to 2 nm) by adsorption of iodine from solution. It is equivalent to surface area of activated carbon between 900 m²/g and 1100 m²/g. It is the standard measure for liquid phase applications.

Molasses

Some carbons are more adept at adsorbing large molecules. Molasses number or molasses efficiency is a measure of the macropore content of the activated carbon (greater than 20 Å, or larger than 2 nm) by adsorption of molasses from solution. A high molasses number indicates a high adsorption of big molecules (range 95–600).

Molasses number is a measure of the degree of decolourisation of a standard molasses solution that has been dilited and standardised against standardised activated carbon. Due to the size of colour bodies, the molasses number represents the potential pore volume available for larger adsorbing species.

Tannin

Tannins are a mixture of large and medium size molecules. Carbons with a combination of macropores and mesopores adsorb tannins. The ability of a carbon to adsorb tannins is reported in parts per million concentration (range 200–362 ppm).

Methylene blue

Some carbons have a mesopore (20 Å to 50 Å, or 2 to 5 nm) structure which adsorbs medium size molecules, such as the dye methylene blue. Methylene blue adsorption is reported in g/100 g (range 11–28 g/100 g).

Dechlorination

Some carbons are evaluated based on the dechlorination half-value length, which measures the chlorine-removal efficiency of activated carbon. The dechlorination half-value length is the depth of carbon required to reduce the chlorine level of a flowing stream from 5 ppm to 3.5 ppm. A lower half-value length indicates superior performance.

Apparent density

Higher density provides greater volume activity and normally indicates better quality activated carbon.

Hardness/abrasion number

It is a measure of the activated carbon's resistance to attrition. It is important indicator of activated carbon to maintain its physical integrity and withstand frictional forces imposed by backwashing, etc. There are large differences in the hardness of activated carbons, depending on the raw material and activity level.

Ash content

It reduces the overall activity of activated carbon. It reduces the efficiency of reactivation. The metals (Fe_2O_3) can leach out of activated carbon resulting in discolouration. Acid/water soluble ash content is more significant than total ash content. Soluble ash content can be very important for aquarists, as ferric oxide can promote algal growths, a carbon with a low soluble ash content should be used for marine, freshwater fish and reef tanks to avoid heavy metal poisoning and excess plant/algal growth.

Carbon tetrachloride activity

Measurement of the porosity of an activated carbon by the adsorption of saturated carbon tetrachloride vapour.

Particle size distribution

The finer the particle size of an activated carbon, the better the access to the surface area and the faster the rate of adsorption kinetics. In vapour phase systems this needs to be considered against pressure drop, which will affect energy cost. Careful consideration of particle size distribution can provide significant operating benefits.

Applications

Activated carbon is used in gas purification, gold purification, metal extraction, water purification, medicine, sewage treatment, air filters in gas masks and filter masks, filters in compressed air and many other applications.

One major industrial application involves use of activated carbon in metal finishing field. It is very widely employed for purification of electroplating solutions. For example, it is a main purification technique for removing organic impurities from bright nickel plating solutions. A variety of organic chemicals are added to plating solutions for improving their deposit qualities and for enhancing properties like brightness, smoothness, ductility, etc. Due to passage of direct current and electrolytic reactions of anodic oxidation and cathodic reduction, organic additives generate unwanted break down products in solution. Their excessive build up can adversely affect the plating quality and physical properties of deposited metal. Activated carbon treatment removes such impurities and restores plating performance to the desired level.

Environmental applications

Carbon adsorption has numerous applications in removing pollutants from air or water streams both in the field and in industrial processes such as: (i) spill cleanup, (ii) groundwater remediation, (iii) drinking water filtration, (iv) air purification, and (v) volatile organic compounds capture from painting, dry cleaning, gasoline dispensing operations, and other processes.

Medical applications

Activated carbon is used to treat poisonings and overdoses following oral ingestion. It is thought to bind to poison and prevent its absorption by the gastrointestinal tract.

In cases of suspected poisoning, medical personnel administer activated charcoal on the scene or at a hospital's emergency department. Dosing is usually empirical at 1 g/kg of body weight, usually given only once, but depending on the drug taken, it may be given more than once. In rare situations activated charcoal is used in intensive care to filter out harmful drugs from the blood stream of poisoned patients. Activated carbon has become the treatment of choice for many poisonings, and other decontamination methods such as ipecac-induced emesis or stomach pumps are now used rarely.

While activated carbon is useful in an acute poisoning situation, it has been shown to not be effective in long term accumulation of toxins, such as with the use of toxic herbicides.

Activated charcoal is also used for bowel preparation by reducing intestinal gas content before abdominal radiography to visualise bile, pancreatic and renal stones.

Gas purification

Filters with activated carbon are usually used in compressed air and gas purification to remove oil vapours, odours, and other hydrocarbons from the air. The most common designs use a 1 stage or 2 stage filtration principle where activated carbon is embedded inside the filter media. Activated charcoal is also used in spacesuit primary life support systems.

Distilled alcoholic beverage purification

Activated carbon filters can be used to filter vodka and whiskey of organic impurities which can affect colour, taste, and odour. Passing an organically impure vodka through an activated carbon filter at the proper flow rate will result in vodka with an identical alcohol content and significantly increased organic purity, as judged by odour and taste.

Mercury scrubbing

Activated carbon, often impregnated with iodine or sulphur, is widely used to trap mercury emissions from coal fired power stations, medical incinerators, and from natural gas at the wellhead.

CARBON BLACK

Carbon black is a material produced by the incomplete combustion of heavy petroleum products such as FCC tar, coal tar, ethylene cracking tar, and a small amount from vegetable oil. Carbon black is a form of amorphous carbon that has a high surface area to volume ratio, and as such it is one of the first nanomaterials to find common use, although its surface area to volume ratio is low compared to activated carbon. It is similar to soot but with a much higher surface area to volume ratio. Carbon black is used as a pigment and reinforcement in rubber and plastic products.

Short-term exposure to high concentrations of carbon black dust may produce discomfort to the upper respiratory tract, through mechanical irritation.

The most common use (70 per cent) of carbon black is as a pigment and reinforcing phase in automobile tyres. Carbon black also helps conduct heat away from the tread and belt area of the tyre, reducing thermal damage and increasing tyre life. Carbon black particles are also employed in some radar absorbent materials and in printer toner.

The highest volume use of carbon black is as a reinforcing filler in rubber products, especially tyres. While a pure gum vulcanisate of styrene-butadiene has a tensile strength of no more than 2.5 MPa, and almost nonexistent abrasion resistance. Carbon black (colour index international, PBL-7) is the name of a common black pigment, traditionally produced from charring organic materials such as wood or bone. It consists of pure elemental carbon, and it appears black because it reflects almost no light in the visible part of the spectrum. It is known by a variety of names, each of which reflects a traditional method for producing carbon black:

1. Ivory black was traditionally produced by charring ivory or bones.

2. Vine black was traditionally produced by charring desiccated grape vines and stems.
3. Lamp black was traditionally produced by collecting soot, also known as lampblack, from oil lamps.

Newer methods of producing carbon black have superseded these traditional sources, although some materials are still produced using traditional methods. For artisanal purposes, it is very useful.

Surface chemistry: All carbon blacks have chemisorbed oxygen complexes (i.e. carboxylic, quinonic, lactonic, phenolic groups and others) on their surfaces to varying degrees depending on the conditions of manufacture. These surface oxygen groups are collectively referred to as volatile content. It is also known to be a nonconductive material due to its volatile content.

The coatings and inks industries prefer grades of carbon black that are acid oxidised. Acid is sprayed in high temperature dryers during the manufacturing process to change the inherent surface chemistry of the black. The amount of chemically-bonded oxygen on the surface area of the black is increased to enhance performance characteristics.

Manufacture of Carbon Black

The manufacture of carbon blacks has evolved from primitive methods in ancient times to continuous, high capacity systems using sophisticated designs and control equipment today. Carbon black is a product of a very carefully controlled manufacturing process. The methods generally adopted for the manufacture of carbon black are based on the incomplete combustion of hydrocarbons in diffuse flames, carbon particles being removed from the flame by impingement on cooled metal surfaces or in cyclone-precipitator-filter systems. Carbon blacks are made by the partial oxidation or thermal decomposition of hydrocarbon gases or liquids. Many different processes have been developed. In general, each process is capable of producing a number of types or grades of carbon black with characteristics somewhat different from those produced by any other process.

Thus, various kinds of carbon blacks are often referred to with respect to the process by which they are made: channel blacks, gas-furnace blacks, oil-furnace blacks, thermal blacks, etc. However, the unique character of carbon blacks from a given process is not absolute. In particular, continuing development has made and is making the oil-furnace process versatile enough to produce reasonable approximations to many of the carbon blacks made by other processes in addition to carbon blacks which can be made only by the oil-furnace process.

Carbon black manufacturing processes may be classified as follows:

1. Impingement processes: In these types of processes, the carbon black is formed by impingement of open flames upon a surface from which

the carbon black is recovered. This category includes the channel and oil impingement processes.

2. Furnace processes: In these processes, combustion (heat generation) and carbon black formation occur simultaneously in a confined reactor or furnace, the carbon black being dispersed in the effluent gases. This category includes the gas-furnace and oil-furnace processes. The lampblack process, although quite different in mechanics, may also be placed in this category.

3. Furnace processes: In this processes,combustion and carbon black formation do not proceed simultaneously. This category comprises the cyclic thermal black processes in which the combustion and carbon black formation are separated in time, and the more specialised acetylene black process (thermal and electrical arc variations). The term furnace process may be a slight misnomer applied to acetylene black process, since although energy generation is involved, direct combustion is not.

Carbon black production processes may also be classified with respect to whether the primary raw material is gas (usually natural gas) or heavy hydrocarbon oil.

1. Gas processes: (i) channel process, (ii) gas-furnace process, (iii) gas-thermal process, and (iv) acetylene black process (acetylene feedstock)

2. Oil processes: (i) lampblack process, (ii) oil-furnace process, (iii) oil-impingement process, and (iv) oil-thermal process.

Apart from these manufacturing processes, a small amount of by-product carbon from the manufacture of synthesis gas from liquid hydrocarbons has found applications in electrically conductive compositions. Table 7.2 lists various production methods and important raw materials used for making carbon blacks.

Table 7.2. Production methods and raw materials.

Chemical process	Production method	Raw materials
Thermal-oxidative decomposition	Oil-furnace/furnace black process, lampblack process	Aromatic oils on basis of coal tar or petrochemical oils, natural gas
Thermal decomposition	Thermal black process Acetylene black process	Natural gas (or oils) Acetylene

The lampblack process is the ancestor of all modern carbon black processes. Until the 1870s, it was the only commercial process, and because of this the word lampblack is occasionally used as a generic term for carbon black.

In the lampblack process, oils are burned in open, shallow pans in a restricted air supply. The heavy carbon-laden smoke is passed through a series of settling chambers and filters from which the flocculated carbon deposits are recovered.

During the latter half of the nineteenth century, the channel process was developed. In the channel process, iron channels are used for the collection of carbon blacks from the impingement of thousands of small luminous flames burning in a restricted atmosphere of air. This process dominated the carbon black industry for over 50 years. In 1926, there were 33 producers in USA. Because of poor carbon yields from natural gas in the range of 1–5 per cent and severe atmospheric pollution this process has become extinct. The last channel black process plant in USA was closed in 1976.

Thus, during the first half of the twentieth century most carbon black was made from natural gas. Before long distance pipelines were built, there was little other use for natural gas and it was available cheaply. This was important consideration for the channel black process because of its low yields. However, development of a natural gas industry fed by high-pressure pipelines has led to a steadily increasing value for natural gas.

In the 1920s two other processes using natural gas were introduced that gave much higher yields with large decreases in atmospheric contamination. One was the cyclic thermal black process. Alternate heating and production cycles in large brick checkered chambers are used to produce a unique large particle size, essentially unaggregated-grade useful for many special rubber and plastic applications. Thermal black is produced in the USA, Canada, England and a few other countries worldwide. The other process, based on natural gas, was the so-called gas-furnace process and is no longer used. This process was continuous and the forerunner of the oil-furnace process. It was discontinued because of the relatively low yield, high raw material cost and limited range of products.

Search for a more efficient process for making reinforcing carbon blacks, led to the development of oil-furnace process in early 1940s, by R.D. Snow and his associate Dr. J.C. Krejci, who was awarded the Charles Goodyear Medal by the Rubber Division of the American Chemical Society in 1974. By 1941, experimental quantities of oil-furnace blacks with good reinforcing properties were available, and by 1943 the first commercial oil-furnace process was fully developed and put into operation by Phillips Petroleum Co. at Borger, Texas, USA.

The early oil-furnace blacks were found to be ideally suited for synthetic rubber, a product also in its infancy at the time but growing rapidly under pressure of World War II needs. Expansion of the synthetic rubber industry and development of the ability of the oil-furnace process to make a multiplicity of carbon blacks of desired characteristics has resulted in a steady growth of oil-furnace black production throughout the world.

The oil-furnace blacks displaced all other types used for the reinforcement of rubber and today account for practically all carbon black production. In the oil-furnace process, heavy aromatic residual oils are atomised into a primary combustion flame where the excess oxygen in the primary zone burns a portion of the residual oil to maintain flame temperatures, and the remaining oil is thermally decomposed into carbon and hydrogen. Yields in this process are in the range of 35 to 50 per cent based on the total carbon input. A broad range of product qualities can be produced. Only brief descriptions of the thermal, gas-furnace and acetylene processes are given and a more detailed description of oil-furnace process is presented.

Thermal process

Production of carbon blacks from thermal process dates from about 1922. The process is cyclic or batch type using natural gas or coke oven gas as feedstock, although a similar process in limited use employs heavy liquid hydrocarbons. The carbon blacks are produced in a non-oxidising atmosphere by heating brick checkerwork by combustion of air and fuel gas, then excluding the air and fuel, and cracking the feedstock using the stored heat in the checkerwork.

Thermal blacks are the lowest surface area (largest particle) members of the carbon black spectrum, and are also characterised by low oxygen content and very low structure. In fact, electron micrographs reveal that they are virtually structureless, consisting essentially of discrete spherical particles. Three grades of thermal blacks are recognised—medium thermal (MT or N990), medium thermal non-staining (MT-NS or N907), and fine thermal (FT or N880). Iodine numbers and nitrogen surface areas if the medium grades are generally 6 to 10, and of fine grades 10 to 15. Thermal blacks are used in both rubber and plastics.

In rubber, these carbon blacks are used where only a modest degree of reinforcement is required but maximum compound volume extension is desirable. Low structure and low reinforcing character allow these carbon blacks to be used at high loadings without attendant high stock viscosity, high hardness, or low vulcanisate elongation. Applications in rubber include mats, hose, sponge, gaskets, tyre innerliners, bead stock and electrical insulation.

A simplified flow sheet of a representative modern thermal black unit is shown in the following Fig. 7.2. The reactor section consists of two refractory lined furnaces filled with firebrick checkerwork. The furnaces may be 12 to 15 ft in diameter and 25 to 35 ft high. Operation of each furnace is cyclic, consisting of a heating cycle followed by a make cycle producing thermal black by cracking natural gas. While one furnace is on the heating cycle, the other is on the make cycle and vice versa, resulting in a reasonably continuous flow of carbon black to downstream equipment.

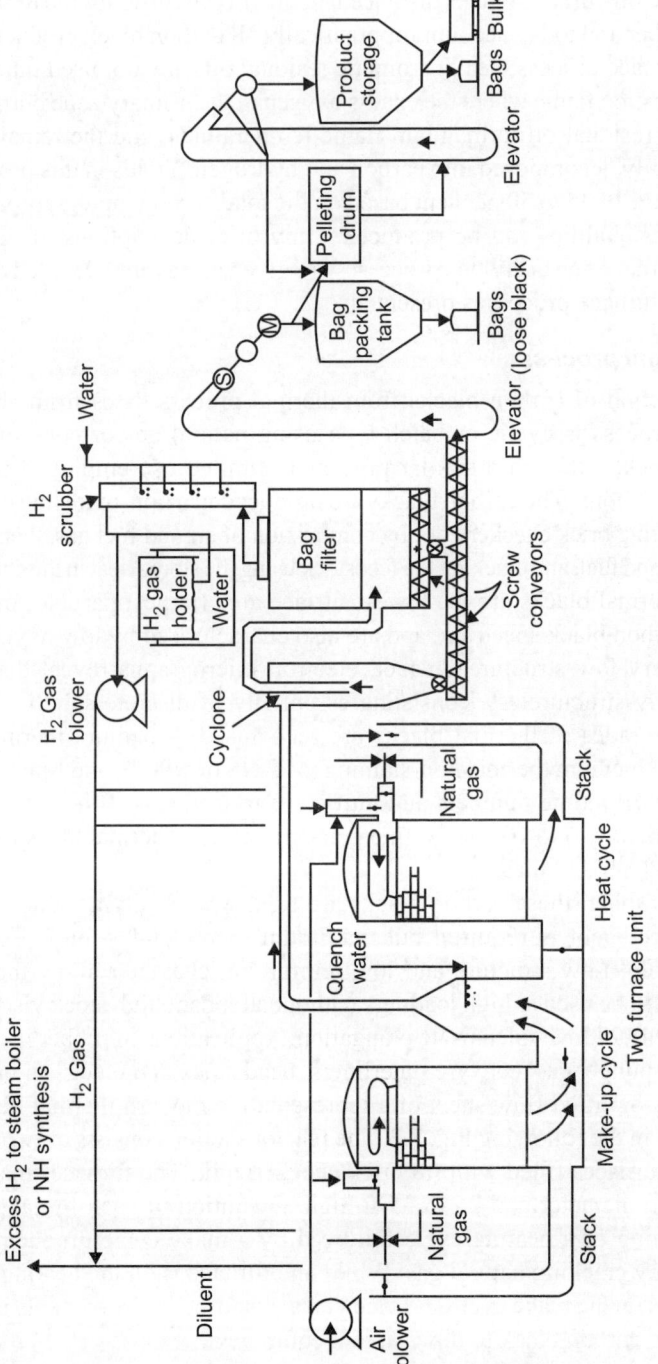

Fig. 7.2. Thermal process (natural gas feedstocks).

The heat required for cracking is supplied during the heating cycle by complete combustion of air and fuel gas in near stochiometric ratio, heat being stored in the checkerwork at 2400° to 2800°F, with the combustion gases passing on to the stack. The fuel gas is the hydrogen-rich gas from the make cycles.

At the conclusion of the heating cycle, fuel gas and air are shut off, natural gas is introduced to initiate the make cycle, and the effluent is routed to the recovery system. At the conclusion of the make cycle, the natural gas feed is shut off and the furnace is briefly purged with fuel gas to sweep out black deposited on the checkerwork so as to improve the yield. Air is then admitted; initiating the ensuing heating cycle, and the other reactor is switched from heating to make cycle. Automatic cycle timers and associated valving alternate each furnace from make to heating cycle about every 5 minutes. The newer units are designed for downflow operation of the furnaces, as indicated in the flow sheet, on both make and heating cycles.

The effluent gas from the make cycle is 85 to 90 per cent hydrogen plus methane and heavier hydrocarbons. The gas carries the carbon black to a quench tower where gas and carbon black are cooled by water sprays, and thence to cyclone separators and a bag filter to collect the carbon black. The effluent gas from the filter is further cooled and dehumidified in a water scrubber, and compressed to about 2.5 lb/in^2 gauge for recycle to the furnaces as fuel or as diluent for the natural gas when making fine thermal black.

A gas holder provides surge and storage for excess gas, which may be used for steam generation or in ammonia synthesis. The collected carbon black is transported by screw conveyors, elevated, screened, passed over a magnetic separator, and fed to a pulveriser. The carbon black is then either pelletised prior to shipping or bagged as loose black using an auger packer. For medium thermal black the natural gas feedstock is fed without dilution. The non-staining type requires slightly higher furnace temperatures and lower rates to more completely eliminate heavy hydrocarbons, which deposit on the carbon black and lead to staining.

To produce fine thermal black, the natural gas feedstock is diluted with two volumes of the hydrogen rich effluent gas per volume of natural gas. This dilution produces the higher surface area, smaller particle size fine thermal black. Yields of medium thermal black are 18 to 20 lb per 1000 std. ft^3 of natural gas; those of fine thermal are somewhat lower.

Gas-furnace process

The gas-furnace process was first commercialised in the 1920s. Today, gas-furnace carbon blacks are supplanted by oil-furnace counterparts. In the gas-furnace process, natural gas played the role of both fuel and feedstock. Reactors were refractory-lined furnaces typically 4×10×14 ft or 5 ft diameter by 30 to 32 ft long. Air and gas were introduced at one end through special slotted or drilled burners, usually 3 to 6 burners per furnace. The carbon black was produced in a rolling mass of flame in the furnace and the smoke (carbon black suspended in the reaction gases) exited at the other end. A unit of 3 to 5 furnaces delivered the smoke to a quench tower where the reaction was quenched and the gases cooled to about 400°F by water sprays.

The carbon black was either pelleted or packaged as densified unpelleted carbon black. Development of gas-furnace process was prompted by the desire to improve the low yields obtained in the channel process and produce a carbon black of intermediate reinforcing character between channel and thermal blacks. Prior to 1942, essentially only one grade of gas-furnace black was made and was termed semireinforcing furnace (SRF). It was similar to the modern carbon black grade N762. A typical furnace processed about one million ft^3 of gas per day making 9000 to 10,000 lb of carbon black. World War II pressures for more reinforcing carbon blacks led to the introduction in 1942 of HMF (high modulus furnace) and later FF (fine furnace). These types required higher air/gas ratios and the yield suffered accordingly, being only 10 to 12 per cent of the carbon in gas for FF.

All the gas-furnace carbon blacks were less reinforcing than the channel blacks and were primarily suitable for tyre carcass and mechanical rubber goods, with some application as pigments. Production of such carbon blacks has declined since about 1958 and presently this process is not used.

Acetylene process

The high carbon content of acetylene (92 per cent) and its property of decomposing exothermically to carbon and hydrogen make it an attractive raw material for conversion to carbon black, which is known as acetylene black. This acetylene black is made by a continuous decomposition process at atmospheric pressure and 800°–1000°C in water-cooled retorts lined with refractory. In the process, acetylene is fed into hot reactors. The exothermic reaction is self-sustaining and requires water-cooling to maintain a constant reaction temperature. The carbon black laden hydrogen stream is then cooled followed by separation of the carbon from the hydrogen tail gas. The tail gas is either flared or used as fuel. After separation from the gas stream, acetylene black is very fluffy with a bulk density of only 19 kg/m^3 (1.2 lb/ft^3). It is difficult to compact and resists pelletisation. Commercial grades are compressed to various bulk densities up to 200 kg/m^3 (12.5 lb/ft^3). Acetylene black is very pure with a carbon content of 99.7 per cent. It has a surface area of about 65 m^2/g, an average particle diameter of 40 nm, and a very high but rather weak structure with a DBPA value of 250 ml/100 g. It is the most crystalline or graphitic of the commercial carbon blacks. These unique features result in high electrical and thermal conductivity, low moisture absorption, and high liquid absorption. Acetylene black is mainly used in dry cell batteries where it contributes low electrical resistance and high capacity. In rubber it gives electrically conductive properties to heater pads, tapes, antistatic belt drives, conveyor belts, and shoe soles. It is also useful in electrically conductive plastics such as electrical magnetic interference (EMI) shielding enclosures. Its contribution to thermal conductivity has been useful in rubber curing bags for tyre manufacture.

Oil-furnace process

The oil-furnace process is overwhelmingly predominant today. This process, based on the partial combustion of liquid aromatic residual hydrocarbons, was first introduced in the USA at the end of World War II. It rapidly displaced the then dominant channel (impingement) and gas-furnace processes because it gave improved yields and better product qualities. It was also independent of the geographical source of raw materials, a limitation on the channel process and other processes dependent on natural gas, making possible the worldwide location of manufacturing plants closer to the tyre customers. Environmentally it favoured elimination of particulate air pollution and was more versatile than all other competing processes. A simplified flow diagram of a modern oil-furnace black production line is shown in the following Fig. 7.3.

The principal pieces of equipment are the air blower, process air and oil preheaters, reactors, quench tower, bag filter, pelletiser and rotary dryer. The basic process consists of atomising the feedstock into the combustion zone of the reactor where the combustion of natural gas and preheated excess air create a high temperature environment of 1200° to 1900°C that almost instantly vapourises the feedstock and decomposes most of it to carbon black and hydrogen. The remaining feedstock reacts with the excess oxygen in the primary combustion stream to maintain the reaction temperature for carbon formation. In some reactors a number of feedstock streams are atomised radially into the high velocity combustion gases. The reaction products must be quenched with water sprays to lower the temperature to prevent loss of the carbon black product through reaction with carbon dioxide and water, products of the combustion reactions.

The hot, heavy carbon black smoke from the reactors enters the air preheater where thermal energy is transferred to preheat the primary combustion air. From the air preheater the lower temperature combustion products are given a secondary quench for a further lowering of temperature in a tower from which they enter the bag filter that separates the fluffy carbon black product from the tail gases. Since the tail gases are composed mainly of water, nitrogen, carbon monoxide, carbon dioxide, and hydrogen, they have heating value as a fuel to supplement the natural gas used to preheat feedstock and for heating the pellet dryers. Unused tail gas is frequently flared prior to venting to the atmosphere after removal of particulate matter. The fluffy carbon black from the bag filter is mechanically agitated to increase its bulk density and is then conveyed to the wet pelletisers where water is added to transform the product into wet granules. Dry pelletisation in rotating drums is practiced for some special applications. The wet pellets are then dried in a rotary dryer after which finished product goes to storage tanks for shipping in bulk or in bags.

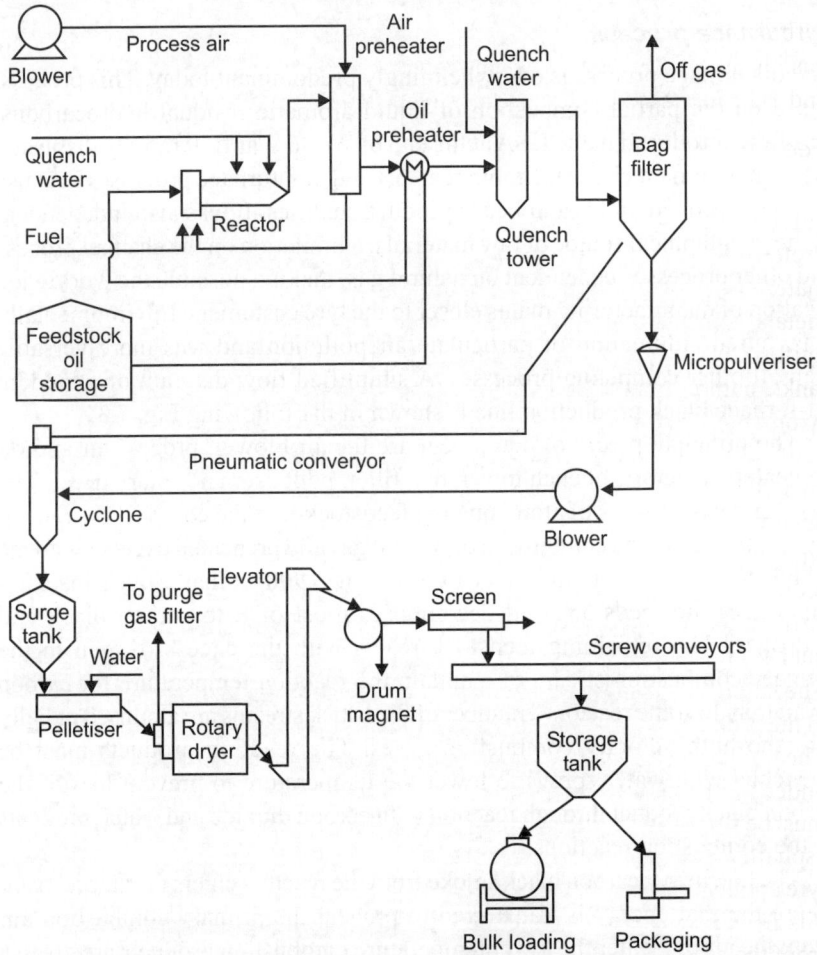

Fig. 7.3. Flow diagram of oil-furnace black process.

A plant may consist of a single line producing several grades of carbon black in a blocked-out operation, or of two or more lines, each capable of producing a different assortment of carbon blacks. A typical line may produce about 140,000 lb/stream day (40 to 50 million lb per year allowing for on stream factor). The reaction section usually includes two to five reactors, each with its own air and feedstock preheaters, the effluents being combined for downstream processing. The energy utilisation in the production of one kg of oil-furnace carbon black is in the range of $9.3–16 \times 10^7$ J ($4–6.9 \times 10^4$ Btu/lb), and the yields are $300–660$ kg/m^3 ($2.5–5.5$ lb/gal) depending on the grade. The energy inputs to the reactor are the heat of combustion of the preheated feedstock, heat of combustion of natural gas, and the thermal energy of the preheated air. The energy output consists of the heat of combustion of the carbon black product, the heat of combustion and the sensible heat of the tail gas, the heat loss from the water quench, heat loss by radiation to atmosphere and the heat transferred to preheat the primary combustion air.

Feedstocks

Feedstocks are viscous aromatic hydrocarbons consisting of branched polynuclear aromatics with smaller quantities of paraffins and unsaturates. Preferred feedstocks are high in aromaticity, free of coke and other gritty materials, and contain low concentrations of asphaltenes, sulphur and alkali metals. Other limitations are the quantities available on a long-term basis, uniformity, ease of transportation, and cost. The ability to handle such oils in tanks, pumps, transfer lines, and spray nozzles are also primary requirements. Aromaticity is the most important property of a carbon black feedstock. It is generally measured by the Bureau of mines correlation index (BMCI) and is an indication of the carbon-to-hydrogen (C/H) ratio. High aromaticity is desirable for successful carbon black operation. For a given production equipment, increased aromaticity results in lower feedstock requirements and reduced gas load to downstream equipment like exchangers, filters, etc. Aromaticity is also a factor controlling carbon black quality, reactor geometry, quenching, and the like, but these are not considered major economic factors. The sulphur content is limited to reduce corrosion, loss of yield, and sulphur in the product. It may be limited in certain locations for environmental reasons. The boiling range must be low enough so that it will be completely volatilised under furnace time-temperature conditions. Alkane insolubles or asphaltenes must be kept below critical levels in order to maintain product quality. Excessive asphaltene content results in a loss of reinforcement and poor treadwear in tyre applications. The principal sources of feedstocks in the USA are the decant oils from petroleum refining operations. These are clarified heavy distillates from the catalytic cracking of gas oils. About 95 per cent of the US feedstock use is decant oil. Another source of feedstock is ethylene process tars obtained as the heavy by-products from the production of ethylene by steam cracking of alkanes, naphthas, and gas oils. There is a wide use of these feedstocks in Europe.

Asian and European operations also use significant quantities of coal tars, creosote oils, and anthracene oils, the distillates from the high temperature coking of coal. European feedstock sources are 50 per cent decant oils and 50 per cent ethylene tars and creosote oils.

The pricing of carbon black feedstocks depends on their alternate market as residual fuel oil, essentially that of high sulphur no. 6 fuel oil. The actual price is determined by the supply-demand relationships for these two markets. Feedstock cost contributes about 60 per cent of the total manufacturing cost. The market price of carbon black is strongly dependent on the feedstock cost and fluctuates along with it.

Contaminants in various feedstocks are important consideration. The deleterious compounds in marginal feedstock fall into two categories—those that are harmful to the reactor and those that adversely affect the properties of the product. Sulphur is not normally considered a problem at this time although it does reduce yield and contribute to air pollution. Inorganic matter such as cracking catalyst, catalyst lubricants, corrosion products, and indigenous metals foul the reactor with deposits and promote slagging. Organic particulate matter, pre-existent coke particles, etc. frequently found in ethylene cracker residuum, contribute to grit and may adversely affect rubber reinforcement and tyre wear. Alkali metals from quench water or seawater contamination depress the structure level of the carbon black and therefore limit flexibility. In addition, certain components of the feedstock, which contribute to high pentane insolubles and high toluene insolubles, are also undesirable since they may contribute to high grit or to the production of unusually large carbon black particles (positive skewness of distribution). Feedstock properties in general are not related to the bulk properties of carbon black. Thus, the successful use of marginal feedstocks depends upon one of two options—clean up the feedstock or adapt the reactor and the process to the use of marginal feedstocks.

Design and operations

Reactors

The reactor is the heart of the oil-furnace process. The diversity of oil-furnace blacks requires different reactor designs and sizes to cover the spectrum, although several different carbon blacks may be produced in any given reactor. Reactors consist of an inlet section, a reaction tunnel, which may be cylindrical or may contain venturi shapes or cylindrical chokes, and provision for quenching the reaction. For hard carbon blacks, reaction zone diameters may range from 6 to 15 inches with lengths up to 15 ft; for soft carbon blacks, larger diameters up to 30 inches or more and lengths up to 30 to 35 ft may be desirable. Hard carbon blacks are the more reinforcing types ranging in surface area from about 75 to 150 m^2/g and soft carbon blacks are the less reinforcing or nonreinforcing types ranging in surface area from about 25 to 50 m^2/g. Each producer or licensor has his own proprietary reactor designs.

The typical reactor is lined with refractories precast in the required shapes from electric furnace mullite, 90 to 99 per cent alumina-silica, or chrome-alumina materials. Depending on cost and service requirements, different parts of a reactor may employ different refractory materials. Properties such as temperature resistance, thermal shock resistance, spalling and slagging resistance and erosion resistance must be weighed in the selection. The liner is backed with high-temperature castable refractory and insulation cement in a steel shell. To maintain precise internal dimensions, the refractory liners, though costly, must usually be replaced at intervals of 1 to 3 years.

Reaction section

Process air is supplied by centrifugal blowers delivering air at 6 to 12 lb/in^2 gauge pressure. Generally, one set of blowers supplies air for all plant reactors, the output being manifolded to the various reactors and metered individually to each. Similarly, feedstock oil from storage is heated by steam to an appropriate temperature for pumpability, typically, 150° to 250°F, and pumped through a manifold at 150 to 300 lb/in^2 gauge with metered takeoffs for each reactor, excess flow returning to storage. Modern air preheaters are usually of the shell and tube type, designed for less than 1 lb/in^2 pressure drop, with fixed tube sheets, disk and donut baffles shell side, and expansion provision in the shell. Flow is concurrent with smoke (reactor effluent) tube-side and air shell side. Attainable air temperature depends on metallurgy; with suitable stainless steels, 1000°F or higher may be reached, requiring 1400° to 1600°F smoke inlet. More typical is 1200° to 1300°F smoke inlet and 500° to 800°F air outlet temperatures. Overall heat transfer rate may be 4 to 5 $Btu/hr/ft^2/°F$. Typical oil preheaters are pipe coil type with oil inside the coil. Overall heat transfer rate may be 10 to 12 $Btu/hr/ft^2/°F$, feedstock oil being normally preheated to 500° to 600°F without vapourisation. In many older lines the oil preheater precedes the air preheater.

Auxiliary fuel is burned in the tangential tunnels of the reactor with the preheated air, the air usually being 40 to 100 per cent in excess of theoretical depending upon the combustion temperature limitations of the refractories and economic considerations (combustion temperature is a function of air preheat and per cent excess air and is maintained at a safe level by establishing and controlling the tangential heat input expressed as air enthalpy plus net heating value of fuel per standard cubic foot of air). In most soft black operations, only a small amount or no auxiliary fuel is used. Feedstock oil enters the reactor axially through a spray nozzle shrouded by a small axial stream of unheated air. Both nozzle spray pattern and longitudinal nozzle position affect carbon black properties, and oil preheat, which affects spraying characteristics must be closely controlled. The excess tangential air and axial air burn a portion of the feedstock oil, providing additional heat for the reactions converting the balance of the oil to carbon black. A feature of the tangential reactor and most other oil-furnace reactors is a rotating flow imparted to the combustion gases. This creates a vortex largely confining the carbon-forming reactions and preventing oil droplets or partially formed carbon black from reaching the reactor walls where they would cause coke decomposition.

Primary quench water sprays appropriately located stop the reaction and adjust the smoke temperature to that required for entry to the preheaters. This may be by two or four side sprays directed radially, but a single spray nozzle at the reactor centreline directed upstream is usually considered superior. Alternate quench locations are frequently provided to allow for earlier or later quenching as required by throughput and grade of carbon black.

Filtration

Smoke leaving the preheaters is combined with that from the other reactors and enters the quench tower. Water sprays (secondary quench) here adjust the smoke temperature as required for the carbon black collection system. The tower also may serve as a grit separator, removing small coke or refractory particles carried from the reactors. In some installations the quench tower is omitted, secondary quench being introduced in the smoke transfer line. Earlier, carbon black collection systems utilised a combination of an electrical precipitator to agglomerate the carbon black to some degree, followed by cyclone separators, and recovered 90 to 97 per cent of the carbon black, later with a bag filter added to improve recovery and decrease carbon black emissions to the atmosphere. In the most modern systems, electrical precipitator and cyclones are omitted and collection is by bag filter only, and loss of carbon black in the off-gas is virtually eliminated.

The filter shell is usually of carbon steel, perhaps with stainless steel in the screw conveyor and hopper parts where corrosion is a problem, and is insulated to maintain proper operating temperature. Filter bags are tubular and may be 5 to 11 inches in diameter and 10 to 25 ft long. The open ends are secured to openings in the cell plate of the filter and the closed ends suspended from the top of the filter with the bag under suitable tension.

Fabrics made of coated glass fibre yarns are most widely employed for bags in carbon black service, with Teflon fabrics now achieving some acceptance. Carbon black filters are divided into compartments, each of which may contain several hundred filter bags.

Filtering temperature must be maintained above the dew point of the gases, which contain 40 to 50 per cent water vapour from water of reaction and vapourised quench water, and below the allowable service temperature of the bags. Aqueous condensate from the gas tends to be acidic and may attack the bag fabric, promote metal corrosion, and impede carbon black flow. Higher operating temperatures are preferable, especially if the off-gas is to be burned, since combustion of this low-heating value gas is assisted by higher gas temperature. With some glass fabrics, permissible operating temperature may reach 600°F. Teflon fabrics are limited to about 500°F, which is one deterrent to their use. Service performance of filter bags varies considerably among carbon black plants. Some filters may operate for two years or more without complete bag replacement, others only six months. Almost all filters require isolated bag replacements between complete changes to avoid leakage and consequent atmospheric pollution.

Filtering rates in carbon black filters tend to be low compared to rates obtained in other industries. With glass fabric bags, typical smoke rates are 1.0 to 1.5 ft^3/min/ft^2 of cloth, whereas in filtering of cement, fly ash, ore, etc. rates from 2 to 6 ft^3/min/ft^2 are reportedly obtained. Teflon bags in carbon black filtration have been found to give at least 50 per cent greater rate than glass bags, but operating problems and higher cost have slowed their acceptance.

Pneumatic conveying

Carbon black from the filter product outlet is usually pneumatically conveyed through a pulveriser to the surge tank feeding the pelletiser. The carrier gas is smoke withdrawn from the filter with the carbon black. A centrifugal blower furnishes the motive power. Typical linear velocity in the conveying line is about 7200 ft/min (120 ft/sec) minimum. The pulveriser, a type of hammer mill, serves only to protect the product from possible inclusion of coarse residue particles (coke or refractory), which may infrequently be carried from the reactor. At the surge tank, a cyclone separator separates the carbon black and delivers it to the surge tank. The cyclone operates with only a few inches of water pressure drop and under the conditions of carrier gas carbon black loading, may recover 90 to 95 per cent of the entering carbon black. The cyclone effluent gas, still carrying a little carbon black, returns to the filter or may be directed to a separate small filter. The pneumatic conveying system is usually considered more satisfactory and economical than the bucket elevators and screw conveyors once used for conveying the lose, fluffy, low-density carbon black.

Pelleting

To facilitate shipping and handling the carbon black is pelleted, giving a free-flowing product with a bulk density of 18 to 30 lb/ft^3 depending on grade of carbon black and pellet character. The preferred size range is such that the majority of pellet diameters are 0.25 to 2.0 mm. Excessive fineness (less than 0.125 mm diameter) may cause handling problems. Pellets must be hard enough to resist breakage in shipping and handling but if too hard the carbon black may be difficult to disperse in the end use. Today, pelleting is primarily by the wet process.

The most commonly used wet pelletiser, constructed of stainless steel and capable of pelleting 5000 to 6000 lb/hr, consists of a horizontal cylindrical shell about 20 inches in diameter and 8 to16 ft long with a rotating shaft extending through the centre. The shaft is fitted with radial pins in a double helical arrangement, the pins extending to a close clearance with the shell. The extremity of each pin is bevelled to form a cutting point.

The pin tips wear in use so must be periodically resharpened and the pins replaced when they become too short. Pin length may be made adjustable to avoid frequent replacement. A short screw conveyor is sometimes used under the carbon black inlet to assist movement of the carbon black into the pin section, and some pelletisers have provision for heating the shell either electrically or by system jacket. The shaft rotating speed may be adjustable in the range of 200 to 800 r/min.

Carbon black is fed from the surge tank through a rotary valve. Water enters the pelletiser through one or more sprays downstream of the black inlet, mixes with, and is absorbed by the carbon black. The mixing and cutting action of the pins converts this damp mass into pellets, rounded to a roughly spherical shape. The pellets exiting the pelletiser, although they contain 40 to 55 per cent water based on dry carbon black, appear dry because the water is completely absorbed. To attain desired pellet properties, pelleting additives are frequently introduced with the pelleting water. Typical additives are molasses or lignin sulphonate, 0 to 3 per cent weight on carbon black, and/or hydrocarbon (e.g. kerosene) up to a few hundred ppm on carbon black. A second pelletiser, termed a polisher, may be operated in series with the first but this practice is declining.

Pelletiser operation is largely an art. Optimum carbon black/water ratio, additive level, and revolutions per minute may vary with type of carbon black, temperature of materials, and pin condition, and must be adjusted by trial and error. Dry pelleting is an older process still employed in relatively few plants primarily because some consumers prefer dry pelleted carbon blacks for certain uses. Dry process pellets are usually softer and more prone to dusting than wet process pellets, and the pelleting process is more troublesome. Dry pelleting of oil-furnace carbon blacks employs large horizontal drums typically 6 to 10 ft in diameter by 48 to 60 ft long, rotating at about 15 r/min, with loose carbon black and recycled pellets fed near one end. The tumbling action produces the pellets assisted by the seeding of the recycle. Close observation is necessary as operating upsets may have serious consequences, even causing loss of pelleting action. Dry pelleted carbon blacks, of course, do not require a drying step.

Drying

A screw conveyor feeds wet pellets from the pelletiser to the dryer where the moisture content is reduced to less than 1 per cent (typically 0.5 per cent). The usual dryer is a large rotating drum 6 to 8 ft in diameter by 48 to 70 ft long, made of suitably corrosion and temperature-resistant alloy steels, mounted on trunnions, and rotated at 3 to 4 r/min. Wet pellets are fed at one end and dry pellets exit at the other. Internal helical flights attached to the drum shell rapidly move the carbon black into the heated section, and agitating flights or lugs tumble the carbon black as it passes through the heated length. The drum is encircled by a refractory lined firebox with a number of fuel burners located in its lower part along its length. The combustion gases pass around the drum, heating it, and thence to a stack. The heating pattern is adjusted by varying the burner-firing pattern. In some newer designs, a separate combustion furnace is employed with the combustion gases distributed along the heated length of the drum by a refractory duct with adjustable openings.

A purge gas blower pulls a portion of the combustion gas into the drum through inlets known as snorkels and located near the outlet end. The purge gas passes through the drum countercurrent to the carbon black flow to sweep out the steam. From the blower the purge gas is usually routed to a small bag filter separate from the main filter.

The majority of plants today employ clean reactor off-gas (from the main bag filter) as dryer fuel and the number is growing, dictated both by conservation and atmospheric pollution considerations. Some of these low heating value off-gases require enrichment with other fuel (e.g. natural gas) for satisfactory combustion, but with suitable burning systems, many can be burned without enrichment.

Dryer product temperature is controlled at specific values, usually in the range of 300° to 500°F at the outlet, often with separate control in different portions of the drum. These temperatures must be high enough to produce suitably dry product, but temperatures too high may promote undesirable oxidation of the carbon black or even create a fire hazard. Therefore, close control, though difficult, is essential. Dryer shell temperatures are usually limited to about 1000°F or less. An 8 × 60 ft dryer may be capable of drying 5000 to 6000 lb/hr of carbon black.

This oil-furnace process of carbon black manufacture is the most important process used worldwide. Therefore, it has been studied in great detail covering process variables, energy conservation and recovery, cost of production, feedstock quality and yields, automation and control, and corrosion of plant equipment. Table 7.3 compares yields of carbon black from various processes. The energy requirements for each process are also listed.

Environmental aspects

Oil-furnace plants are potential sources of contaminants emitted to the land, atmosphere, and surface waters. Fuel and feedstock are handled in large quantities in carbon black plants. Carbon black is produced in dilute form (smoke) collected in filters, handled in conveyors, densified (pelletised), stored in tanks and transferred to packages or bulk containers for shipment.

Table 7.3. Comparison of yields from different processes.

Process	Raw material	Commercial yields, g/m^3	Yield % of theoretical carbon	Energy utilisation, J/kg content
Channel	Natural gas	8–32	1.6–6.0	$1.2–2.3 \times 10^9$
Gas-furnace	Natural gas	144–192	27–36	$2.3–3.0 \times 10^8$
Thermal	Natural gas	160–240	30–45	$2.0–2.8 \times 10^8$
Oil-furnace	Liquid aromatic hydrocarbons	300–660 kg/m³ (2.5–5.5 lb/gal)	23–70	$9.3–16 \times 10^7$

Gases from the combustion and cracking processes are converted by various heat recovering equipment to an oxidised state; but ultimately oxides of carbon, sulphur and nitrogen must be discharged to the atmosphere or given further special treatment (stack gas scrubbing).

Also, the different types of equipment must be cleaned properly to remove carbon black and washed with water. Such complexity of processing and solids handling equipment is a source of many contaminants. However, over the years, the carbon black industry has made an extreme effort to contain this contamination within the confines of the equipment and facilities specially developed for the purpose. Modern carbon black plants contribute little in the form of particulate environmental contaminants. Figure 7.4 shows a modern oil-furnace plant with emphasis on environment protection facilities.

Storage and shipping

Typical carbon black storage tanks may have a capacity of around one million lb of pelleted carbon black. They are usually divided into two to four compartments for storage of different grades of carbon black, and are lined or otherwise constructed to minimise corrosion. The tanks are elevated so that loading of bulk shipments and delivery to packaging equipment can be by gravity flow. Product leaving the dryer is lifted by a bucket elevator, passes over a drum magnet and usually a scalper screen, and is delivered to the proper storage compartment by a screw conveyor system.

The drum magnet guards against inclusion of magnetic material, infrequently found, in the product. Since any magnetic material probably results from steel corrosion, its appearance calls for corrective action.

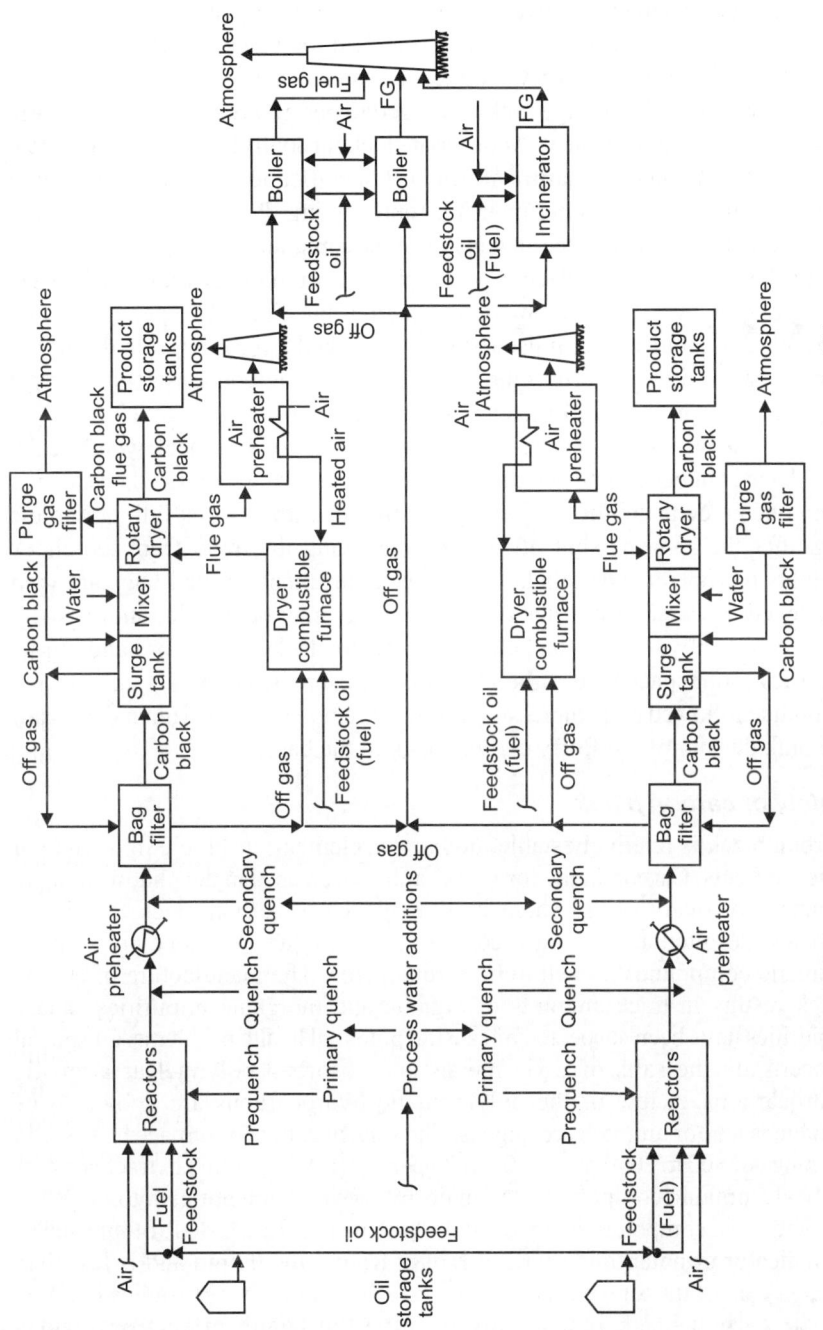

Fig. 7.4. Oil furnace process showing environmental protection facilities.

The scalper screen removes oversize pellets (over, say, 2 mm diameter) and is particularly needed for some types of carbon black for which pelleting control is difficult. The screw conveyor system consists of a set of screw conveyors above the storage tanks arranged to permit product delivery to any desired storage compartment. Appropriate lockout controls prevent inadvertent delivery to a wrong compartment. Any off-specification product is delivered to a separate tank or compartment for reprocessing. Pelleted carbon black is shipped in bulk or packaged in bags or other containers.

Packaging of carbon black is predominantly in multiply kraft paper bags, using one or two spout automatic packaging machines. The pellets are fed through the spout in a fluidised state established by passing air through a porous pad and weighed into the bag. Usually, a bag contains 50 lb (25 kg) carbon black product and modern machines can package 5 to 10 bags per minute. The bags from the packaging machine pass over a check weigh scale.

Changes in pellet type and quality, usually occurring mainly when shifting from one grade to another, affect the quantity of carbon black falling into the bag after the spout is shut off, but by observing the trend from the check weights the operator can make necessary adjustments. Filled bags are then conveyed to warehouse, truck or rail car. Bags may be handled individually but are generally unitised on pellets bearing 40-stacked bags for handling by lift truck. Some packaged carbon black is sold unpelleted. A vacuum packing technique followed by compression is used to remove occluded air and increase the bulk density of the fluffy carbon black in the bag.

Safety of carbon black

Carbon black is relatively stable, unreactive element, insoluble in organic or other solvents. Carbon in the form of char has been used in the pharmaceutical industry for many years. There is no evidence of carbon black toxicity in humans, despite the fact that it contains trace amounts of some polynuclear aromatic compounds known to be carcinogenic. The manufacture of carbon black results in trace amounts of organic and inorganic impurities. These impurities have been suspected of causing potential health problems. Of special concern are the salts of toxic metals and absorbed polynuclear aromatic hydrocarbons. A few of the polyaromatic hydrocarbons are known to be mutagens and/or animal carcinogens. The solvent extract of furnace blacks is in the range of 300 to 2000 ppm (0.03–0.20 per cent). Most of this extract consists of 10–15 organic compounds, the majority of which are not genotoxic.

One toxic compound is benzo[alpha]pyrene [50-32-8], BαP, often used as an indicator of potential hazard. It ranges from 0 to 50 ppm and is less than one per cent of the total extract. There have been a number of studies initiated by the carbon black industry to examine the health effects of various commercial carbon blacks and their benzene extracts.

Although the solvent extracts of carbon black do show toxic properties, the aqueous systems of concern in humans show no elution of BαP and no toxic properties. The BαP is believed to be so strongly absorbed on the surface of carbon black and in such high dilution that it is inactive in animal testing for carcinogenicity.

There is no evidence of increased cancer risk from exposure to industrial carbon blacks. Most studies of carbon black dust inhalation and intratracheal administration with animals indicate that carbon black is not carcinogenic. OSHA regulations for carbon black dust concentrations call for an average exposure level over a given time period of not more than 3.5 mg/m^3.

Carbon blacks, which according to their application properties may be used for food products, cosmetics, drinking water pipes, food packaging materials, and toys, must comply with local regulations. The testing methods for the approval of carbon blacks for such applications are different in different countries. The general aim, however, is to give limitations for the content of polyaromatic hydrocarbons and of heavy metals.

INDUSTRIAL DIAMOND INDUSTRY

The market for industrial-grade diamonds operates much differently from its gem-grade counterpart. Industrial diamonds are valued mostly for their hardness and heat conductivity, making many of the gemological characteristics of diamonds, such as clarity and colour, irrelevant for most applications.

The dominant industrial use of diamond is in cutting, drilling, grinding, and polishing. Most uses of diamonds in these technologies do not require large diamonds; in fact, most diamonds that are gem-quality except for their small size, can find an industrial use. Diamonds are embedded in drill tips or saw blades, or ground into a powder for use in grinding and polishing applications. Specialised applications include use in laboratories as containment for high pressure experiments, high-performance bearings, and limited use in specialised windows.

With the continuing advances being made in the production of synthetic diamonds, future applications are beginning to become feasible. Garnering much excitement is the possible use of diamond as a semiconductor suitable to build microchips form, or the use of diamond as a heat sink in electronics.

The boundary between gem-quality diamonds and industrial diamonds is poorly-defined and partly depends on market conditions (for example, if demand for polished diamonds is high, some suitable stones will be polished into low-quality or small gemstones rather than being sold for industrial use). Within the category of industrial diamonds, there is a sub-category comprising the lowest-quality, mostly opaque stones, which are known as bort or boart.

Three natural types are employed: (i) crystalline and cleavable diamonds, more or less off-grade and off-colour, (ii) bort, translucent to opaque, gray or dark brown, with a radiated or confused crystalline structure, and (iii) carbonado, frequently known as black diamond or carbon, which occurs in an opaque, tough, crystalline aggregate without cleavage. Bort is the most widely used of industrial diamond stones. Synthetic diamond has become the standard of performance for grinding wheels, saws, and drills, and is a major factor in the economy of the industry.

An extensive use of industrial diamonds is in the grinding of shaped carbide or ceramic tips for turning and boring tools. Bonded-diamond wheels for grinding hard, abrasive materials such as cemented carbide tools, ceramics, and glass are made from synthesised diamond powder or crushed bort, held by a matrix of resin, vitreous material, or sintered metal. Tools set with industrial diamond are a practical necessity in truing and dressing worn abrasive wheels in all industries where precision grinding is performed. Diamond lathe tools, because of the hardness and high heat conductivity of diamond, allow greatly increased machining speeds in turning nonferrous metal parts to close tolerances and fine finishes. Since the continuous drawing of wire through reduction dies causes much wear, dies for accurate wire drawing are made of diamonds. Such dies are generally made from fine-quality single-crystal stones with an appropriate hole drilled through them. Recently, successful wire-drawing dies have been made from diamond compacts which consist of fine diamond powder grains thoroughly bonded together. The holes are started by using sharp diamond points, or by a fine laser beam, and are finished by lapping with fine diamond dust. When worn, dies must be redressed to the next larger wire size. Diamond dies are generally used only for drawing wire smaller than 2.5 mm in diameter. Diamond bits for drilling through hard rock formations for oil, gas, or water, for boring blast-charge holes in mines and quarries, and for drilling holes in and taking cores from concrete structures are set with small whole diamonds in a sintered metal bond.

Here the diamonds do the actual work, the steel heads simply serving to hold them in the cutting position. Diamond drill crowns allow actual cores of rock to be taken from great depths and examined to yield information on what minerals or deposits are present. The increasing trend toward automation is causing the domestic consumption of industrial diamonds to be the fastest growing in the abrasives industry.

In mounting diamond crystals in tools they should be oriented in the crystal direction which provides the greatest abrasion resistance. The most favourable orientation can be done by eye if the shape of the crystal is such that the crystal directions can be recognised. Otherwise X-ray diffraction must be used. Proper orientation of the diamonds can improve the performance of a tool many fold over that of one having a random diamonds orientation.

General Electric Co. commercially introduced synthetic diamond grit for resin-bonded grinding wheels based on a process which requires pressures and temperatures in the region of thermodynamic stability for diamond (versus graphite) and a molten catalyst-solvent metal consisting of a group VIII metal or alloy. Special ultra high-pressure apparatus is used, the moving members of which are forced together by large hydraulic presses. It is a batch process requiring a period of a number of minutes. Different types and sizes of diamond particles, or crystals, require different conditions of pressure, temperature, catalyst-solvent, and reaction time. The crude diamonds are chemically cleaned and sized by sieving. Many different types are now currently available, ranging from the finest polishing powders up to 25/35 mesh blocky crystals which yield excellent serve in saws and drills for ceramic, stone, and concrete cutting. Fine diamond polishing powders can be made directly from certain graphites by shock-compression processes. A shock wave of the order of 300 to 1200 kbar intensity and 10^{-6} s duration is passed through the graphite. In the Du Pont process the graphite particles are dispersed in a metal matrix, which helps to transmit the pressure wave and to thermally quench the extremely hot carbon components of the system immediately after compression. The thermal quench helps prevent graphitisation of the diamond formed during the compression. The metal is removed by acids, and the extraneous graphite by selective oxidation processes.

Techniques have been developed for bonding diamond powder into strong, relatively tough, compacts. These, when sharpened and mounted properly, serve as effective cutting tips for tools to machine very hard workpieces. These diamond compacts also serve well as cores for small wire drawing dies.

The new diamond like abrasive material, cubic boron nitride, is now synthesised, marketed, and used commercially. It is made from graphitic boron nitride at pressures and temperatures similar to those used in diamond synthesis, but with different catalyst-solvent materials. Compacts of cubic boron nitride are also made. Cubic boron nitride abrasive grain is particularly effective in grinding hard ferrous metals with which diamond tends to react. In general, diamond abrasive works best on hard nonmetallic materials, while cubic boron nitride abrasive does best on hard metallic materials.

Gem-quality diamonds can be grown in the laboratory, but the process is slow and expensive and is not considered economically feasible in the usual market. However, these diamonds have the highest thermal conductivity of any known material, more than matching that of type II natural diamond crystals, and they may become useful as heat sinks for high-power-density microelectric solid-state devices and for high-intensity laser beam windows.

CARBON FIBRE

Carbon fibre or (alternatively called graphite fibre, graphite fibre or carbon graphite) is a material consisting of extremely thin fibres about 0.005–0.010 mm in diameter and composed mostly of carbon atoms. The carbon atoms are bonded together in microscopic crystals that are more or less aligned parallel to the long axis of the fibre. The crystal alignment makes the fibre very strong for its size. Several thousand carbon fibres are twisted together to form a yarn, which may be used by itself or woven into a fabric. Carbon fibre has many different weave patterns and can be combined with a plastic resin and wound or moulded to form composite materials such as carbon fibre reinforced plastic (also referenced as carbon fibre) to provide a high strength-to-weight ratio material. The density of carbon fibre is also considerably lower than the density of steel, making it ideal for applications requiring low weight. The properties of carbon fibre such as high tensile strength, low weight, and low thermal expansion make it very popular in aerospace, civil engineering, military, and motorsports, along with other competition sports.

Properties

Carbon fibres are the closest to asbestos in a number of properties. Each carbon filament thread is a bundle of many thousand carbon filaments. A single such filament is a thin tube with a diameter of 5–8 micrometers and consists almost exclusively of carbon.

The atomic structure of carbon fibre is similar to that of graphite, consisting of sheets of carbon atoms (graphene sheets) arranged in a regular hexagonal pattern. The difference lies in the way these sheets interlock. Graphite is a crystalline material in which the sheets are stacked parallel to one another in regular fashion. The intermolecular forces between the sheets are relatively weak van der Waals forces, giving graphite its soft and brittle characteristics. Depending upon the precursor to make the fibre, carbon fibre may be turbostratic or graphitic, or have a hybrid structure with both graphitic and turbostratic parts present. In turbostratic carbon fibre the sheets of carbon atoms are haphazardly folded, or crumpled, together. Carbon fibres derived from polyacrylonitrile (PAN) are turbostratic, whereas carbon fibres derived from mesophase pitch are graphitic after heat treatment at temperatures exceeding 2200°C. Turbostratic carbon fibres tend to have high tensile strength, whereas heat-treated mesophase-pitch-derived carbon fibres have high Young's modulus and high thermal conductivity.

Applications

Carbon fibre is most notably used to reinforce composite materials, particularly the class of materials known as carbon fibre or graphite reinforced polymers. Non-polymer materials can also be used as the matrix for carbon fibres. Due to the formation of metal carbides and corrosion considerations, carbon has seen limited success in metal matrix composite applications. Reinforced carbon-carbon (RCC) consists of carbon fibre-reinforced graphite, and is used structurally in high-temperature applications. The fibre also finds use in filtration of high-temperature gases, as an electrode with high surface area and impeccable corrosion resistance, and as an anti-static component.

Synthesis

Each carbon filament is made out of long, thin filaments of carbon sometimes transformed to graphite. A common method of making carbon filaments is the oxidation and thermal pyrolysis of polyacrylonitrile (PAN), a polymer based on acrylonitrile used in the creation of synthetic materials. Like all polymers, polyacrylonitrile molecules are long chains, which are aligned in the process of drawing continuous filaments. A common method of manufacture involves heating the PAN to approximately 300°C in air, which breaks many of the hydrogen bonds and oxidises the material.

The oxidised PAN is then placed into a furnace having an inert atmosphere of a gas such as argon, and heated to approximately 2000°C, which induces graphitisation of the material, changing the molecular bond structure. When heated in the correct conditions, these chains bond side-to-side (ladder polymers), forming narrow graphene sheets which eventually merge to form a single, jelly roll-shaped or round filament. The result is usually 93–95 per cent carbon.

Lower-quality fibre can be manufactured using pitch or rayon as the precursor instead of PAN. The carbon can become further enhanced, as high modulus, or high strength carbon, by heat treatment processes. Carbon heated in the range of 1500°–2000°C (carbonisation) exhibits the highest tensile strength (8,20,000 psi or 5650 MPa or 5650 N/mm^2), while carbon fibre heated from 2500° to 3000°C (graphitising) exhibits a higher modulus of elasticity (77,000,000 psi or 531 GPa or 531 kN/mm^2).

Textile

Precursors for carbon fibres are PAN, rayon and pitch. Carbon fibre filament yarns are used in several processing techniques: the direct uses are for prepregging, filament winding, pultrusion, weaving, braiding, etc. Carbon fibre yarn is rated by the linear density (weight per unit length = 1 g/1000 m = tex) or by number of filaments per yarn count, in thousands. For example, 200 tex for 3000 filaments of carbon fibre is three times as strong as 1000 carbon fibres but is also three times as heavy.

This thread can then be used to weave a carbon fibre filament fabric or cloth. The appearance of this fabric generally depends on the linear density of the yarn and the weave chosen. Some commonly used types of weave are twill, satin and plain.

PYROLYTIC CARBON

Pyrolytic carbon is a material similar to graphite, but with some covalent bonding between its graphene sheets as a result of imperfections in its production. Generally it is produced by heating a hydrocarbon nearly to its decomposition temperature, and permitting the graphite to crystallise (pyrolysis). One method is to heat synthetic fibres in a vacuum. Another method is to place seeds or a plate in the very hot gas to collect the graphite coating.

Physical Properties

Pyrolytic carbon samples usually have a single cleavage plane, similar to mica, because the graphene sheets crystalise in a planar order, as opposed to graphite, which forms microscopic randomly-oriented zones. Because of this, pyrolytic carbon exhibits several unusual anisotropic properties. It is more thermally conductive along the cleavage plane than graphite, making it one of the best planar thermal conductors available. It is also more diamagnetic against the cleavage plane, exhibiting the greatest diamagnetism of any room temperature (by weight) diamagnet. It is even possible to levitate reasonably pure and sufficiently ordered samples over rare earth permanent magnets.

Applications

Pyrolytic carbon applications are following:
1. It is used nonreinforced for missile nose cones, and ablative (boiloff-cooled) rocket motors.
2. In fibre form, it is used to reinforce plastics and metals
3. Pebble bed nuclear reactors use a coating of pyrolytic carbon as a neutron moderator for the individual pebbles.
4. Used to coat graphite cuvettes (tubes) in graphite furnace atomic absorption furnaces to decrease heat stress, thus increasing cuvette lifetimes.
5. Pyrolytic carbon is used for several applications in electronic thermal management: thermal interface material, heat spreaders (sheets) and heat sinks (fins).
6. It is used to fabricate grid structures in some high power vacuum tubes.

Biomedical applications

Because blood clots do not easily form on it, it is often advisable to line a blood-contacting prosthesis with this material in order to reduce the risk of thrombosis. For example, it finds use in artificial hearts and artificial heart valves. Blood vessel stents, by contrast, are often lined with a polymer that has heparin as a pendant group, relying on drug action to prevent clotting. This is at least partly because of pyrolytic carbon's brittleness and the large amount of permanent deformation which a stent undergoes during expansion. Pyrolytic carbon is also in medical use to coat anatomically correct orthopaedic implants, and replacement joint. In this application it is currently marketed under the name pyrocarbon.

Graphite Electrodes

Graphite electrodes are used mainly in electric arc furnace steel production. They are presently the only products available that have the high levels of electrical conductivity and the capability of sustaining the extremely high levels of heat generated in this demanding environment. Graphite electrodes are also used to refine steel in ladle furnaces and in other smelting processes.

Chapter 8

Ceramic Industries

INTRODUCTION

The traditional ceramic industries, sometimes referred to as the clay products or silicate industries, have as their finished materials a variety of products that are essentially silicates. In recent years new products have been developed as a result of the demand for materials that withstand higher temperatures, resist greater pressures, have superior mechanical properties, possess special electrical characteristics or can protect against corrosive chemicals.

CLAY

Clay is a naturally occurring material composed primarily of fine-grained minerals, which show plasticity through a variable range of water content, and which can be hardened when dried and/or fired. Clay deposits are mostly composed of clay minerals (phyllosilicate minerals), which impart plasticity and harden when fired and/or dried, and variable amounts of water trapped in the mineral structure by polar attraction. Organic materials which do not impart plasticity may also be a part of clay deposits.

Clay minerals are typically formed over long periods of time by the gradual chemical weathering of rocks (usually silicate-bearing) by low concentrations of carbonic acid and other diluted solvents. These solvents (usually acidic) migrate through the weathering rock after leaching through upper weathered layers. In addition to the weathering process, some clay minerals are formed by hydrothermal activity. Clay deposits may be formed in place as residual deposits, but thick deposits usually are formed as the result of a secondary sedimentary deposition process after they have been eroded and transported from their original location of formation. Clay deposits are typically associated with very low energy depositional environments such as large lake and marine deposits.

Clays are distinguished from other fine-grained soils by various differences in composition. Silts, which are fine-grained soils which do not include clay minerals, tend to have larger particle sizes than clays, but there is some overlap in both particle size and other physical properties, and there are many naturally occurring deposits which include both silts and clays. The distinction between silt and clay varies by discipline. Geologists and soil scientists usually consider the separation to occur at a particle size of 2 μm (clays being finer than silts), sedimentologists often use 4–5 μm, and colloid chemists use 1 μm. Primary clays, also known as kaolins, are located at the site of formation. Secondary clay deposits have been moved by erosion and water from their primary location.

Depending upon academic source, there are three or four main groups of clays: kaolinite, mont-morillonite-smectite, illite, and chlorite. Chlorites are not always considered a clay, sometimes being classified as a separate group within the phyllosilicates. There are approximately thirty different types of 'pure' clays in these categories, but most natural clays are mixtures of these different types, along with other weathered minerals. Varve (or varved clay) is clay with visible annual layers, formed by seasonal differences in erosion and organic content. This type of deposit is common in former glacial lakes.

Quick clay is a unique type of marine clay indigenous to the glaciated terrains of Norway, Canada, Northern Ireland and Sweden. It is a highly sensitive clay, prone to liquefaction, which has been involved in several deadly landslides.

Dry clay is normally much more stable than sand with regard to excavations. Clays exhibit plasticity when mixed with water in certain proportions. When dry, clay becomes firm and when fired in a kiln, permanent physical and chemical reactions occur which, amongst other changes, causes the clay to be converted into a ceramic material. It is because of these properties that clay is used for making pottery items, both practical and decorative. Different types of clay, when used with different minerals and firing conditions, are used to produce earthenware, stoneware, and porcelain. Depending on the content of the soil, clay can appear in various colours, from a dull gray to a deep orange–red.

Clay is also used in many industrial processes, such as paper making, cement production and chemical filtering. Clay is also often used in the manufacture of pipes for smoking tobacco. Clay, being relatively impermeable to water, is also used where natural seals are needed, such as in the cores of dams, or as a barrier in landfills against toxic seepage (lining the landfill, preferably in combination with geotextiles).

Recent studies have been carried out to investigate clay's adsorption capacities in various applications, such as the removal of heavy metals from waste-water and air purification.

Industrial Plasticine

It is a modelling material based on wax and typically contains sulphur which gives a characteristic smell to most clays. Often, the styled object will be used to create moulds from this.

Clay animation

Clay animation is one of many forms of stop motion animation. Each animated piece, either character or background, is deformable—made of a malleable substance, usually plasticine clay.

In clay animation, which is one of the many forms of stop motion animation, each object is sculpted in clay or a similarly pliable material such as plasticine, usually around a wire skeleton called an armature. As in other forms of object animation, the object is arranged on the set (background), a film frame is taken and the object or character is then moved slightly by hand. Another frame is taken and the object moved slightly again. This cycle is repeated until the animator has achieved the desired amount of film.

Clay minerals

Clay minerals are hydrous aluminium phyllosilicates, sometimes with variable amounts of iron, magnesium, alkali metals, alkaline earths and other cations. Clays have structures similar to the micas and therefore form flat hexagonal sheets. Clay minerals are common weathering products (including weathering of feldspar) and low temperature hydrothermal alteration products. Clay minerals are very common in fine grained sedimentary rocks such as shale, mudstone and siltstone and in fine grained metamorphic slate and phyllite.

Clay minerals include the following groups:
1. Kaolin group which includes the minerals kaolinite, dickite, halloysite and nacrite.
2. Some sources include the serpentine group due to structural similarities.
3. Smectite group which includes dioctahedral smectites such as montmorillonite and nontronite and trioctahedral smectites for example saponite.
4. Illite group which includes the clay-micas. Illite is the only common mineral. Chlorite group includes a wide variety of similar minerals with considerable chemical variation.

Clay pit

A clay pit is a quarry or mine for the extraction of clay, which is generally used for manufacturing pottery, bricks or Portland cement.

The brick factory is often located alongside the clay pit to reduce the transport costs of the raw material. These days pottery producers are often not sited near the source of their clay and usually do not own the clay deposits. The other essential raw material is fuel for firing and potteries may be located near to fuel deposits rather than the clay.

Bentonite

Bentonite is an absorbent aluminium phyllosilicate, generally impure clay consisting mostly of montmorillonite. There are a few types of bentonites and their names depend on the dominant elements, such as K, Na, Ca, and Al.

Sodium bentonite

Sodium bentonite expands when wet, possibly absorbing several times its dry mass in water. Because of its excellent colloidal properties it is often used in drilling mud for oil and gas wells and for geotechnical and environmental investigations. The property of swelling also makes sodium bentonite useful as a sealant, especially for the sealing of subsurface disposal systems for spent nuclear fuel and for quarantining metal pollutants of groundwater.

Sodium bentonite can also be 'sandwiched' between synthetic materials to create geosynthetic clay liners (GCL) for the aforementioned purposes. This technique allows for more convenient transport and installation and it greatly reduces the volume of sodium bentonite required.

Calcium bentonite

Calcium bentonite is a useful adsorbent of ions in solution as well as fats and oils, being a main active ingredient of Fuller's earth, probably one of humankind's first industrial cleaning agents. Calcium bentonite may be converted to sodium bentonite (termed sodium beneficiation or sodium activation) to exhibit many of sodium bentonite's properties by a process known as 'ion exchange'.

Commonly this means adding 5–10 per cent of a soluble sodium salt such as sodium carbonate to wet bentonite, mixing well, and allowing time for the ion exchange to take place and water to remove the exchanged calcium. Some properties, such as viscosity and fluid loss of suspensions, of sodium beneficiated calcium bentonite (or sodium activated bentonite) may not be fully equivalent to natural sodium bentonite. For example, residual calcium carbonates (formed if exchanged cations are insufficiently removed) may result in inferior performance of the bentonite in geosynthetic liners. Pascalite is a commercial name for the calcium bentonite clay. Potassium bentonite also known as potash bentonite or K-bentonite, potassium bentonite is a potassium rich illitic clay formed from alteration of volcanic ash.

CERAMIC

Ceramics is the art and science of making things from inorganic, non-metallic materials by the action of heat. The term refers to the study of those materials, their purification and combination, and their formation into useful objects. Ceramic materials may have a crystalline or partly crystalline structure, or may be of glass. They are formed by the action of heat and subsequent cooling.

The earliest ceramics were pottery objects made from clay, either by itself or mixed with other materials. Ceramics now includes domestic, industrial and building products and art objects. In the 20th century new ceramic materials were developed for use in advanced ceramic engineering, for example, in semiconductors.

Types of Ceramic Products

For convenience ceramic products are usually divided into four sectors, and these are shown below with some examples:
1. Structural, including bricks, pipes, floor and roof tiles.
2. Refractories, such as kiln linings, gas fire radiants, steel and glass making crucibles.
3. Whitewares, including tableware, wall tiles, pottery products, and sanitary ware.
4. Technical, is also known as engineering, advanced, special, and in Japan, fine ceramics. Such items include tiles used in the space shuttle programme, gas burner nozzles, ballistic protection, nuclear fuel uranium oxide pellets, bio-medical implants, jet engine turbine blades, and missile nose cones. Frequently the raw materials do not include clays.

Technical ceramics

Technical ceramics can also be classified into three distinct material categories:
1. Oxides: Alumina, zirconia.
2. Non-oxides: Carbides, borides, nitrides, silicides.
3. Composites: Particulate reinforced, combinations of oxides and non-oxides.

Each one of these classes can develop unique material properties.

Applications of ceramics

1. Ceramics are used in the manufacture of knives.
2. Ceramics such as alumina and boron carbide have been used in ballistic armoured vests to repel large-caliber rifle fire.
3. Ceramic balls can be used to replace steel in ball bearings.
4. Work is being done in developing ceramic parts for gas turbine engines.
5. Recently, there have been advances in ceramics which include bio-ceramics, such as dental implants and synthetic bones. Hydroxyapatite, the natural mineral component of bone, has been made synthetically from a number of biological and chemical sources and can be formed into ceramic materials.
6. High-tech ceramic is used in watchmaking for producing watch cases.

Ceramic Materials

Ceramic materials are inorganic, non-metallic raw materials and things made from them. They may be crystalline or partly crystalline. They are formed by the action of heat and subsequent cooling. Clay was one of the earliest ceramic materials in human use, but many different ceramic materials are now used in domestic, industrial and building products.

A ceramic material may be defined as any inorganic crystalline oxide material. It is solid and inert. Ceramic materials are brittle, hard, strong in compression, weak in shearing and tension. They withstand chemical erosion that occurs in an acidic or caustic environment. Ceramics generally can withstand very high temperatures such as temperatures that range from 1000° to 1600°C (1800° to 3000°F). Exceptions include inorganic materials that do not have oxygen such as silicon carbide. Glass by definition is not a ceramic because it is an amorphous solid (noncrystalline). However, glass involves several steps of the ceramic process and its mechanical properties behave similarly to ceramic materials.

Traditional ceramic raw materials include clay minerals such as kaolinite, whereas more recent materials include aluminium oxide, more commonly known as alumina. The modern ceramic materials, which are classified as advanced ceramics, include silicon carbide and tungsten carbide. Both are valued for their abrasion resistance, and hence find use in applications such as the wear plates of crushing equipment in mining operations. Advanced ceramics are also used in the medicine, electrical and electronics industries.

Crystalline ceramics

Crystalline ceramic materials are not amenable to a great range of processing. Methods for dealing with them tend to fall into one of two categories — either make the ceramic in the desired shape, by reaction *in situ*, or by forming powders into the desired shape, and then sintering to form a solid body. Ceramic forming techniques include shaping by hand (sometimes including a rotation process called throwing), slip casting, tape casting (used for making very thin ceramic capacitors, etc.), injection moulding, dry pressing, and other variations. A few methods use a hybrid between the two approaches.

Non-crystalline ceramics

Non-crystalline ceramics, being glasses, tend to be formed from melts. The glass is shaped when either fully molten, by casting, or when in a state of toffee-like viscosity, by methods such as blowing to a mould. If later heat treatments cause this class to become partly crystalline, the resulting material is known as a glass-ceramic.

Properties of ceramics

The physical properties of any ceramic substance are a direct result of its crystalline structure and chemical composition.

Physical properties of chemical compounds which provide evidence of chemical composition include odour, colour, volume, density (mass/volume), melting point, boiling point, heat capacity, physical form at room temperature (solid, liquid or gas), hardness, porosity, and index of refraction. Mechanical properties are important in structural and building materials as well as textile fabrics. They include the many properties used to describe the strength of materials such as—elasticity/plasticity, tensile strength, compressive strength, shear strength, fracture toughness and ductility (low in brittle materials), and indentation hardness.

In modern materials science, fracture mechanics is an important tool in improving the mechanical performance of materials and components.

Some ceramics are semiconductors. Most of these are transition metal oxides that are II–VI semiconductors, such as zinc oxide. Under some conditions, such as extremely low temperature, some ceramics exhibit high temperature superconductivity.

Ceramics Processing

Techniques of ceramics processing and materials science engineering include the science of creating high performance ceramic components from inorganic, non-metallic materials—typically in the form of fine powders. The most common methods utilised are based on the application of high temperatures. But more recently, wet chemical methodologies have been developed for obtaining similar microstructures and properties at much lower temperatures. The method includes the purification and/or chemical synthesis of raw materials, the study and production of the chemical compounds concerned, their formation into technologically valuable components and the study of their structural features (on various spatial scales), chemical composition and physical properties. Traditional ceramic raw materials include the most common chemical compounds found in the earth's outer crust. These include silica (SiO_2), alumina (Al_2O_3), magnesia (MgO), calcia (CaO), and iron oxide (Fe_2O_3, or rust). Clay minerals composed of several common oxides which are combined in specific whole number ratios include kaolinite, boehmite or mullite (silica-alumina composite). Other materials used in the development of advanced ceramics include silicon carbide and silicon nitride, boron carbide and titanium carbide.

Chemical processing

Microstructural uniformity

In the processing of fine ceramics, the irregular particle sizes and shapes in a typical powder often lead to non-uniform packing morphologies that result in packing density variations in the powder compact. Uncontrolled agglomeration of powders due to attractive van der Waals forces can also give rise to in microstructural inhomogeneities.

Self-assembly

Self-assembly is the most common term in use in the modern scientific community to describe the spontaneous aggregation of particles (atoms, molecules, colloids, micelles, etc.) without the influence of any external forces. Large groups of such particles are known to assemble themselves into thermodynamically stable, structurally well-defined arrays, quite reminiscent of one of the 7 crystal systems found in metallurgy and mineralogy (e.g. face-centered cubic, body-centered cubic, etc.).

Sol-gel processing

The sol-gel process is a wet-chemical technique for the fabrication of materials (typically a metal oxide) starting either from a chemical solution which acts as the precursor for an integrated network (gel) of either discrete particles or network polymers. Typical precursors are metal alkoxides and metal chlorides, which undergo hydrolysis and polycondensation reactions to form either a network elastic solid or a colloidal suspension (or dispersion)—a system composed of discrete (often amorphous) submicron particles dispersed to various degrees in a host solvent. Formation of a metal oxide involves connecting the metal centres with oxo (M-O-M) or hydroxo (M-OH-M) bridges, therefore generating metal-oxo or metal-hydroxo polymers in solution. Thus, the sol evolves towards the formation of an gel-like diphasic system containing both a liquid phase and solid phase whose morphologies range from discrete particles to continuous polymer networks.

Alternative methods

Melt processing

Several methods for making ceramics either use no powders at all or do not directly make the ceramic from powder. The most extensive of these are the various melt processing methods, of which melt casting produces by far the largest volumes and individual sizes. Arc melting using graphite electrodes predominates, but some induction melting is used, directly coupling to the ceramic to be melted, usually after heating part of the mass by other means to about 1000°C for sufficient coupling.

Chemical vapour deposition

Another nonpowder-base method is that of chemical vapour deposition (CVD). While used for making ceramic powders, this method is also used quite extensively for making ceramic coatings, as well as monolithic components.

Typically, inorganic and related precursors, which are substantially less expensive, are utilised for bulk, as in structural bodies, with processing temperatures commonly in the 1000°–1500°C range. Organometallic precursors, typically much more expensive, can be used, for example, for coatings for electronics, with depositions at temperatures from a few to several 100°C.

Physical vapour deposition

Various methods of physical vapour deposition (PVD) are also used. These include evaporation (e.g. electron beam), sputtering, and reaction process (e.g. reactive sputtering). Since the deposition rates are quite low, such processes are restricted to thin coatings. Coatings for wear applications, especially for many cutting and related tools—for example, with tin by reactive deposition—are now widely done on an industrial scale. A number of ceramics can also be deposited by electrochemical means which, in principle, could be used for producing monolithic components as well as coatings as with CVD.

Polmer pyrolysis

One of the newest nonpowder-based methods of preparing ceramics or ceramic coatings is polymer pyrolysis. Here, one obtains a ceramic by the pyrolysis of an appropriate metal organic polymer, in direct analogy with the fabrication of glassy carbon or graphite fibres by pyrolysis of appropriate organic polymers. While this method is applicable to some oxides, it is predominantly for non-oxides.

Ceramic composites

Substantial interest has arisen in recent years in fabricating ceramic composites. While there is considerable interest in composites with one or more non-ceramic constituents, the greatest attention is on composites in which all constituents are ceramic. These typically comprise two ceramic constituents: a continuous matrix, and a dispersed phase of ceramic particles, whiskers, or short (chopped) or continuous ceramic fibres. The challenge, as in wet checmical processing, is to obtain a uniform or homogeneous distribution of the dispersed particle or fibre phase.

REFRACTORY

A refractory material is one that retains its strength at high temperatures. ASTM C71 defines refractories as non-metallic materials having those chemical and physical properties that made them applicable for structures, or as components of systems, that are exposed to environments above 1000°F (538°C).

Refractory materials are used in linings for furnaces, kilns, incinerators and reactors. They are also used to make crucibles.

Refractory materials must be chemically and physically stable at high temperatures. Depending on the operating environment, they need to be resistant to thermal shock, be chemically inert, and/or have specific ranges of thermal conductivity and of the coefficient of thermal expansion.

The oxides of aluminium (alumina), silicon (silica) and magnesium (magnesia) are the most important materials used in the manufacturing of refractories. Another oxide usually found in refractories is the oxide of calcium (lime). Fireclays are also widely used in the manufacture of refractories.

Refractories must be chosen according to the conditions they will face. Some applications require special refractory materials. Zirconia is used when the material must withstand extremely high temperatures. Silicon carbide and carbon are two other refractory materials used in some very severe temperature conditions, but they cannot be used in contact with oxygen, as they will oxidise and burn.

Types of Refractories

Acidic refractories cannot be used in a basic environment and basic refractories cannot be used in an acidic environment because they will be corroded. Zircon, fireclay and silica are acidic, dolomite and magnesite are basic and alumina, chromite, silicon carbide, carbon and mullite are neutral.

Refractory materials are used extensively in the metal industries, along with glass melting and other heat treatment operations.

There are two common forms of refractories, bricks and monolithics. Bricks (also known as firebrick) are pre-sintered forms which can hold their shape. Monolithics are loose material which can be formed into complex shapes, or sprayed into place, and have to be sintered before use. Castable refractory cement is also commonly used.

Manufacture of Ceramic

Refractories

Initial processing may include an extensive survey of the raw material deposit, selective mining, stockpiling by grade, and beneficiation techniques such as weathering, grinding, washing, heavy-media separation, froth flotation, etc. Further steps include crushing and grinding, screening, mixing, forming, drying, and burning.

Speciality refractories

Bulk refractory products include gunning, ramming, or plastic mixes, granular materials, hydraulic-setting castables, and mortars. These products are generally made from the same raw materials as their brick counterparts.

Selection and Uses

Any manufacturing process requiring refractories depends upon proper selection and installation. When selecting refractories, the environmental conditions are evaluated first, then the functions to be served, and finally the expected length of service. All factors pertaining to the operation service, design, and construction of equipment must be related to the physical and chemical properties of the various classes of refractories.

By far the most common industrial refractories are those composed at single or mixed oxides of Al, Ca, Cr, Mg, Si and Zr. These oxides exhibit relatively high degrees of stability under both reducing and oxidising conditions. Carbon, graphite, and silicon carbide have been used both alone and in combination with these oxides. Refractories made from the above materials are used in ton-lot quantities in industrial applications. Other refractory oxides, nitrides, borides, and silicides are used in relatively small quantities for speciality applications in the nuclear, electronic, and aerospace industries.

Silica refractories: Silica refractories are used in open-hearth roof linings, refractories for coke ovens, coreless-induction foundry furnaces, and fused-silica technical ceramic products.

Semisilica refractories: Semisilica refractories are used in shapes for open-hearth stoves and checkers.

Fireclay refractories: Fireclay refractories are used in kilns, ladles, and heat regenerators, acid-slag-resistant applications, boilers, blast furnances, and rotary kilns.

High alumina refractories: These refractories are used in kilns, ladles, and furnaces that operate at temperatures or under conditions for which fireclay refractories are not suited.

Chrome refractories: These refractories are used in non-ferrous metallurgical furnaces, rotary-kiln linings, secondary refining vessels, such as argon-oxygen decarborisers (AODs) and glass-tank regenerators.

Magnesite refractories: These refractories are used in lining and maintenance of steelmaking and refining vessels and checkers.

Dolomite refractories: Dolomite refractories are primarily used in refining vessels, and in ladles and cement kilns.

Spinel refractories: Spinel refractories are used in cement kilns and steel-ladle linings.

Forsterite refractories: Forsterite refractories are used include non-ferrous-metal-furnace roofs and glass-tank refractories not in contact with the melt, i.e. checkers, ports, and uptakes.

Silicon carbide refractories: Silicon carbide has a wide range of refractory uses including chemical tanks and drains, kiln furniture, abrasion-resistance linings, blast-furnace linings, and non-ferrous metallurgical crucibles and furnace linings.

Zirconia refractories: The most common zirconia-containing refractories are made from zircon sand and are used mostly for glass-tank paver brick.

Properties of Refractories

In making the refractory best suited for a definite operation it is necessary to consider the materials, the working temperature of the furnace where the refractory is needed, the rate of temperature change, the load applied during heats, and the chemical reactions encountered. Generally, several types of refractories are required for the construction of any one furnace, because usually no single refractory can withstand all the different conditions that prevail in the various parts of furnaces.

Chemical properties

The usual classification of commercial refractories divides them into acid, basic, and neutral groups, although in many cases a sharp distinction cannot be made. Silica bricks are decidedly acid, and magnesite bricks are strongly basic; however, fire-clay bricks are generally placed in the neutral group, though they may belong to either of these classes, depending upon the relative silica-alumina content. It is usually inadvisable to employ an acid brick in contact with an alkaline product, or vice versa. Neither chemical reactions nor physical properties are the only criteria of acceptable behaviour; both should be considered. Chemical action may be due to contact with slags, fuel ashes, and furnace gases, as well as with products such as glass or steel.

Porosity

Porosity is directly related to many other physical properties of brick, including resistance to chemical attack. The higher the porosity of the brick, the more easily it is penetrated by molten fluxes and gases. For a given class of brick, those with the lowest porosity have the greatest strength, thermal conductivity, and heat capacity.

Fusion points

Fusion points are found by the use of pyrometric cones of predetermined softening points. Most commercial refractories soften gradually over a wide range and do not have sharp melting points because they are composed of several different minerals, both amorphous and crystalline.

Spalling

A fracturing, or a flaking off, of a refractory brick, or block, due to uneven heat stresses or compression caused by heat is known as spalling. Refractories usually expand when heated. Bricks that undergo the greatest expansion at the least uniform rate are the most susceptible to spalling when subjected to rapid heating and cooling.

Strength

Cold strength usually has only a slight bearing on strength at high temperatures. Resistance to abrasion or erosion is also important for many furnace constructions, such as coproduct coke-oven walls and linings of the discharge end of rotary cement kilns.

Resistance to temperature changes

Bricks with the lowest thermal expansion and coarsest texture are the most resistant to rapid thermal change; also, less strain develops. Bricks that have been used for a long time are usually altered properties, and are often melted to glassy, slags on the outside surface or even more or less corroded away.

Thermal conductivity

The densest and least porous bricks have the highest thermal conductivity because of the absence of air in voids. Though heat conductivity is wanted in some furnace constructions, as in muffle walls, it is not so desirable as some other properties of refractories, such as resistance to firing conditions. Insulation is desired in special refractories.

Heat capacity

Furnace heat capacity depends upon the thermal conductivity, the specific heat, and the specific gravity of the refractory. The low quantity of heat absorbed by lightweight brick works as an advantage when furnaces are operated intermittently, because the working temperature of the furnace can be obtained in less time with less fuel. Conversely, dense, heavy fire-clay brick is best for regenerator checker work, as in coke ovens, glass furnaces, and stoves for blast furnaces.

Manufacture of Refractories

Included are these physical operations and chemical conversions: grinding and screening, mixing, pressing or moulding and repressing, drying, and burning or vitrification. Usually, the most important single property to produce in manufacture is high bulk density, which affects many of the other important properties, such as strength, volume stability, slag and spalling resistance, as well as heat capacity. On the other hand, for insulating refractories, a porous structure is required, which means low density.

Grinding

Obviously, one of the most important factors is the size of the particles in the batch. It is known that a mixture in which the proportion of coarse and fine particles is about 55:45, with only few intermediate particles, gives the densest mixtures. Careful screening, separation, and recycling are necessary for close control. This works very well on highly crystalline materials but is difficult to obtain in mixes of high plasticity.

Mixing

The real function of mixing is the distribution of the plastic material so as to coat thoroughly the nonplastic constituents. This serves the purpose of providing a lubricant during the moulding operation and permits the bonding of the mass with a minimum number of voids.

Moulding

The great demand for refractory bricks of greater density, strength, volume, and uniformity has resulted in the adoption of the dry-press method of moulding with mechanically operated presses. The dry-press method is particularly suited for batches that consist primarily of nonplastic materials. In order to use high-pressure forming, it is necessary to deair the bricks during pressing to avoid laminations and cracking when the pressure is released. When pressure is applied, the gas is absorbed by the clay or condensed. The use of a vacuum is applied through vents in the mould box. Large special shapes are not easily adapted to machine moulding. Drying is used to remove the moisture added before moulding to develop plasticity. It should be noted that the elimination of water leaves voids and causes high shrinkage and internal strains. In some cases drying is omitted entirely, and the small amount necessary is accomplished during the heating stage of the firing cycle.

Burning may be carried out in typical round, downdraft kilns or continuous-tunnel kilns. Two important things take place during burning: the development of a permanent bond by partial vitrification of the mix, and the development of stable mineral forms for future service. The changes that take place are removal of the water of hydration, followed by calcination of carbonates and oxidation of ferrous iron. During these changes the volume may shrink as much as 30 per cent, and severe strains are set up in the refractory. This shrinkage may be eliminated by prestabilisation of the materials used.

POTTERY

Pottery is the ceramic ware made from clay by potters. Major types of pottery include earthenware, stoneware, and porcelain. The places where such wares are made are called potteries. Pottery is one of the oldest human technologies and art-forms, and remains a major industry today. Ceramic art covers the art of pottery, whether in items made for use or purely for decoration.

Methods of Shaping

The potter's most basic tools are the hand, but many additional tools have been developed over the long history of pottery manufacture, including the potter's wheel and turntable, shaping tools (paddles, anvils, ribs), rolling tools (roulettes, slab rollers, rolling pins), cutting/piercing tools (knives, fluting tools, wires) and finishing tools (burnishing stones, rasps, chamois).

Handwork or hand building

This is the earliest and the most individualised and direct forming method. Wares can be constructed by hand from coils of clay, from flat slabs of clay, from solid balls of clay—or some combination of these. Parts of hand-built vessels are often joined together with the aid of slurry or slip, a runny mixture of clay and water. Hand building is slower and more gradual than wheel-throwing, but it offers the potter a high degree of control over the size and shape of wares. While it isn't difficult for an experienced potter to make identical pieces of hand-built pottery, the speed and repetitiveness of wheel-throwing is more suitable for making precisely matched sets of wares such as table wares. Some studio potters find hand building more conducive to fully using the imagination to create one-of-a-kind works of art, while others find this with the wheel.

The potter's wheel. A ball of clay is placed in the center of a turntable, called the wheel-head, which the potter rotates with a stick, or with foot power (a kick wheel or treadle wheel) or with a variable speed electric motor. (Often, a disk of plastic, wood or plaster—called a bat—is first set on the wheel head, and the ball of clay is thrown on the bat rather than the wheel head so that the finished piece can be removed intact with its bat, without distortion.)

During the process of throwing the wheel rotates rapidly while the solid ball of soft clay is pressed, squeezed, and pulled gently upwards and outwards into a hollow shape. The first step, of pressing the rough ball of clay downward and inward into perfect rotational symmetry, is called centering the clay, a most important (and often most difficult) skill to master before the next steps: opening (making a centered hollow into the solid ball of clay), flooring (making the flat or rounded bottom inside the pot), throwing or pulling (drawing up and shaping the walls to an even thickness), and trimming or turning (removing excess clay to refine the shape or to create a foot).

The potter's wheel can be used for mass production, although it is often employed to make individual pieces. Wheel-work makes great demands on the skill of the potter, but an accomplished operator can make many near-identical plates, vases, or bowls in the course of a day's work. Because of its inherent limitations, wheel work can only be used to create wares with radial symmetry on a vertical axis. These can then be altered by impressing, bulging, carving, fluting, faceting, incising, and by other methods making the wares more visually interesting. Often, thrown pieces are further modified by having handles, lids, feet, spouts, and other functional aspects added using the techniques of handworking.

Jiggering and jolleying

These operations are carried out on the potter's wheel and allow the time taken to bring wares to a standardised form to be reduced. Jiggering is the operation of bringing a shaped tool into contact with the plastic clay of a piece under construction, the piece itself being set on a rotating plaster mould on the wheel. The jigger tool shapes one face whilst the mould shapes the other. Jiggering is used only in the production of flat wares, such as plates, but a similar operation, jolleying, is used in the production of hollow-wares, such as cups.

Jiggering and jolleying have been used in the production of pottery since at least the 18th century. In large-scale factory production jiggering and jolleying are usually automated, which allows the operations to be carried out by semi-skilled labour.

Roller-head machine

This machine is for shaping wares on a rotating mould, as in jiggering and jolleying, but with a rotary shaping tool replacing the fixed profile. The rotary shaping tool is a shallow cone having the same diameter as the ware being formed and shaped to the desired form of the back of the article being made. Wares may in this way be shaped, using relatively unskilled labour, in one operation at a rate of about twelve pieces per minute, though this varies with the size of the articles being produced. The roller-head machine is now used in factories worldwide.

RAM pressing

A factory process for shaping table wares and decorative ware by pressing a bat of prepared clay body into a required shape between two porous moulding plates. After pressing, compressed air is blown through the porous mould plates to release the shaped wares.

Granulate pressing

As the name suggests, this is the operation of shaping pottery by pressing clay in a semi-dry and granulated condition in a mould. The clay is pressed into the mould by a porous die through which water is pumped at high pressure. The granulated clay is prepared by spray-drying to produce a fine and free flowing material having a moisture content of between about five and six per cent. Granulate pressing, also known as dust pressing, is widely used in the manufacture of ceramic tiles and, increasingly, of plates.

Slipcasting

Slipcasting is often used in the mass-production of ceramics and is ideally suited to the making of wares that cannot be formed by other methods of shaping. A slip, made by mixing clay body with water, is poured into a highly absorbent plaster mould. Water from the slip is absorbed into the mould leaving a layer of clay body covering its internal surfaces and taking its internal shape. Excess slip is poured out of the mould, which is then split open and the moulded object removed. Slipcasting is widely used in the production of sanitary wares and is also used for making smaller articles, such as intricately-detailed figurines.

Glazing

Glaze is a glassy coating applied to pottery, the primary purposes of which include decoration and protection. Glazes are highly variable in composition but usually comprise a mixture of ingredients that generally, but not always, mature at kiln temperatures lower than that of the pottery that it coats. One important use of glaze is in rendering pottery vessels impermeable to water and other liquids. Glaze may be applied by dusting it over the clay, spraying, dipping, trailing or brushing on a thin slurry composed of glaze minerals and water. Brushing tends not to give an even covering but can be effective as a decorative technique. The colour of a glaze before it has been fired may be significantly different than afterwards. To prevent glazed wares sticking to kiln furniture during firing, either a small part of the object being fired (for example, the foot) is left unglazed or, alternatively, special refractory spurs are used as supports. These are removed and discarded after the firing. Special methods of glazing are sometimes carried out in the kiln. One example is salt-glazing, where common salt is introduced to the kiln to produce a glaze of mottled, orange peel texture. Materials other than salt are also used to glaze wares in the kiln, including sulphur. In wood-fired kilns fly-ash from the fuel can produce ash-glazing on the surface of wares.

Decorating

Pottery may be decorated in a number of ways, including:
1. In the clay body; by, for example, incising patterns on its surface.
2. Underglaze decoration, in the manner of many blue and white wares.
3. In-glaze decoration.
4. On-glaze decoration.
5. Enamel.

Additives can be worked into the clay body prior to forming, to produce desired effects in the fired wares. Coarse additives, such as sand and grog (fired clay which has been finely ground) are sometimes used to give the final product a required texture. Contrasting coloured clays and grogs are sometimes used to produce patterns in the finished wares. Colourants, usually metal oxides and carbonates, are added singly or in combination to achieve a desired colour. Combustible particles can be mixed with the body or pressed into the surface to produce texture.

Agateware

So-named after its resemblance to the quartz mineral agate which has bands or layers of colour that are blended together. Agatewares are made by blending clays of differing colours together, but not mixing them to the extent that they lose their individual identities. The wares have a distinctive veined or mottled appearance. The term 'agateware' is used to describe such wares in the United Kingdom; in Japan the term neriage is used and in China, where such things have been made since at least the Tang Dynasty, they are called marbled wares. Great care is required in the selection of clays to be used for making agatewares as the clays used must have matching thermal movement characteristics.

Banding

This is the application, by hand or by machine, of a band of colour to the edge of a plate or cup. Also known as lining, this operation is often carried out on a potter's wheel.

Burnishing

The surface of pottery wares may be burnished prior to firing by rubbing with a suitable instrument of wood, steel or stone, to produce a polished finish that survives firing. It is possible to produce very highly polished wares when fine clays are used, or when the polishing is carried out on wares that have been partially dried and contain little water, though wares in this condition are extremely fragile and the risk of breakage is high.

Engobe

This is a clay slip, often white or cream in colour that is used to coat the surface of pottery, usually before firing. Its purpose is often decorative, though it can also be used to mask undesirable features in the clay to which it is applied. Engobe slip may be applied by painting or by dipping, to provide a uniform, smooth, coating. Engobe has been used by potters from pre-historic times until the present day, and is sometimes combined with sgraffito decoration, where a layer of engobe is scratched through to reveal the colour of the underlying clay. With care it is possible to apply a second coat of engobe of a different colour to the first and to incise decoration through the second coat to expose the colour of the underlying coat. Engobes used in this way often contain substantial amounts of silica, sometimes approaching the composition of a glaze.

Litho

This is a commonly used abbreviation for lithography, although the alternative names of transfer print or decal are also common. These are used to apply designs to articles. The litho comprises three layers: the colour, or image, layer which comprises the decorative design; the cover coat, a clear protective layer, which may incorporate a low-melting glass; and the backing paper on which the design is printed by screen printing or lithography.

There are various methods of transferring the design while removing the backing-paper, some of which are suited to machine application.

Gold

Decoration with gold is used on some high quality ware. Different methods exist for its application, including:

1. Best gold—a suspension of gold powder in essential oils mixed with a flux and a mercury salt extended. This can be applied by a painting technique. From the kiln the decoration is dull and requires burnishing to reveal the full colour.
2. Acid gold—a form of gold decoration developed in the early 1860s. The glazed surface is etched with diluted hydrofluoric acid prior to application of the gold. The process demands great skill and is used for the decoration only of ware of the highest class.
3. Bright gold—consists of a solution of gold sulphoresinate together with other metal resonates and a flux. The name derives from the appearance of the decoration immediately after removal from the kiln as it requires no burnishing
4. Mussel gold—an old method of gold decoration. It was made by rubbing together gold leaf, sugar and salt, followed by washing to remove solubles.

Firing

Firing produces irreversible changes in the body. It is only after firing that the article can be called pottery. In lower-fired pottery the changes include sintering, the fusing together of coarser particles in the body at their points of contact with each other. In the case of porcelain, where different materials and higher firing-temperatures are used the physical, chemical and mineralogical properties of the constituents in the body are greatly altered. In all cases the object of firing is to permanently harden the wares and the firing regime must be appropriate to the materials used to make them. As a rough guide, earthenwares are normally fired at temperatures in the range of about 1000° to 1200°C; stonewares at between about 1100° to 1300°C; and porcelains at between about 1200° to 1400°C. However, the way that ceramics mature in the kiln is influenced not only by the peak temperature achieved, but also by the duration of the period of firing. Thus, the maximum temperature within a kiln is often held constant for a period of time to soak the wares, to produce the maturity required in the body of the wares.

The atmosphere within a kiln during firing can affect the appearance of the finished wares. An oxidising atmosphere, produced by allowing air to enter the kiln, can cause the oxidation of clays and glazes. A reducing atmosphere, produced by limiting the flow of air into the kiln, can strip oxygen from the surface of clays and glazes. This can affect the appearance of the wares being fired and, for example, some glazes containing iron fire brown in an oxidising atmosphere, but green in a reducing atmosphere. The atmosphere within a kiln can be adjusted to produce complex effects in glaze.

Kilns may be heated by burning wood, coal and gas, or by electricity. When used as fuels, coal and wood can introduce smoke, soot and ash into the kiln which can affect the appearance of unprotected wares. For this reason wares fired in wood or coal fired kilns are often placed in the kiln in saggars; lidded ceramic boxes, to protect them. Modern kilns powered by gas or electricity are cleaner and more easily controlled than older wood or coal fired kilns and often allow shorter firing times to be used. In a Western adaptation of traditional Japanese Raku ware firing, wares are removed from the kiln while hot and smothered in ashes, paper or woodchips, which produces a distinctive, carbonised, appearance. This technique is also used in Malaysia in creating traditional labu sayung.

FELDSPAR

Feldspar is the name of a group of rock-forming minerals which make up as much as 60 per cent of the earth's crust. Feldspars crystallise from magma in both intrusive and extrusive igneous rocks, and they can also occur as compact minerals, as veins, and are also present in many types of metamorphic rock. Rock formed entirely of plagioclase feldspar is known as anorthosite. Feldspars are also found in many types of sedimentary rock.

Uses

Feldspar is a common raw material in the production of ceramics and geopolymers. It is used for thermoluminescence dating and optical dating in earth sciences and archaeology. It is an ingredient in Bon Ami US brand household cleaner. It is often an anti-caking agent used in powdered forms of non-dairy creamer.

SAND

Sand is a naturally occurring granular material composed of finely divided rock and mineral particles. As the term is used by geologists, sand particles range in diameter from 0.0625 (or 1/16 mm, or 62.5 micrometers) to 2 millimeters. An individual particle in this range size is termed a sand grain. The next smaller size class in geology is silt: particles smaller than 0.0625 mm down to 0.004 mm in diameter. The next larger size class above sand is gravel, with particles ranging from 2 mm up to 64 mm. Sand feels gritty when rubbed between the fingers (silt, by comparison, feels like flour).

The most common constituent of sand, in inland continental settings and non-tropical coastal settings, is silica (silicon dioxide, or SiO_2), usually in the form of quartz, which, because of its chemical inertness and considerable hardness, is resistant to weathering. The composition of sand is highly variable, depending on the local rock sources and conditions. The bright white sands found in tropical and subtropical coastal settings are eroded limestone and may contain coral and shell fragments in addition to other organic or organically derived fragmental material. Some sands contain magnetite, chlorite, glauconite or gypsum. Sands rich in magnetite are dark to black in colour, as are sands derived from volcanic basalts and obsidian. Chlorite-glauconite bearing sands are typically green in colour, as are sands derived from basalt (lava) with a high olivine content.

Uses of Sand

Uses of sand are the following:
1. Sand is often a principal component of concrete.
2. Moulding sand, also known as foundry sand, is moistened or oiled and then shaped into moulds for sand casting. This type of sand must be able to withstand high temperatures and pressure, allow gases to escape, have a uniform, small grain size and be non-reactive with metals.
3. It is the principal component in glass production.
4. Sand is spread on roads to improve traction (and thus traffic safety) in icy or snowy conditions.
5. Graded sand is used as an abrasive in sandblasting and is also used in media filters for filtering water.
6. Brick manufacturing plants use sand as an additive with a mixture of clay and other materials for manufacturing bricks.
7. Sand is sometimes mixed with paint to create a textured finish for walls and ceilings or a non-slip floor surface.

While sand is generally harmless, one must take care with some activities involving sand such as sandblasting. Bags of silica sand used for sandblasting now carry labels warning the user to wear respiratory protection and avoid breathing the fine silica dust. There have been a number of lawsuits in recent years where workers have developed silicosis, a lung disease caused by inhalation of fine silica particles over long periods of time. Material safety data sheets (MSDS) for silica sand state that 'excessive inhalation of crystalline silica is a serious health concern'. In areas of high pore water pressure sand can partially liquefy to form quicksand. Quicksand, once dried, produces a considerable barrier to escape for creatures caught within, who often die from exposure as a result.

BRICK

A brick is a block of ceramic material used in masonry construction, usually laid using mortar. Bricks may be made from clay, shale, soft slate, calcium silicate, concrete, or shaped from quarried stone. Clay is the most common material, with modern clay bricks formed in one of three processes—soft mud, dry press, or extruded. A new type of brick, based on fly ash, a by-product of coal power plants.

Mud Bricks

The soft mud method is the most common, as it is the most economical. It starts with the raw clay, preferably in a mix with 25–30 per cent sand to reduce shrinkage. The clay is first ground and mixed with water to the desired consistency. The clay is then pressed into steel moulds with a hydraulic press. The shaped clay is then fired (burned) at 900°–1000°C to achieve strength.

Rail kilns

In modern brickworks, this is usually done in a continuously fired tunnel kiln, in which the bricks move slowly through the kiln on conveyors, rails, or kiln cars to achieve consistency for all bricks. The bricks often have added lime, ash, and organic matter to speed the burning.

Dry pressed bricks

The dry press method is similar to mud brick but starts with a much thicker clay mix, so it forms more accurate, sharper-edged bricks. The greater force in pressing and the longer burn make this method more expensive.

Extruded bricks

With extruded bricks the clay is mixed with 10–15 per cent water (stiff extrusion) or 20–25 per cent water (soft extrusion). This is forced through a die to create a long cable of material of the proper width and depth. This is then cut into bricks of the desired length by a wall of wires. Most structural bricks are made by this method, as hard dense bricks result, and holes or other perforations can be produced by the die. The introduction of holes reduces the needed volume of clay through the whole process, with the consequent reduction in cost. The bricks are lighter and easier to handle, and have thermal properties different from solid bricks. The cut bricks are hardened by drying for between 20 and 40 hours at 50°–150°C before being fired. The heat for drying is often waste heat from the kiln.

Calcium silicate bricks

The raw materials for calcium silicate bricks include lime mixed with quartz, crushed flint or crushed siliceous rock together with mineral colourants. The materials are mixed and left until the lime is completely hydrated, the mixture is then pressed into moulds and cured in an autoclave for two or three hours to speed the chemical hardening. The finished bricks are very accurate and uniform, although the sharp arrises need careful handling to avoid damage to brick (and brick-layer). The bricks can be made in a variety of colours, white is common but pastel shades can be achieved.

Fly ash bricks

These are produced when fly ash and water are compressed at 4000 psi (27,939 kPa) for two weeks. Owing to the high concentration of calcium oxide in fly ash, the brick is considered self-cementing. The brick is toughened using an air entrainment agent, which traps microscopic bubbles inside the brick so that it resists penetration by water, allowing it to withstand up to 100 freeze-thaw cycles. Since the manufacturing method uses a waste by-product rather than clay, and solidification takes place under pressure rather than heat, it has several important environmental benefits. It saves energy, reduces mercury pollution, alleviates the need for landfill disposal of fly ash, and costs 20 per cent less than traditional clay brick manufacture.

Uses

Bricks are used for building and pavement. Bricks are also used in the metallurgy and glass industries for lining furnaces. They have various uses, especially refractory bricks such as silica, magnesia, chamotte and neutral (chromomagnesite) refractory bricks. This type of brick must have good thermal shock resistance, refractoriness under load, high melting point, and satisfactory porosity.

KILN

Kilns are thermally insulated chambers, or ovens, in which controlled temperature regimes are produced. They are used to harden, burn or dry materials. Specific uses include:

1. To dry green lumber so that the lumber can be used immediately.
2. Drying wood for use as firewood.
3. Heating wood to the point of pyrolysis to produce charcoal.
4. For annealing, fusing and deforming glass, or fusing metallic oxide paints to the surface of glass.
5. For cremation (at high temperature).
6. Drying of tobacco leaves.

7. Firing of material, such as clay, to form ceramics.
8. Drying malted barley for brewing.
9. Smelting ore to extract metal.
10. Heating limestone with clay to make cement.

Ceramic Kilns

Kilns are an essential part of the manufacture of all ceramics, which, by definition, require heat treatment, often at high temperature. During this process, chemical and physical reactions occur which cause the material to be permanently altered. In the case of pottery, clay materials are shaped, dried and then fired in a kiln. The final characteristics are determined by the composition and preparation of the clay body, the temperature at which it is fired, and by the glazes that may be used. While modern kilns often have sophisticated electrical systems to control the firing temperatures, pyrometric devices have been used to provide visual indication of the firing regime.

Clay consists of fine-grained particles, that are relatively weak and porous. Clay is combined with other minerals to create a workable clay body. Part of the firing process includes sintering. This process heats the clay until the particles partially melt and flow together, creating a strong, single mass, composed of a glassy phase interspersed with pores and crystalline material. Through firing, the pores are reduced in size, causing the material to shrink slightly. This crystalline material is a matrix of predominantly silicon and aluminium oxides, and is very hard and strong, although usually somewhat brittle.

Types of kilns

In the broadest terms there are two types of kiln, both sharing the same basic characteristics of being an insulated box with controlled inner temperature and atmosphere.

Intermittent kiln

The ware to be fired, is loaded into the kiln. The kiln is sealed and the internal temperature increased according to a schedule. After the firing process is completed, both the kiln and ware are cooled.

Continuous kiln

Continuous, or sometimes called tunnel. These are long structures in which only the central portion is directly heated. From the cool entrance, ware is slowly transported through the kiln, and its temperature is increased steadily as it approaches the central, hottest part of the kiln. From there, its transportation continues and the temperature is reduced until it exits the kiln at near room temperature. A continuous kiln is the most energy efficient because heat given off during cooling is recycled to pre-heat the incoming ware. A specialty type of kiln, common in tableware and tile manufacture, is the Roller-hearth Kiln, in which ware placed on bats is carried through the kiln on rollers.

Kiln technology is very old. The development of the kiln from a simple earthen trench filled with pots and fuel, pit firing, to modern methods happened in stages. One improvement was to build a firing chamber around pots with baffles and a stoking hole, this allowed heat to be conserved and used more efficiently. The use of a chimney stack improves the air flow or draw of the kiln, thus burning the fuel more completely.

Early examples of kilns found in Britain include those made for the making of roof-tiles during the Roman occupation. These kilns were built up the side of a slope, such that a fire could be lit at the bottom and the heat would rise up into the kiln.

Anagama kiln

The Asian anagama kiln has been used since medieval times and is the oldest style of kiln in Japan. This kiln usually consists of one long firing chamber, pierced with smaller stacking ports on one side, with a firebox at one end and a flue at the other. Firing time can vary from one day to several weeks. Traditional anagama kilns are also built on a slope to allow for a better draft.

Noborigama kiln

The Noborigama is an evolution from Anagama design as a multi-chamber kiln, usually built on a slope, where wood is stacked from the front firebox at first, then only through the side-stoking holes with the benefit of having air heated up to 600°C from the front firebox, enabling more efficient firings.

Bottle kiln

A type of intermittent kiln, usually coal-fired, formerly used in the firing of pottery; such a kiln was surrounded by a tall brick hovel or cone, of typical bottle shape.

Catenary arch kiln

Catenary arch kiln, typically used for the firing of pottery using salt, these by their form (a catenary arch) tend to retain their shape over repeated heating and cooling cycles, whereas other types require extensive metalwork supports.

Sevres kiln

Sevres kiln was invented in sevres, France and enabled to reach efficiently high-temperature (1280°C) in order to have fully water-proof ceramic bodies and easy to obtain glazes. It features a down-draft design that enabled to reach high temperature in shorter time, even with wood-firing.

Top-hat kiln

It is an intermittent kiln of a type sometimes used in the firing of pottery. The ware is set on a refractory hearth, or plinth, over which a box-shaped cover is then lowered.

Feller kiln

Feller kiln brought contemporary design to wood firing by reusing the unburnt gas from the chimney in order to heat the air up before entering the firebox. This leads to an even shorter firing cycle and less wood consumption. This design requires external ventilation to prevent the in chimney radiator from melting, being typically in metal. The result is a very efficient wood kiln firing one cubic meter of ceramics with one cubic meter of wood.

Electric kilns

Kilns is operated by electricity were developed in the 20th century, primarily for smaller scale use such as in schools, universities, and hobby centers. The atmosphere in most designs of electric kiln is rich in oxygen, as there is no open flame to consume oxygen molecules, however reducing conditions can be created with appropriate gas input.

Modern kilns

Modern kilns with the advent of the industrial age, kilns were designed to utilise electricity and more refined fuels, including natural gas and propane. The majority of large, industrial pottery kilns now use natural gas, as it is generally clean, efficient and easy to control. Modern kilns can be fitted with computerised controls, allowing for refined adjustments during the firing cycle. A user may choose to control the rate of temperature climb or ramp, hold or soak the temperature at any given point, or control the rate of cooling. Both electric and gas kilns are common for smaller scale production in industry and craft, handmade and sculptural work.

Microwave assisted firing

This technique combine microwave energy with more conventional energy sources such as radiant gas or electric heating in order to process ceramic materials to the required high temperatures.

Microwave-assisted firing offers significant economic benefits.

Lime Kiln

A lime kiln is a kiln used to produce quicklime by the calcination of limestone (calcium carbonate). The chemical equation for this reaction is:

$$CaCO_3 + heat \rightarrow CaO + CO_2$$

This reaction takes place at 900°C (at which temperature the partial pressure of CO_2 is 1 atmosphere), but a temperature around 1000°C (at which temperature the partial pressure of CO_2 is 3.8 atmospheres) is usually used to make the reaction proceed quickly. Excessive temperature is avoided because it produces unreactive, dead-burned lime (Fig. 8.1).

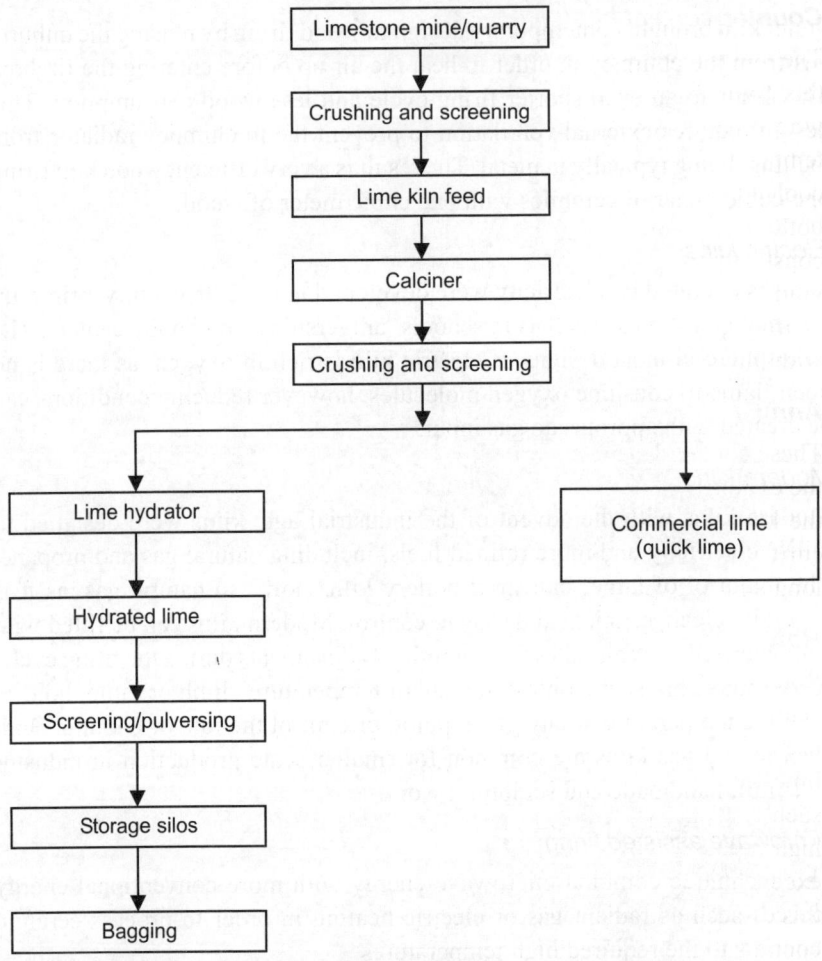

Fig. 8.1. Flow sheet of lime manufacturing operations.

Shaft Kilns

The theoretical heat (the standard enthalpy) of reaction required to make high-calcium lime is around 3.15 MJ per kg of lime, so the batch kilns were only around 20 per cent efficient. The key to development in efficiency was the invention of continuous kilns, avoiding the wasteful heat-up and cool-down cycles of the batch kilns.

The first were simple shaft kilns, similar in construction to blast furnaces. These are counter-current shaft kilns. Modern variants include regenerative and annular kilns. Output is usually in the range 100–500 tons per day.

Counter-current shaft kilns

The fuel is injected part way up the shaft, producing maximum temperature at this point. The fresh feed fed in at the top is first dried then heated to 800°C, where decarbonation begins, and proceeds progressively faster as the temperature rises. Below the burner, the hot lime transfers heat to, and is cooled by, the combustion air. A mechanical grate withdraws the lime at the bottom. A fan draws the gases through the kiln, and the level in the kiln is kept constant by adding feed through an airlock. As with batch kilns, only large, graded stone can be used, in order to ensure uniform gas flows through the charge. The degree of burning can be adjusted by changing the rate of withdrawal of lime. Heat consumption as low as 4 MJ/kg is possible.

Annular kilns

These contain a concentric internal cylinder. This gathers preheated air from the cooling zone, which is then used to pressurise the middle annular zone of the kiln. Air spreading outward from the pressurised zone causes counter-current flow upwards, and co-current flow downwards. This again produces a long, relatively cool calcining zone.

Rotary kilns

Rotary kilns started to be used for lime manufacture at the start of the 20th century and now account for a large proportion of new installations. The early use of simple rotary kilns had the advantages that a much wider range of limestone size could be used, from fines upwards, and undesirable elements such as sulphur can be removed. On the other hand, fuel consumption was relatively high because of poor heat exchange compared with shaft kilns, leading to excessive heat loss in exhaust gases. Modern installations partially overcome this disadvantage by adding a preheater, which has the same good solids/gas contact as a shaft kiln, but fuel consumption is still somewhat higher. In the design shown, a circle of shafts (typically 8–15) is arranged around the kiln riser duct. Hot limestone is discharged from the shafts in sequence, by the action of a hydraulic 'pusher plate'. Kilns of 1000 tons per day output are typical.

Gas cleaning

All the above kiln designs produce exhaust gas that carries an appreciable amount of dust. Lime dust is particularly corrosive. Equipment is installed to trap this dust, typically in the form of electrostatic precipitators or bag filters. The dust usually contains a high concentration of elements such as alkali metals, halogens and sulphur.

Carbon Dioxide Emissions

The lime industry is a significant carbon dioxide emitter. The manufacture of one ton of calcium oxide involves decomposing calcium carbonate, with the formation of 785 kg of CO_2. In addition, the heat supplied to form the lime.

Health and Safety Factors

Because industrial refractories are by their very nature stable materials, they usually do not constitute a physiological hazard. This statement does not apply, however, to unusual refractories that might contain heavy metals or radioactive oxides such as thoria, urania, and plutonia, or binders or additives that may be toxic. Inhalation of certain fine dusts may constitute a health hazard.

Chapter 9

Portland Cements, Calcium and Magnesium Compounds

INTRODUCTION

In the most general sense of the word, a cement is a binder, a substance which sets and hardens independently, and can bind other materials together. The volcanic ash and pulverised brick additives which were added to the burnt lime to obtain a hydraulic binder were later referred to as cementum, cimentum, cäment and cement. Cements used in construction are characterised as hydraulic or non-hydraulic. The most important use of cement is the production of mortar and concrete — the bonding of natural or artificial aggregates to form a strong building material which is durable in the face of normal environmental effects.

Cement should not be confused with concrete as the term cement explicitly refers to the dry powder substance. Upon the addition of water and/or additives the cement mixture is referred to as concrete, especially if aggregates have been added. Cement is made by heating limestone with small quantities of other materials (such as clay) to 1450°C in a kiln. The resulting hard substance, called clinker, is then ground with a small amount of gypsum into a powder to make ordinary portland cement, the most commonly used type of cement (often referred to as OPC). Portland cement is a basic ingredient of concrete, mortar and most nonspeciality grout. The most common use for portland cement is in the production of concrete. Concrete is a composite material consisting of aggregate (gravel and sand), cement, and water. As a construction material, concrete can be cast in almost any shape desired, and once hardened, can become a structural (load bearing) element. Portland cement may be gray or white.

MANUFACTURE OF PORTLAND CEMENT

Raw Materials

The materials from which portland cement is manufactured may be divided into two classes: those which supply the lime and those which supply the silica, iron oxide and alumina. The first are termed calcareous and the second argillaceous. The following groups show the principal materials used in the manufacture of portland cement.

Calcareous materials	*Argillaceous materials*
Limestone	Clay
Marl	Shale
Chalk	Slate
Alkali waste	Blast furnace slag

The manufacture of cement consists of mixing the calcareous and argillaceous materials together intimately and heating them to the point of incipient fusion. The intimate mixing of the two materials is accomplished by finely grinding them together. The powder is then subjected to a temperature of from 1400°–1600°C, when a sintering or semi-fusion takes place and the mixture rolls up into little balls varying in size from that of a walnut down to that of wheat, with an occasional larger piece and some fine sand. After cooling, these lumps or clinkers are mixed with a small amount (2–3 per cent) of gypsum and finely pulverised. The resulting powder is portland cement.

Manufacturing Process

The two processes employed for the manufacture of cement are known respectively as the wet process and the dry process. The wet process is the older of the two and is used almost universally. The dry process originated in America and is employed to the greater extent. The two processes differ only in the treatment of the raw materials and very much the same equipment is used in each. The treatment of the burned clinker is the same in both cases (Fig. 9.1).

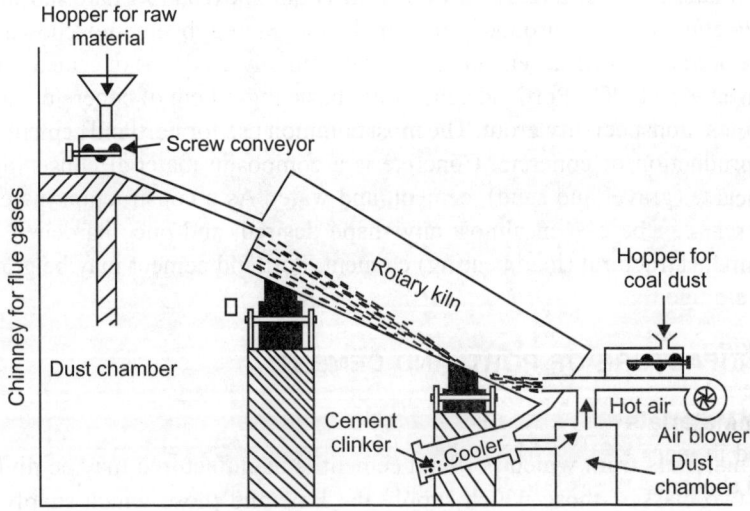

Fig. 9.1. Flow diagram for manufacture of portland cement.

The wet process is always used for marl and clay and the dry process for cement rock and for blast furnace slag. Both processes are used for limestone and clay or shale. Where applicable, the dry process is the more economical but it is easier to control the composition of the cement by the wet process. This later is also better where the materials cannot easily be dried.

Where the wet process is employed for limestone and shale, the two materials are crushed and stored without drying just as in the dry process. They are then mixed in proper proportions and fed to the grinding machinery, at which point water is added and the materials ground wet. The result is a thin mud or slurry, as it is called, which is made just fluid enough to flow easily. This slurry, containing from 35 to 40 per cent water, is fed directly into the kilns and burned. In the process, the limestone, cement rock, and shale are usually crushed to about two inches or smaller and then dried. The crushing is nearly always done in large jaw or gyratory crushers which are followed by hammer mills. The dried materials are stored in separate bins or piles and are drawn out of these as desired and mixed in proper proportions by automatic scales. The mixture is then ground and burned. Sometimes the storage and mixing precede the drying. Drying is done by means of rotary driers. These are cylinders of sheet steel from 6 to 8 feet in diameter and from 60 to 100 feet in length. They are unlined, and are usually provided with channel irons bolted to the inside to act as shelves, to carry the rock up and drop it through the hot gases. The driers are heated by a coal fire at the lower end or else by the waste gases from the rotary kilns. They are similar in construction to the rotary kilns, except that they are smaller and are not lined with firebrick.

Storage

In order to provide a constant supply of materials in case of a cessation of quarry operations, due to bad weather, etc. a storage is usually provided for the crushed raw material. This is often sufficiently large to hold a week or ten days supply of material. The most common form of storage is large covered concrete bins provided with belt conveyors—one overhead for bringing in the stone and other in tunnels underneath the bin for taking it out. Another form of storage is provided with a travelling crane and grab bucket, similar to the storage for clinker.

Mixing

The materials are usually mixed after leaving the storage, various methods being employed for proportioning the two different kinds. Hopper scales are used in many of the older mills, while in most of the newer ones limestone and shale are proportioned just as they are fed to the grinding mills by means of some type of adjustable feeder such as a poidometer or a rotating table feeder. Clay is often first worked up with water and the thin slip formed is proportioned by volume to the limestone just before the latter goes to the ball mill or compeb mill. A revolving wheel with wickets, known as a Ferris wheel, is usually employed for this purpose.

Where the wet process is employed for limestone and shale or clay, no drier is employed and sufficient water to make the ground material flow is added just before the mill.

Grinding

The grinding of the raw materials may be done in one or in two stages. The combinations now most often employed are: (i) ball mill and tube mill, (ii) kominuter and tube mill, (iii) hercules mill and tube mill, and (iv) griffin mill and tube mill.

In place of the combination of ball and tube mill, a single mill combining the elements of these two mills is now most generally used in the newer cement plants. These combination mills go under different names according to the maker such as compeb mill, combination mill, etc. These mills are somewhat similar to a tube mill except that they are made longer and there are usually three or more compartments, separated by perforated or grid partitions.

The degree of fineness to which the raw material is to be ground depends entirely upon conditions. It is stated as a general rule that it should be sufficiently fine so 90 per cent will pass through 100-mesh sieve, and in most cases 95 to 98 per cent is required to produce a sound cement. The finer the grinding the more perfect the combination between the silica, the alumina, the iron and the lime during the burning operation. If the raw materials are not finely enough ground, the cement will be unsound—that is, some of the lime will not combine. This yields a cement which disintegrates rapidly.

Flotation

The fine grinding of the cement materials affords the possibility of the use of a unique method for adjusting the composition of the mix. As has been indicated above, the various oxides must be proportioned within fairly narrow limits to give a satisfactory product. This means that, many deposits of natural materials cannot be used because they cannot be proportioned in any way to give the desired chemical composition. Recently the operation of flotation has been used with great success to separate the finely ground mineral constituents from each other. It is now being used to correct mixtures and to make many raw materials usable which were formerly considered unusable. By removing impurities from cheap materials they can be used to replace the more costly ones. Proper control of the flotation operation produces a better mix before burning. This has been one of the principal developments in the cement industry in recent years.

Storage of ground material

In the wet process, the ground materials containing from 33 to 40 per cent water, are stored in tanks or basins which are agitated with either mechanical or compressed air agitators. The materials as ground is usually passed into one set of vats called correction basins from which samples of the slurry are drawn and analysed. When a basin is full, if the composition is not correct, it is adjusted either by stirring in more clay or by mixing the contents of two or more basins. The slurry whose composition has been satisfactorily adjusted is then passed on to a second set of basins known as 'kiln feed basins'. These are also provided with agitators.

In the dry process, it was quite usual at the older plants to send the material directly from the grinding mills to the kilns. At the newer plants, however, the ground material is sent into large storage tanks where its composition is checked and if found unsatisfactory is adjusted by blending the contents of two or more tanks, etc. When this is done the dry process will give fully as uniform cement as the wet.

Conveying materials

The material is usually carried from one stage of manufacture to another by various types of conveyors. The product of the crushers is conveyed to the granulating mills on belt conveyors, and the product of these latter mills and the tube mills is transported by screw-conveyors. The elevating is done by means of bucket elevators of the link-belt type. Finely ground raw material and cement may also be conveyed by means of the Fuller-Kinyon system through a pipe line. Slurry and marl are pumped through pipe lines by means of either plunger or centrifugal pumps, or by a compressed air system.

Burning

In the early days of the American portland cement industry, the burning was done in intermittent upright kilns, similar to those used for burning lime. These were soon improved, by making them continuous in action in order to economise on fuel. This allowed the charge to receive the waste heat from the clinkering of the cement, and the air for combustion to be preheated, by passing through the fully-burned materials.

Rotary kiln

The rotary kiln, in its usual form consists of a cylinder, from 6 to 12 ft in diameter and from 60 to 350 ft long, made of sheet steel and lined with firebrick. The burning of cement is essentially an application of the unit operation of heat transfer, by radiation, and therefore it is necessary to have equipment and conditions which will bring about a large amount of heat transfer from the flame to the solid material.

The steel sheets of the kiln are held together by single-strap butt joints. This long cylinder is supported at a very slight pitch (1/2 to 3/4 inch to the foot) from the horizontal, on two or more tyres made of rolled steel, which in turn revolve one heavy friction rollers. The kiln is driven at a speed of from one-half to one revoluation per minute by a girth-gear situated near its middle, and a train of reducing gears. The power is supplied by either a line shaft or a motor. The upper end of the kiln projects into a brick flue, which is surmounted by a steel stack, also lined with firebrick for its entire height. The flue is provide with a door at the bottom, which serves not only to allow the flue to be cleared of the dust which accumulates in it, but also a damper to control the draft of the kiln. The material to be burned is usually fed into the kiln through a horizontal water-jacketed screw-conveyor or else spouted into it through an inclined cast-iron pipe. The raw material feeding device is usually attached to the driving gear of the kiln, so that when the kiln stops the feed also stops.

The lower end of the kiln is closed by a firebrick hood. This is usually mounted on rollers, so it can be moved away from the kiln when the latter has to be relined. The hood is provided with two openings; one for the entrance and support of the fuel-burning apparatus, and the other for observing the operation, temperature, etc. of the kiln, and through which bars may be inserted to break up the rings of material which is formed, and to patch and repair the lining. The lower part of the hood is left partly open. Through this opening the clinker falls out and most of the air for combustion enters.

Reaction in the kiln

1. 1/4 of the kiln is drying zone and water is driven out.
2. In the next zone, i.e. in the calcination zone organic matters burn away and $CaCO_3$ breaks to CaO and CO_2. Temperature is about 700°C. When reaction is complete and material has travelled 3/4 of the of the kiln decarbonation is complete. After decarbonation the material goes to the hottest zone. The temperature is 1400°–1450°C. Here 20–30 per cent material is converted to lime and the material reacts with the silica and sintering takes place in the last zone and clinkers are formed. The clinkers fall out from the outlet of the kiln to clinker cooler.

All these reactions take place in the kiln:

1. Up to 100°C evaporation of water take place.
2. At 500°C the evolution of combined water from clay starts. The reaction is endothermic.
3. Between 900°–1200°C, the main reaction takes place between clay and lime.
4. Between 1200°–1250°C there is liquid formation and the reaction is endothermic.

5. At 1200°C and beyond up to 1450°C further formation of liquid and then sintering takes place. The reaction is endothermic. The clinker is formed.

Heating the kiln

The kiln is heated by a jet of burning fuel, usually powdered coal, but sometime natural gas or fuel oil are used. The necessary temperature of the hottest part of the kiln is about 1400°C, and is rarely less then 1300°C. To maintain this temperature, about 80 to 160 lb of fuel are required per barrel of cement, the actual amount depending on the coal itself, the material to be burned and the dimensions of the kiln. The longer the kiln, the greater the fuel economy. Dry materials require much less coal than slurry. With limestone and shale mixture, in a kiln 100 ft long by 7 ft in diameter, the coal consumption will amount to about 90 lb of good gas slack per barrel. A kiln 60 ft long by 6 ft in diameter will, on the other hand, require about 110 lb of coal per barrel. Wet materials require about 30 per cent more fuel.

Grinding of coal

When coal is used for burning, this is pulverised in mills similar to those used for grinding the raw materials. It is, however, first crushed by passing through rolls or roll-jaw crushers, and then dried in rotary driers of special type. The mills mostly used for coal pulverising are the Fuller mill and the Raymond mill. The coal should be pulverised so that 90 per cent of it will pass a sieve having 100 meshes to the linear inch, and should contain from 30 to 45 per cent volatile matter.

Thermal efficiency

Of the heat supplied to the kiln by the burning of the coal, by far the larger proportion is wasted. About 50 to 75 per cent of it is carried off by the waste gases of the stack, and form 10 to 15 per cent by the hot clinker falling from the lower end of the kiln. The gases enter the stack at from 600°–800°C, and the clinker leaves the kiln at not much under 1200°C. If the kiln could be made to show the same economy as is common in good boiler practice, a barrel of cement could be burned with 25 lbs of coal.

The gases leave the dry process kiln at about 800°C. In many plants the waste gases are led through waste heat boilers located at the end of the kilns. By so doing about 4 to 5 lb of steam are generated per pound of coal burned. The flow of gases through the boilers at a high velocity is one of the requisites for successful employment of the waste gases for steam generation and this is obtained by means of an inducted draft fan. By employing modern turning engines, directly connected to electric generators, enough power may be obtained from the waste gases to operate the entire plant. The gases are sometimes purified and the carbon dioxide reclaimed.

Dust losses

Normally from 3 to 5 per cent of the raw material is carried away in the exit gases of the kiln as dust. Various schemes have been tried with a view to eliminating the dust, such as settling chambers, water sprays and electrical precipitation. The later, Cottrell precipitator, is the only method, however, which is used to any extent. Several installations of this system are now in operation. The dust collected by the latter is found to contain considerable potash for about half the potash in the raw materials is volatilised in the kiln. Some of this potash is water soluble and may be recovered and used for fertiliser.

Forming the clinker

The raw material as it enters the kiln contains about 33 per cent carbon dioxide. For the first 30 ft of its journey through a 100 ft kiln, it is merely heated up, and whatever water it contains is driven off. In the next 40 ft it loses all its carbon dioxide and sticks together, forming small, soft, lemon-yellow balls, which, as they reach the hottest part of the kiln—the last 30 ft, practically vitrify, become rough and hard. Properly burned portland cement clinker is greenish-black in colour, of vitreous lustre, and usually, when just cooled, sparkles with small bright glistening specks. It forms in lumps from the size of a walnut to hardly more than dust, with here and there a larger lump. Under-burned clinker is more or less soft, is irregular in shape, and not as black as the well-burned material. It usually show soft-brown centres. Hard brown centres are due to very hard burning.

The cement is brought into the stock house by an overhead conveyor or the Fuller-Kinyon system pipe line and dropped into any desired bin. A screw-conveyor also runs under the floor of the stock house. The bins are provided with gates, and when it is desired to pack from any bin, these gates are opened and the cement is allowed to run into the screw-conveyor. The screw-conveyors then carry it to the packing machines.

Cooling of clinkers

The clinkers through clinker conveyor falls inside the cooler. In the cooler, atmospheric air passes over hot clinker, cooling the clinkers and itself getting heated. The common type of coolers are rotary coolers.

The quality of cement largely depends on the rate of cooling.
1. If the rate of cooling is slow no glass is formed but there will be dusty clinkers. Dusting means conversion of dicalcium silicate to powder

form and not crystalline form. This is caused by the conversion of SiO_2 which has no binding property and hydrates very slowly.

2. If the rate of cooling is high then the alumina and iron solidifies into glass and there is any formation of crystalline component of alumina and iron. This non-formation of any crystalline component is adverse to the process.

3. If the rate of cooling is such that the melt liquid in the clinker crystallises—there is formation of a large quantity of $3CaOSiO_2$ which gives higher strength to cement.

4. If the rate of cooling is so high that the melted liquid in the clinker turns into glass and if there is magnesia also in the clinker—that will be converted to glass and will not crystallise out as periclase (MgO). This will give beneficial effect to the properties of cement as the crystalline periclase MgO hydrates very slowly—when it is exposed to water and this causes large expansion. If the cement after several years, when magnesia is in the form of glass, it does not expand to a high degree when exposed to water. So, by cooling rapidly the clinker containing magnesia—there will not be excessive expansion of cement after several years.

It is very necessary that the cooling of clinker should be controlled to produce a definite degree of crystallisation of the melted clinker.

Mixing of additives

1. Retarder: Addition of retarder is necessary to prevent quick settling of cement plaster. Such a retarder is gypsum ($CaSO_4 \cdot 2H_2O$) or plaster of Paris, $CaSO_4 \cdot 1/2H_2O$. The quantity required is 2.5–3.0 per cent.

2. Dispersing agent: Small quantity of dispersing agent, e.g. sodium salt or naphthalene or sulphuric acid is added to the cement. These dispersing agents prevent formation of lumps and cakes in the cement. The power cost of cement manufacture is reduced by 30–35 per cent.

3. Additives which cause air entrainment improves durability of concrete, particularly against the alternate freezing and thawing. For this vinyl resins or Darex may be added. These agents have the property of imparting air into the cement paste.

4. Water porosity or dispersing agents are also added to cement to make then resistant to water porosity.

Grading of clinkers

The clinkers are pulverised to fine grains in tube mill. During grinding, additives are added. The grind powder is packed into bags through automatic packaging machine.

Fineness of cement

The portland cement is so fine that 98–99 per cent passes through 200 mesh and 9 per cent through 325 mesh screen.

LIME

Lime is nearly pure calcium oxide, CaO; or a mixture of calcium and magnesium oxides, CaO + MgO; sometimes called quicklime. High calcium limes are stronger than those containing considerable percentages of magnesia. They are also better suited for mortar work, as they slake more readily. Magnesium limes, on the other hand, are better for plaster finishing because they work more smoothly under the trowel. Pure lime whether magnesium or not, is snow white. However, a very small percentage of certain impurities such as iron or manganese may give the lime a gray or yellow colour. Through certain methods of burning the ash of the fuel may be introduced into the lime, causing discolouration.

Lime is made by burning limestone in suitable furnaces at a temperature sufficient to drive off all of its carbon dioxide, the reaction being:

$$CaCO_3 + 21,900 \text{ cals} \rightleftarrows CaO + CO_2$$

Theoretically, 806 g calories per g of calcium oxide and 733 g calories per gram of magnesium oxide are required to produce this change. At atmospheric pressure the temperature at which calcium carbonate decomposes is stated as 898°C, while magnesium carbonate decomposes at 575°C (Fig. 9.2).

Kilns

Lime kilns are ordinarily operated at from 900° to 1100°C, in the hottest or burning zone. If a temperature much above 1200°C is employed, the lime will be partially vitrified on the outside of the lumps, due to combination of the CaO with impurities SiO_2 and Al_2O_3, always present in small quantities in even the purest limestone. This causes the lime to be very slow in slaking, which is undesirable, as some of it may escape hydration in the mortar box and later will expand, or blow or pop in the wall. This manifests itself in small blisters in the finished plaster work.

Intermittent Kilns

The type of kilns ordinarily employed in burning lime may be divided into two classes—intermittent and continuous. The intermittent kilns are primitive and uneconomical, though they are frequently used by farmers and other small producers of lime. There is a great waste of heat and time in such a kiln, owing to the fact that it must be cooled and reheated each time it is charged. They are seldom, if ever, used in large scale commercial operations.

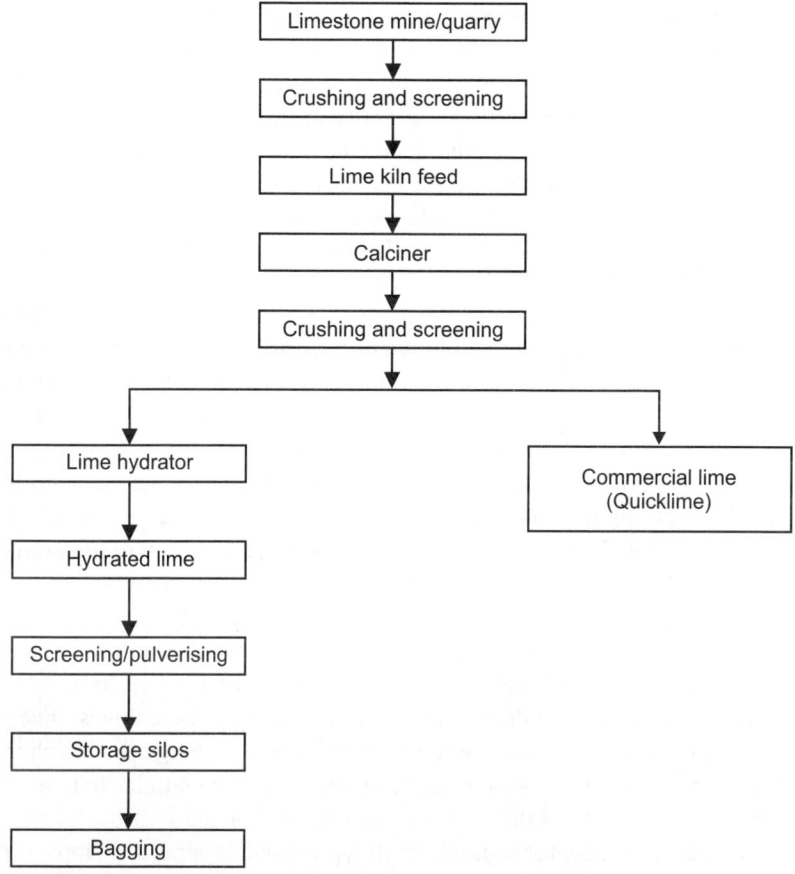

Fig. 9.2. Flow sheet of lime manufacturing operations.

Continuous Kilns

Three different types of continuous kilns are employed: these are:
1. The vertical kiln with mixed feed, in which the limestone and fuel are charged in alternate layers.
2. The vertical kiln with separate feed, in which the limestone and fuel are not brought in contact.
3. The rotary kiln.

Vertical kiln, mixed feed

Vertical kilns with mixed feed are similar to the intermittent ones, except that they are provided with an arrangement whereby the lime may be drawn at regular intervals from below. They are built on the side of a hill, usually of limestone blocks, and are sometimes lined with firebrick. In charging them, first a layer of anthracite coal or coke and then a layer of limestone is fed into the top. Fire is started at the bottom and works its way up. The process of charging and drawing the lime is continuous. These kilns are economical and, for the same size kiln, yield a larger quantity of product than do the vertical kilns with separate feed. On the other hand, the lime is contaminated by ash of the fuel, and the lime burned in these kilns must be carefully sorted in order to discard those lumps to which the fuel ash has adhered.

Vertical kiln, separated feed

The vertical kiln with separate feed usually consists of a steel cylinder lined with firebrick. This is equipped with two or four fireplaces for the burning of the fuel, which are built into the sides of the kiln, so that the fuel is not mixed with the stone. The hot gases of combustion pass from the fire-box into the kiln, while the ash of the fuel drops through the grate bars into an ash-pit below, and does not mix with the lime. The kilns are often constructed with a hopper-shaped cooling chamber, set below the fire-box, which is closed by doors at the bottom. The cooling chamber holds about one draw of lime. They are from 6 to 10 ft in cross-section, and from 40 to 50 ft in height. They are usually charged by employing an incline and a cable hoist, by means of which the cars of limestone are dawn from the quarry to the top of the kilns. They are sometimes provided with steel stacks in order to induce a better draft.

An improved system of draft employs an exhauster. Where this is done, it is of course necessary to close the top of the kiln and to charge the limestone into the latter through a door or a charging bell somewhat similar to that of a blast furnace. When the kiln gases are drawn off for their carbon dioxide (as in the Solvay process for soda) this bell type of seal is generally employed. A properly installed induced draft will often increase the capacity of a kiln 50 per cent.

Wood, oil, and coal are employed for burning the wood. Wood is the best fuel, as it burns with a long flame of comparatively low temperature. This is an advantage, as it is essential that the heat should be dispersed at a considerable distance throughout the kiln without having excessive temperatures at the mouth of the fire-box. The steam, which the wood introduces, also seems to be beneficial, and indeed some manufacturers prefer to use green wood because of the greater quantity of steam which it introduces. Wood-burned lime is whiter than that burned with coal.

Where coal is employed as a fuel, it is customarily used wet. A steam jet is also often employed, being inserted below the fire-box. The steam passing through the hot bed of coal is decomposed into hydrogen and carbon monoxide as follows:

$$H_2O + C = H_2 + CO$$

These gases are burned in the kiln itself, and hence carry the heating zone further up the shaft.

With hand-fired kilns the diameter cannot be increased beyond the limit to which the flames from the fire-box can reach effectively or the limestone in the centre of the kiln will not be burned. The limiting diameter for such a kiln is about 6 to 7 ft. With gas, on the other hand, the kiln may be made much larger because the gas may be made to burn in all parts of the kiln. The use of gas also saves the labour of stoking the grates. On the other hand, more skill is required to burn lime with gas than with any other fuel and considerable experimenting is usually required before satisfactory results are obtained. Natural gas is used and producer gas is now being used for kilns of large capacity.

Usually the producer gas enters through openings around the side of the kiln and the air through holes in the bottom of the kiln. The air is thus preheated by passing up through the hot lime, which it cools. It then comes in contact with the producer gas in the centre of the kiln, where combustion takes place, spreading back to the gas openings. Gas-fired lime kilns are now built having capacities of from 40 to 60 tons per day.

Oil also makes an excellent fuel. When it is used in the kiln described, the burner is placed in the door openings of the fire-boxes and these are bricked in with firebrick leaving openings for the burner and for observing the lime.

The shaft kiln (Fig. 9.3) described above when heated with wood, coal, or oil will produce from 8 to 20 tons of lime per day, depending on the kiln and the limestone burned and whether natural or induced draft is employed. When fired with gas the capacity will range from 17 to 60 tons per day.

Rotary kilns

Lime is also burned in rotary kilns similar to those described in the section on portland cement. The limestone is first crushed to pieces ranging in size for $2\frac{1}{2}$ inches down to dust and fed into the kiln, which is heated by producer gas, oil or powdered coal. These kilns are peculiarly adapted to burning highly crystalline stone, which would crumble when subjected to heat and so stop the draft of the vertical kiln, and to supplying lime for chemical and metallurgical purposes.

There was at one time considerable objection to the use of rotary kiln lime, due to small particle size produced. The mason, being accustomed to lime in large lumps, assumed the rotary kiln lime was air-slaked. Now, however, this prejudice has been overcome and much lime is sold in the powdered or granular condition. Powdered lime, made by grinding lump or granular lime which results by decrepitation when certain crystalline limestones are burned, slakes much quicker and is less likely to pit than are most lump limes. Pebble lime is a rotary kiln lime made by burning carefully sized stone usually ranging between 1 to $2\frac{1}{2}$ inches. Better results are obtained as regards the uniformity of the lime if material finer than $\frac{1}{4}$ inch is screened from the stone before burning.

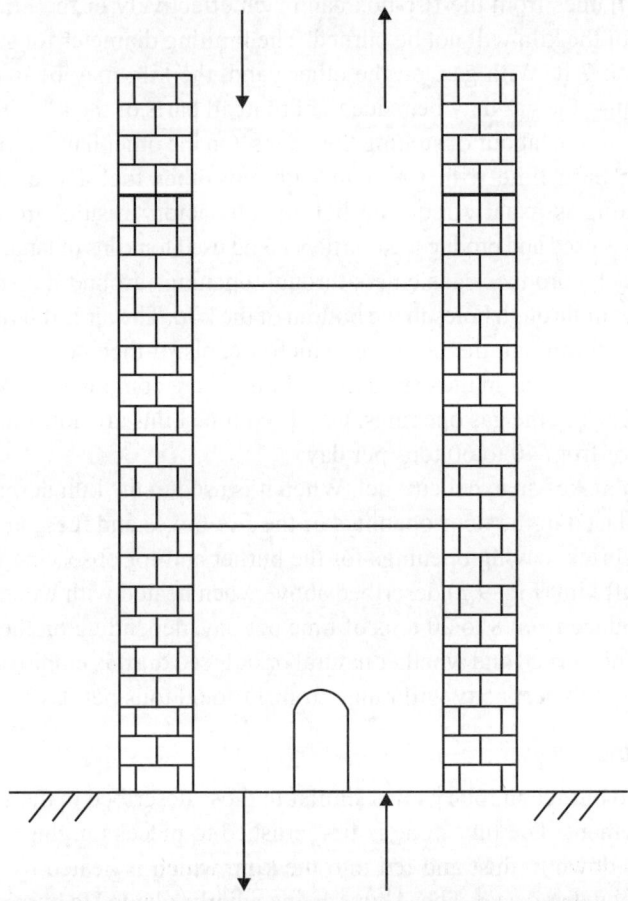

Fig. 9.3. Schematic diagram of a mixed feed shaft kiln.

The rotary kiln is more economical of labour than is the shaft kiln. This applies to both the quarry, where stone for the latter must be carefully sized and all small stones discarded, and to the attention required by the kiln itself. It requires more fuel then the best type of shaft kiln, but the heat in the stack gases may be recovered by installing waste-heat boilers after the kilns, passing the waste gases through the boilers and utilising the steam for power, etc.

Rotary kilns are now being quite generally employed for burning building lime where a large output is required. A kiln 8 ft in diameter and 150 ft long will produced 100 tons or more per day. Rotary kilns are also used to burn lime form various wastes, such as lime-sludge from paper mills and sugar purification.

Lime is now burned from the calcium carbonate waste from caustic soda manufacture at a number of plants by first passing the waste through some form of dewatering device such as a continuous rotary, drum or disc filter, the waste being obtained from the latter in the form of a wet mud containing about 50 per cent water. The recovered lime is very pure and usually contains some alkali otherwise lost, the percentage varying from 3 to 8 per cent of the lime recovered. This practice not only can be carried out generally for less than the cost of purchasing lump lime but it also disposes of a troublesome waste product. The rotary kiln is particularly well adapted to burning chemical lime and is to be preferred where a very well-burned lime, free from carbon dioxide, is desired as in the manufacture of carbide and in metallurgy.

A refractory product is now made by grinding together doomite, $CaMg(CO_3)_2$, and a small amount of iron ore, and burning the mixture in a rotary kiln at a somewhat higher temperature than is required for lime. The product so obtained consists of small roughly rounded modulus, hard and dark brown to black, the latter compound is to make the slake less readily on exposure to the air and so improve its keeping qualities. This material is known in the trade as dead burned dolomite and under various trade names.

Fuel Requirements for Shaft Kilns

The quantity of fuel required depends on many things, among which may be mentioned the kind and quality of fuel, skill of the operator, and the limestone itself. Magnesium limestone burns more easily than high calcium stone and an impure stone more easily than a pure one. The amount of fuel actually required is about as follows:

1 ton of good bituminous coal, hand-fired, will burn from 3 to $4\frac{1}{2}$ tons of lime.

1 barrel of fuel oil will burn 2/3 to 1 ton of line.

1 cord of seasoned hardwood will burn from $1\frac{1}{2}$ to 2 1/2 tons of lime.

Hydrated Lime

When quicklime is treated with water it combines with the water to form calcium hydroxide:

$$CaO + H_2O = Ca(OH)_2$$

If the lime is free from impurities, it will take up 32 per cent of its own weight of water. This quantity, however, is reduced somewhat because of the impurities that are always, found to a greater or less extent in all commercial limes. When lime slakes, heat is generated. One gram of CaO converted to $Ca(OH)_2$ liberates 270 g-calories, or enough heat to evaporate 0.5 g of water at 100°C. The chemical reaction itself requires 0.31 g of water. Thus, assuming the water and lime both to be at 100°C, 1 g of CaO could satisfy 0.81 g of water. In practice appreciably less is employed due to radiation losses, and heat required to bring the materials up to 100°C. Formerly, lime was hydrated, or slaked, by the mason just preparatory to its use. An excess of water was always used, and the calcium hydroxide formed a wet mass called lime putty. Now, mechanical means of hydration have been introduced whereby the lime is hydrated by the manufacturer with just sufficient water to form the hydrate, leaving none in excess. This hydrated lime is a fine dry powder, practically all of which will pass through a 100-mesh screen. It is packed in paper bags or cloth sacks and will keep indefinitely. It can be stored without danger of causing fire, which is not true of caustic lime. When added to cement, it makes it waterproof to some extent and more easy to trowel.

In manufacturing hydrated lime the lump lime is first ground to small size. It is then mixed with predetermined amount of water, when it falls to a fine powder. The slaked lime is then sieved to separate out the unhydrated lumps or siliceous cores from the latter, or else these cores are ground so fine that they will cause no popping. The plan adopted in grinding the quicklime is the most successful hydrating plants consists in crushing the lime by means of a swing hammer mill or a sturtevant open-door crusher. This reduces it to pieces about $\frac{1}{2}$ inch and under. Lime which is to be hydrated should not be burned at as high a temperature as is ordinarily used. Fresh lime hydrates much more readily than that which has been allowed to remain for some time in the air. There are a number of processes and machines for mixing the lime with water which have been successfully used in hydrating.

GYPSUM PRODUCTS

Gypsum occurs in nature as flattened and often twinned crystals and transparent cleavable masses called selenite. It may also occur silky and fibrous, in which case it is commonly called satin spar. Finally it may also be granular or quite compact. In hand-sized samples, it can be anywhere from transparent to opaque. A very fine-grained white or lightly-tinted variety of gypsum is called alabaster, which is prized for ornamental work of various sorts. In arid areas, gypsum can occur in a flower-like form typically opaque with embedded sand grains called desert rose. The most visually striking variety, however, is the giant crystals from Naica Mine. Up to the size of 11 m long, these megacrystals are among the largest crystals found in nature.

Gypsum is a common mineral, with thick and extensive evaporite beds in association with sedimentary rocks. Deposits are known to occur in strata from as early as the Permian age. Gypsum is deposited in lake and sea water, as well as in hot springs, from volcanic vapours, and sulphate solutions in veins. Hydrothermal anhydrite in veins is commonly hydrated to gypsum by groundwater in near surface exposures. It is often associated with the minerals halite and sulphur. The word gypsum is derived from the Greek word, chalk or plaster. Because the gypsum from the quarries of the Montmartre district of Paris has long furnished burnt gypsum used for various purposes, this material has been called plaster of Paris. It is also used in foot creams, shampoos and many other hair products. It is water-soluble.

Uses

There are a large number of uses for gypsum. Some of these uses are: (i) dry wall, (ii) plaster ingredient, (iii) fertiliser and soil conditioner, (iv) a binder in fast-dry tennis court clay, (v) plaster of paris, (vi) wood substitute, (vii) blackboard chalk, (viii) a component of portland cement used to prevent flash setting of concrete, and (ix) soil/water potential monitoring.

Calcination of Gypsum

The usual method of calcination of gypsum consists in grinding the mineral and placing it in large calciners holding 10 to 25 tons. The temperature is raised to about 120° to 150°C, with constant agitation to maintain a uniform temperature. The material in the kettle, known to the public as plaster of paris and to the manufacturer as first-settle plaster, may be withdrawn and sold at this point, or it may be heated further to 190°C to give a material known as second-settle plaster. First-settle plaster is approximately the half hydrate $CaSO_4 \cdot \frac{1}{2} H_2O$, and the second form is anhydrous. Practically all gypsum plaster sold is in the form of first-settle plaster mixed with sand or wood pulp. The second form is used in the manufacture of plasterboard and other gypsum products. Gypsum may also be calcined in rotary kilns similar to those used for limestone.

Plaster of Paris

The term plaster can refer to plaster of paris, lime plaster, or cement plaster. Plaster of paris is a type of building material based on calcium sulphate hemihydrate, nominally $CaSO_4 \cdot 0.5H_2O$. It is produced by heating gypsum to about 150°C. A large gypsum deposit at Montmartre in Paris is the source of the name. When the dry plaster powder is mixed with water, it reforms into gypsum. Plaster is used as a building material similar to mortar or cement. Like those materials plaster starts as a dry powder that is mixed with water to form a paste which liberates heat and then hardens. Unlike mortar and cement, plaster remains quite soft after drying, and can be easily manipulated with metal tools or even sandpaper. These characteristics make plaster suitable for a finishing, rather than a load-bearing material.

Plaster was a common building material for wall surfaces in a process known as lath and plaster, whereby a series of wooden strips on a studwork frame was covered with a semi-dry plaster that hardened into a surface. The plaster is used in most lath and plaster construction was mainly lime plaster, with a cure time of about a month. To stabilise the lime plaster during curing, small amounts of plaster of paris were incorporated into the mix. Because plaster of paris sets quickly, retardants were used to slow setting time enough to allow workers to mix large working quantities of lime putty plaster. A modern form of this method uses expanded metal mesh over wood or metal structures, which allows a great freedom of design as it is adaptable to both simple and compound curves. Today this building method has been partly replaced with drywall, also composed mostly of gypsum plaster. In both these methods a primary advantage of the material is that it is resistant to a fire within a room and so can assist in reducing or eliminating structural damage or destruction provided the fire is promptly extinguished.

One of the skills used in movie and theatrical sets is that of plasterer, and the material is often used to simulate the appearance of surfaces of wood, stone, or metal. Now-a-days, plasterers are just as likely to use expanded polystyrene, although the job title remains unchanged.

Uses

It is used in architecture for use in room interiors. It is used in medicine as a support for broken bones; a bandage impregnated with plaster is moistened and then wrapped around the damaged limb, setting into a close-fitting yet easily removed tube, known as an orthopedic cast; however, this is slowly being replaced by a fibreglass variety.

Safety issues

The chemical reaction that occurs when plaster is mixed with water is exothermic in nature. Some variations of plaster that contain powdered silica or asbestos may present health hazards if inhaled. Special cleanup methods should be used with of plaster products, as they can interfere with the flow of plumbing systems downstream of the disposal area. The residue of these products will often solidify underwater and plug up drains, stain gutters and sidewalks and spoil planting areas.

MISCELLANEOUS CALCIUM COMPOUNDS

Calcium Carbonate

Calcium carbonate is a chemical compound with the chemical formula $CaCO_3$. It is a common substance found as rock in all parts of the world, and is the main component of shells of marine organisms, snails, and eggshells. Calcium carbonate is the active ingredient in agricultural lime, and is usually the principal cause of hard water. It is commonly used medicinally as a calcium supplement or as an antacid, but high consumption can be hazardous.

Calcium carbonate is found naturally as the minerals and rocks: (i) aragonite, (ii) calcite, (iii) vaterite or (μ-$CaCO_3$), (iv) chalk (blackboard chalk is calcium sulphate, $CaSO_4$), (v) limestone, (vi) marble, and (vii) travertine. To test whether a mineral or rock contains carbonate, strong acids such as hydrochloric acid or sulphuric acid can be added to it; if the sample does contain carbonate, it will fizz and produce carbon dioxide and water. (Although sulphuric acid reacts, the reaction soon ceases because the calcium sulphate produced is rather insoluble in water and limits the reaction.)

Weak acids such as acetic acid will react, albeit less vigorously. All of the rocks/minerals mentioned above will react with acid.

To test for calcium, prepare a platinum or nichrome wire and dip it into some hydrochloric acid. Then dip the wire into some crushed sample to be tested. Place the wire in a bunsen burner flame; if calcium is presented in the sample a brick-red flame will be produced. If a sample gives positive results for both of the two tests above, the presence of calcium carbonate is indicated.

Chemical properties

Calcium carbonate shares the typical properties of other carbonates. Notably:

1. It reacts with strong acids, releasing carbon dioxide:
$$CaCO_{3(s)} + 2HCl_{(aq)} \rightarrow CaCl_{2(aq)} + CO_{2(g)} + H_2O_{(l)}$$

2. It releases carbon dioxide on heating (to above 840°C in the case of $CaCO_3$), to form calcium oxide, commonly called quicklime, with reaction enthalpy 178 kJ/mole:
$$CaCO_3 \rightarrow CaO + CO_2$$

3. Calcium carbonate will react with water that is saturated with carbon dioxide to form the soluble calcium bicarbonate.
$$CaCO_3 + CO_2 + H_2O \rightarrow Ca(HCO_3)_2$$

This reaction is important in the erosion of carbonate rocks, forming caverns, and leads to hard water in many regions.

Preparation

The vast majority of calcium carbonate used in industry is extracted by mining or quarrying. Pure calcium carbonate (e.g. for food or pharmaceutical use), can be produced from a pure quarried source (usually marble). Alternatively, calcium oxide is prepared by calcining crude calcium carbonate. Water is added to give calcium hydroxide, and carbon dioxide is passed through this solution to precipitate the desired calcium carbonate, referred to in the industry as precipitated calcium carbonate (PCC):

$$CaCO_3 \rightarrow CaO + CO_2$$
$$CaO + H_2O \rightarrow Ca(OH)_2$$
$$Ca(OH)_2 + CO_2 \rightarrow CaCO_3 + H_2O$$

Uses

The main use of calcium carbonate is in the construction industry, either as a building material in its own right (e.g. marble) or limestone aggregate for road building or as an ingredient of cement or as the starting material for the preparation of builder's lime by burning in a kiln.

Calcium carbonate is also used in the purification of iron from iron ore in a blast furnace. Calcium carbonate is calcined *in situ* to give calcium oxide, which forms a slag with various impurities present, and separates from the purified iron. Calcium carbonate is also used in the oil industry in drilling fluids as a formation bridging and filtercake sealing agent and may also be used as a weighing material to increase the density of drilling fluids to control downhole pressures.

Calcium carbonate is also one of the main sources used in growing Seacrete, or Biorock. The growing of marijuana is also attributed to high calcium carbonate deposits throughout the southern United States.

Precipitated calcium carbonate, pre-dispersed in slurry form, is also now widely used as filler material for latex gloves with the aim of achieving maximum saving in material and production costs.

Calcium carbonate is widely used as an extender in paints, in particular matte emulsion paint where typically 30 per cent by weight of the paint is either chalk or marble. Calcium carbonate is also widely used as a filler in plastics.

Fine ground calcium carbonate is an essential ingredient in the microporous film used in babies' diapers and some building films as the pores are nucleated around the calcium carbonate particles during the manufacture of the film by biaxial stretching.

Calcium carbonate is known as whiting in ceramics/glazing applications, where it is used as a common ingredient for many glazes in its white powdered form. When a glaze containing this material is fired in a kiln, the whiting acts as a flux material in the glaze. It is used in swimming pools as a pH corrector for maintaining alkalinity buffer to offset the acidic properties of the disinfectant agent. It is commonly called chalk as it has been a major component of blackboard chalk. Chalk may consist of either calcium carbonate or gypsum, hydrated calcium sulphate $CaSO_4 \cdot 2H_2O$. Ground calcium carbonate is further used as an abrasive (both as scouring powder and as an ingredient of household scouring creams), in particular in its calcite form, which has the relatively low hardness level of 3 on the Mohs scale of mineral hardness, and will therefore not scratch glass and most other ceramics, enamel, bronze, iron, and steel, and have a moderate effect on softer metals like aluminium and copper.

Health and dietary applications

Calcium carbonate is widely used medicinally as an inexpensive dietary calcium supplement or antacid. It may be used as a phosphate binder for the treatment of hyperphosphatemia (primarily in patients with chronic renal failure). It is also used in the pharmaceutical industry as an inert filler for tablets and other pharmaceuticals.

Calcium carbonate is used in the production of toothpaste and is also used in homeopathy as one of the constitutional remedies. Excess calcium from supplements, fortified food and high-calcium diets, can cause the 'milk alkali syndrome', which has serious toxicity and can be fatal.

Calcium Sulphide

Calcium sulphide is the chemical compound with the formula CaS. This white material crystallises in cubes like rock salt. CaS has been studied as a component in a process that would recycle gypsum, a product of flue gas desulphurisation. Like many salts containing sulphide ions, CaS typically has an odour of H_2S, which results from small amount of this gas formed by hydrolysis of the salt.

In terms of its atomic structure, CaS crystallises in the same motif as sodium chloride indicating that the bonding in this material is highly ionic. The high melting point is also consistent with its description as an ionic solid. In the crystal, each S^{2-} ion is surrounded by an octahedron of six Ca^{2+} ions, and complementarily, each Ca^{2+} ion surrounded by six S^{2-} ions.

CaS is produced by 'carbothermic reduction' of calcium sulphate, which entails the conversion of carbon, usually as charcoal, to carbon dioxide:

$$CaSO_4 + 2C \rightarrow CaS + 2CO_2$$

and can react further:

$$3CaSO_4 + CaS \rightarrow 4CaO + 4SO_2$$

Calcium sulphide decomposes upon contact with water, including moist air, giving a mixture of $Ca(SH)_2$, $Ca(OH)_2$, and $Ca(SH)(OH)$.

$$CaS + H_2O \rightarrow Ca(SH)(OH)$$

$$Ca(SH)(OH) + H_2O \rightarrow Ca(OH)_2 + H_2S$$

Milk of lime, $Ca(OH)_2$, reacts with elemental sulphur to give a 'lime-sulphur', which has been used as an insecticide. The active ingredient is probably a calcium polysulphide, not CaS.

It is yellow to light-gray powder with odour of hydrogen sulphide in moist air, having unpleasant alkaline taste. Gradually decomposes in moist air or in weak acids. Decomposed by acidx, slightly soluble in water with partial decomposition, insoluble in alcohol, sp. gr. 2.6.

It is prepared by strong heating of pulverised calcium sulphate and charcoal. It is irritant to skin and mucous membranes. It is used in luminous paint; depilatory; preparation of arsenic-free hydrogen sulphide; lubricant additive; ore dressing and flotation agent.

Calcium Sulphite

Calcium sulphite, is a chemical compound which is the salt of calcium cation and sulphite anion with the molecular formula $CaSO_3$. As a food additive it is used as a preservative under the E number E226. It is commonly used in preserving wine, cider, fruit juice, canned fruit and vegetables.

Like other metal sulphites, calcium sulphite reacts with acids to produce the respective salt, sulphur dioxide gas and water. It is in the form of white powder, loses water at 100°C. Soluble in sulphurous acid, slightly soluble in water. Low toxicity. It is prepared by the action of sulphurous acid on calcium carbonate. It is used in textiles (antichlor), disinfectant in sugar in dustry, brewing, biological cleansing; food preservative and discolouration retarder, paper manufacture.

Halide

A halide is a binary compound, of which one part is a halogen atom and the other part is an element or radical that is less electronegative than the halogen, to make a fluoride, chloride, bromide, iodide, or astatide compound. Many salts are halides. All Group 1 metals form halides with the halogens and they are white solids. A halide ion is a halogen atom bearing a negative charge. The halide anions are fluoride (F^-), chloride (Cl^-), bromide (Br^-), iodide (I^-) and astatide (At^-). Such ions are present in all ionic halide salts. In organic chemistry halides represent a functional group. Any organic compound that contains a halogen atom can be considered a halide. Alkyl halides are organic compounds of the type R-X, containing an alkyl group R covalently bonded to a halogen X.

Pseudohalides resemble halides in their charge and reactivity; common examples are azides NNN, isocyanate –NCO, isocyanide, CN–, etc. Calcium chloride is obtained commercially as a by-product of chemical manufacture and from natural brines which contain more or less magnesium chloride. For this reason, and since large tonnages are available, it is a very cheap chemical.

Its main applications are in solutions used to lay dust on highways (because it is deliquescent and remains moist) and in low-temperature refrigeration. Calcium bromide and iodide have properties similar to those of the chloride. They are prepared by the action of the halogen acids on calcium oxide or calcium carbonate. They are sold as hexahydrates for use in medicine and photography. Calcium fluoride occurs naturally as a fluorspar.

MAGNESIUM OXYCHLORIDE CEMENT

This cement, sometimes called Sorel's cement, is produced by the exothermic action of a 20 per cent solution of magnesium chloride on a blend of magnesia obtained by calcining magnesite and magnesia obtained from brines:

$$3MgO + MgCl_2 + 11H_2O \longrightarrow 3MgO \cdot MgCl_2 \cdot 11H_2O$$

The resulting crystalline oxychloride ($3MgO \cdot MgCl_2 \cdot 11H_2O$) contributes the cementing action to the commercial cements. The product is hard and strong, but is attacked by water; which leaches out the magnesium chloride. Its main applications are as a flooring cement with an inert filler and a colouring pigment, and as a base for such interior floorings as tile and terrazzo. It is strongly corrosive to iron pipes in contact with it. Sand and wood pulp may be added as fillers. The expense of these cements restricts their use to special purposes. They do not reflect sound. They can be made spark-proof and as such have been widely employed in ordnance plants. The magnesia used may contain small amounts of calcium oxide, calcium hydroxide, or calcium silicates, which during the setting process increase the volume changes, thus decreasing strength and durability. To eliminate this time effect, hydrated magnesium sulphate ($MgSO_4 \cdot 7H_2O$) or 10 per cent finely divided metallic copper is added to the mixture. The use of copper powder not only prevents excessive expansion, but greatly increases water resistance, adhesion, and dry and wet strength over that of ordinary magnesium oxychloride cement. Such a product even adheres in thin layers to concrete and serves to seal cracks therein.

Magnesium Compounds

Magnesium is one of the most widely distributed elements, occupying 1.9 per cent of the earth's crust. It occurs usually in the chloride, silicate, hydrated oxide, sulphate, or carbonate, in either a complex or in simple salts.

Raw materials and uses

Important domestic sources of magnesium salts are sea water, certain salt wells, bitterns from sea brine, salines, dolomite, and magnesite ($MgCO_3$). Magnesium compounds are used extensively for refractories and insulating compounds, as well as in the manufacture of rubber, printing inks, pharmaceuticals, and toilet goods. Magnesium oxide is finding new important uses in air pollution control systems for the removal of sulphur dioxide from stack gases. New uses for magnesium may mean major growth for this metal in the future.

Manufacture

The manufacture of magnesium compounds from salines has long been successful in Germany. As a result of thorough physical and chemical study, the International Minerals and Chemical Corp. is making magnesium chloride from langbeinite, crystallising out carnallite ($KCl \cdot MgCl_2 \cdot 6H_2O$). This double salt is decomposed to furnish magnesium chloride.

1. Manufacture from sea water without evaporation, using sea water and lime as the principal raw material. The Dow Chemical Co. of Freeport and Velasco, Tex., manufactures magnesium hydrate, which is dissolved in 10 per cent hydrochloric acid to furnish a solution of magnesium chloride. This is concentrated in direct-fired evaporators, followed by shelf dryers, producing 76 per cent magnesium chloride ready to be delivered to electrolytic cells to make metallic magnesium.

2. Manufacture from bitterns or mother liquors from the solar evaporation of sea water for salt.

3. Manufacture from dolomite and sea water.

4. Manufacture from deep-well brines. The small amount of bromine is freed by chlorine and following its removal, the $Mg(OH)_2$ is precipitated by pure slaked dolime (calcined dolomite). The $Mg(OH)_2$ produced is settled, filtered, and washed to provide a slurry containing 45 per cent $Mg(OH)_2$ of high purity. This is calcined at high temperatures to produce periclase, a sintered MgO nodule used in making refractory brick.

The production of magnesium compounds from sea water is made possible by the almost complete insolubility of magnesium hydroxide in water. Success in obtaining magnesium compounds by such a process depends upon the following factors:

1. Means to soften the sea water cheaply, generally with lime or calcined dolomite.

2. Preparation of a purified lime or calcined dolomite slurry of proper characteristics.

3. Economical removal of the precipitated hydroxide from the large volume of water.

4. Inexpensive purification of the hydrous precipitates.

5. Development of means to filter the slimes.

The reactions are:

$$MgCl_2(aq) + Ca(OH)_2(c) \longrightarrow Mg(OH)_2(c) + CaCl_2(aq) \qquad \Delta H = +2.260 \text{ kcal}$$

$$MgSO_4(aq) + Ca(OH)_2(c) + 2H_2O(l) \longrightarrow Mg(OH)_2(c) + CaSO_4 \cdot 2H_2O(c)$$
$$\Delta H = -3.170 \text{ kcal}$$

Magnesium products from sea water producing such fine chemical and pharmaceuticals as milk of magnesia and several basic magnesium carbonates such as $3MgCO_3 \cdot Mg(OH)_2 \cdot 4H_2O$ for tooth powders and antacid remedies, for coating table salt to render it noncaking, and for paint fillers. Certain of these basic magnesium compounds are also employed with rubber accelerators. Where calcined dolomite is used instead of calcium carbonate, only about one-half of the magnesia must come from the magnesium salts in the sea water. Consequently, the size of the plant is much smaller and the cost of production probably lower.

The $Mg(OH)_2$ may be calcined at about 1300°F to active, chemical MgO, or at about 2700° to 3000°F to periclase MgO. This $Mg(OH)_2$ is quite different from the slow-settling $Mg(OH)_2$ precipitated by a soluble alkali or by milk of lime. The reactions, as illustrated are principally the following:

Calcination:

$$2CaMg(CO_3)_2(c) \longrightarrow 2CaO(c) + 2MgO(c) + 4CO_2(g) \qquad \Delta H = +145.9 \, kcal$$

Slacking:

$$2CaO(c) + 2MgO(c) + 4H_2O(l) \longrightarrow 2Ca(OH)_2(c) + 2Mg(OH)_2(c)$$
$$\Delta H = -40.3 \, kcal$$

Precipitation:

$$2Ca(OH)_2(c) + 2Mg(OH)_2(c) + MgO_2(aq) + MgSO_4(aq) + 2H_2O(l) \longrightarrow$$
$$4Mg(OH)_2(c) + CaCl_2(aq) + CaSO_4 \cdot 2H_2O(s)$$
$$\Delta H = -5.4 \, kcal$$

Calcination:

$$4Mg(OH)_2(c) \longrightarrow 4MgO(c) + 4H_2O(g) \qquad \Delta H = +59.3 \, kcal$$

Hydrochlorination:

$$Mg(OH)_2(c) + 2HO(aq) \longrightarrow MgCl_2(aq) + 2H_2O(l) \qquad \Delta H = +11.4 \, kcal$$

Only about 7 per cent of the slaked calcined dolomite is needed for softening the sea water, the rest precipitating crystalline $Mg(OH)_2$ which is settled, filtered, and washed. This hydroxide is converted to other products.

Magnesium Carbonate

The term magnesium carbonate is generally reserved for the synthetic, pure variety. The naturally occurring material is called magnesite. It is light, bulky white powder; bulk density about 4 lb per cubic foot, actual sp. gr. about 3.0, decomposes 350°C. Refractive index about 1.52, soluble in acids, very slightly soluble in water; insoluble in alcohol, nontoxic, and noncombustible.

It is extracted from mines as natural material. It is also prepared by carbonation of MgO or $Mg(OH)_2$ with CO_2. It is also prepared by reaction of a soluble Mg salt solution with sodium carbonate or bicarbonate. It is used in magnesium salts, heat insulation and refractory, rubber reinforcing agent, inks, glass, pharmaceuticals, free-running table salts, antacid, making magnesium citrate, filtering medium. Used in foods as drying agent, colour retention agent, anticaking agent, and carrier.

Magnesium Oxide (Magnesia)

Two forms are produced, one a light fluffy material prepared by a relatively low-temperature dehydration of the hydroxide, the other a dense material made by high-temperature furnacing of the oxide after it has been formed from the carbonate of hydroxide.

It is in the form of white powder, either light or heavy depending upon whether it is prepared by heating magnesium carbonate or the basic magnesium carbonate, sp. gr. about 3.6 (varies), m.p. 2800°C, b.p. 3600°C, slightly soluble in water, soluble in acids and ammonium salt solution. Noncombustible.

It is prepared by calcining magnesium carbonate or magnesium hydroxide. It is also prepared by treating magnesium chloride with lime and heating, or by heating it in air. It is also prepared from sea water via the hydroxide. It is toxic by inhalation of fume. Tolerance (as Mg), 10 mg per cubic meter of air. It is used in refractories, especially for steel furnace linings; polycrystalline ceramic for aircraft wind-shields; electrical insulation; pharmaceuticals and cosmetics; inorganic rubber accelerator; oxychloride and oxysulphate cements; paper manufacture; fertilisers; removal of sulphur dioxide from stack gases; adsorption and catalysis; semiconductors; pharmaceuticals; food and feed additive.

Magnesium Chloride

It is colourless or in the form of white crystals, deliquescent, sp. gr. 2.32, m.p. 708°C, loses $2H_2O$ at 100°C, if heated rapidly, melts at 116°–118°C, b.p. 1412°C, decomposes to oxychloride, soluble in water and alcohol, combustible. It is prepared by the action of hydrochloric acid on magnesium oxide or hydroxide, especially the latter when precipitated from sea water or brines (great salt lake). It is purified by recrystallisation.

It is moderately toxic by ingestion. It is used as source of magnesium metal; disinfectants; fire extinguishers; fireproofing wood; magnesium oxychloride cement; refrigerating brines; ceramics; cooling drilling tools; textiles (size, dressing and filling of cotton and woolen fabrics, thread lubricant; carbonisation of wool); paper manufacture; road dust-laying compounds; floor sweeping compounds; flocculating agent; catalyst.

Magnesium Hydroxide

It occurs naturally as brucite. It is in the form of white powder, odourless, soluble in solutions of ammonium salts and dilute acids, almost insoluble in water and alcohol, sp. gr. 2.36, m.p. decomposes at 350°C, nontoxic, and noncombustible. It is prepared by precipitation from a solution of a magnesium salt by sodium hydroxide and also precipitation from sea water with lime.

It is used as an intermediate for obtaining magnesium metal, sugar refining, medicine (antacid, laxative), residual fuel oil additive, sulphite pulp, uranium processing dentifrices, in foods as drying agent, colour retention agent, and frozen desserts.

Magnesium Silicate

It is in the form of fine, white powder, insoluble in water or alcohol, as an absorbent. Noncombustible, may be toxic by inhalation. It is prepared by interaction of a magnesium salt and a soluble silicate. It is used in foods restricted to 2 per cent. It is used in rubber filler, ceramics, glass, refractories, absorbent for crude oil spills, manufacture of permanently dry resins and resinous compositions, paints, varnishes, and paper (filler), animal and vegetable oils (bleaching agent), odour absorbent, filter medium, catalyst and catalyst carrier, anticaking agent in foods.

Magnesium Sulphate

It is in the form of colourless crystals; saline, bitter taste, neutral to litmus, sp. gr. 2.65, decomposes $1124°C$, loses $6H_2O$ at $150°C$, $7H_2O$ at $200°C$, soluble in glycerol, very soluble in water, sparingly soluble in alcohol, low toxicity, and noncombustible. It is prepared by the action of sulphuric acid on magnesium oxide, hydroxide or carbonate. It is used in fireproofing, textiles, mineral waters, catalyst carrier, ceramics, fertilisers, paper (sizing), cosmetic lotions, and dietary supplement.

Calcium Arsenate

Calcium arsenate $(Ca_3 \cdot AsO_4)_2)$ is an extremely poisonous chemical compound. It was originally used as a pesticide. Its high solubility in water, as compared with lead arsenate, makes it more toxic.

Calcium arsenate is produced by the reaction of calcium chloride, calcium hydroxide, and sodium arsenate or lime and arsenic acid:

$$2CaCl_2(aq) + Ca(OH)_2(c) + 2Na_2H\,AsO_4(aq) \longrightarrow Ca_3(AsO_4)_2(c) + 4NaCl(aq) + 2H_2O(l)$$

$$\Delta H = -6.64 \text{ kcal}$$

Some free lime is usually present. Calcium arsenate is used extensively as an insecticide and as a fungicide. It is especially useful on cotton plants to poison boll weevils.

Calcium organic compounds

Calcium acetate and lactate are prepared by the reaction of calcium carbonate of hydroxide with acetic or lactic acid. The acetate was formerly pyrolysed in large amounts to produce acetone, but now it is employed largely in the dyeing of textiles. The lactate is sold for use in medicines and in foods as a source of calcium; it is an intermediate in the purification of fermentation lactic acid. Calcium soaps such as stearate, palmitate, and the abietate are made by the action of the sodium salts of the acids on a soluble calcium salt such as the chloride. These soaps are insoluble in water, but are soluble in hydrocarbons. Many of them form jellylike masses, which are constituents of greases. These soaps are used mainly as waterproofing agents.

Environmental and Social Impacts

Cement manufacture causes environmental impacts at all stages of the process. These include emissions of airborne pollution in the form of dust, gases, noise and vibration when operating machinery and during blasting in quarries, and damage to countryside from quarrying. Equipment to reduce dust emissions during quarrying and manufacture of cement is widely used, and equipment to trap and separate exhaust gases are coming into increased use. Environmental protection also includes the re-integration of quarries into the countryside after they have been closed down by returning them to nature or re-cultivating them.

Climate

Cement manufacture contributes greenhouse gases both directly through the production of carbon dioxide when calcium carbonate is heated, producing lime and carbon dioxide, and also indirectly through the use of energy, particularly if the energy is sourced from fossil fuels. The cement industry produces 5 per cent of global man-made CO_2 emissions, of which 50 per cent is from the chemical process, and 40 per cent from burning fuel. The amount of CO_2 emitted by the cement industry is nearly 900 kg of CO_2 for every 1000 kg of cement produced.

Chapter 10

Glass Industries

INTRODUCTION

Glass in the common sense refers to a hard, brittle, transparent amorphous solid, such as that used for windows, many bottles, or eyewear, including, but not limited to, soda-lime glass, borosilicate glass, acrylic glass, sugar glass, isinglass (Muscovy-glass), or aluminium oxynitride.

In the technical sense, glass is an inorganic product of fusion which has been cooled to a rigid condition without crystallising. Many glasses contain silica as their main component and glass former.

In the scientific sense the term glass is often extended to all amorphous solids (and melts that easily form amorphous solids), including plastics, resins, or other silica-free amorphous solids. In addition, besides traditional melting techniques, any other means of preparation are considered, such as ion implantation, and the sol-gel method. However, glass science commonly includes only inorganic amorphous solids, while plastics and similar organics are covered by polymer science, biology and further scientific disciplines.

Glass plays an essential role in various scientific fields and in industry. The optical and physical properties of glass make it suitable for applications such as flat glass, container glass, optics and optoelectronics material, laboratory equipment, thermal insulator (glass wool), reinforcement fibre (glass-reinforced plastic, glass fibre reinforced concrete), and art.

MANUFACTURE OF GLASS

Ingredients

Ingredients used in manufacture of glass are: (i) barium carbonate, (ii) borax, (iii) boric acid, (iv) dolomite; (v) feldspar, (vi) fluorspar, (vii) lime stone, (viii) litharge, (ix) potash, (x) sand, (xi) soda ash, and (xii) tin oxide.

Pure silica (SiO_2) has a 'glass melting point'— at a viscosity of 10 Pa·s (100 P)— of over 2300°C (4200°F). While pure silica can be made into glass for special applications, other substances are added to common glass to simplify processing. One is sodium carbonate (Na_2CO_3), which lowers the melting point to about 1500°C (2700°F) in soda-lime glass; 'soda' refers to the original source of sodium carbonate in the soda ash obtained from certain plants. However, the soda makes the glass water soluble, which is usually undesirable, so lime (calcium oxide, generally obtained from limestone), some magnesium oxide (MgO) and aluminium oxide are added to provide for a better chemical durability. The resulting glass contains about 70 to 74 per cent silica by weight and is called a soda-lime glass. Soda-lime glasses account for about 90 per cent of manufactured glass.

As well as soda and lime, most common glass has other ingredients added to change its properties. Lead glass, such as lead crystal or flint glass, is more 'brilliant' because the increased refractive index causes noticeably more 'sparkles', while boron may be added to change the thermal and electrical properties. Adding barium also increases the refractive index. Thorium oxide gives glass a high refractive index and low dispersion, and was formerly used in producing high quality lenses, but due to its radioactivity has been replaced by lanthanum oxide in modern eye glasses. Large amounts of iron are used in glass that absorbs infrared energy, such as heat absorbing filters for movie projectors, while cerium(IV) oxide can be used for glass that absorbs UV wavelengths (biologically damaging ionising radiation).

Two other common glass ingredients are calumite (an iron industry by-product) and 'cullet' (recycled glass). Calumite contains mainly silica, calcium oxide, alumina, magnesium oxide, and traces of iron oxide. The recycled glass saves on raw materials and energy. However, impurities in the cullet can lead to product and equipment failure.

Finally, fining agents such as sodium sulphate, sodium chloride, or antimony oxide are added to reduce the bubble content in the glass. Glass batch calculation is the method by which the correct raw material mixture is determined to achieve the desired glass composition. Typical flow diagram highlighting the various manufacturing steps involved is given in Fig. 10.1

Equipment and Handling in Glass Manufacture

Equipment

Glass is made either by an intermittent process in which the individual containers, or pots, are heated in suitable furnaces, or by a continuous process in a tank. Pot melting has been used since the beginning of glass manufacture, and is so especially suited for many types of small-scale or discontinuous operation that it probably never will be entirely displaced in spite of the commercial advantages of large-scale production in the continuous tank process.

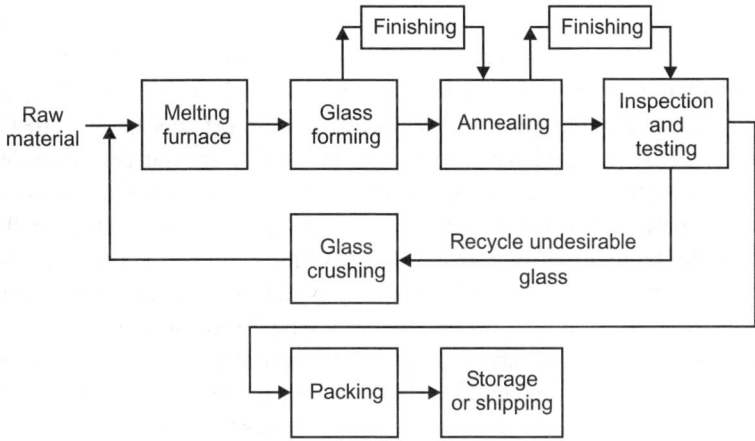

Fig. 10.1. Typical glass manufacturing process

Glass pots

Glass-making pots are made of clay which is carefully chosen and blended to obtain high plasticity, a minimum of shrinkage on heating, and a maximum of refractoriness and resistance to attack by the glass. They were formerly built up by hand, but are now also made by the slip-casting process. The hand-made pots are more porous and require a long time, from three to six months, to dry, while the dense slip-cast pots are made and dried more quickly. The pots may be made open, or sometimes closed by a hood like top with an opening for gathering. In size they range from small pots used for experimental meltings or for small lots of glass to be used in ornamentation, to large pots for plate glass. These small pots hold from 5 to 25 pounds while the large pots hold one or two tons. Most pots are almost cylindrical, with a small taper, and some are oval in cross-section.

In the manufacture of plate glass they are removed from the furnace by overhead cranes. The glass is poured onto the casting-table and the pot is retuned to the furnace. For making ordinary blown and pressed ware, the glass is dipped or gathered from the pot and these pots continue in service until they are worn out. In the manufacture of optical glass the pots are used but once. The pot of glass is removed from the melting furnace and cooled in an auxiliary furnace called a pot-arch, after which melt and pot are broken up. The furnaces used in pot melting are of a wide variety of types, ranging from crude single-pot furnaces without regeneration to regenerative furnaces, either rectangular or circular in cross-section, and holding from sixteen to twenty pots. Plate glass furnaces are usually rectangular, holding as many as twenty pots in two rows. Other furnaces are circular, holding up to sixteen pots, each of which can be set, removed, or worked from, without interfering with the others.

Tanks

Tanks are large furnaces so constructed that the furnace walls themselves are the containers for the glass. The lower part of the tank is built of tank built of tank blocks. These are special refractories of appropriate shapes so that the furnace may be assembled without mortar, and of special composition to offer the greatest resistance to attack by the glass. In recent years the introduction of high alumina and electrocast refractories has so improved the quality of tank blocks as to increase the life of the tank by several fold, which is now often more than a year. Tanks vary widely in size, but a moderately large tank for manufacture of containers may be 16 by 24 feet, and delivers about 90 tons of glass per day. Such tanks are separated into two sections by a bridge-wall in which a channel, called the throat, allows the glass to flow from the large melting end to the working end. Tanks for float glass are usually longer in order to give the glass more time to become free from bubbles (fined or planed), and do not have a bridge-wall. When glass is worked from tanks having no bridge-wall, the gather is taken from within annular clay floaters which provide a glass surface free from scum. In all furnaces the top, or crown, is made of silica brick and the structure is reinforced by steel beams and girders. With regenerative firing the checkerwork is below the firing level. The fuel enters the furnace in pots which are arranged to be symmertical with the exit pots, thereby insuring uniformity of heating when the ports are interchanged on the reverse cycle. The firing is either from the side or the rear.

Fuel

The fuel used in glass melting is either oil or gas, with the greater number of gas-fired furnaces using natural gas. The temperature of a tank at the melting end is usually from 1425° to 1550°C, depending on the type of glass and rate of pull, and at the working end about 1200° to 1300°C. The fuel cost is about one-fifth of the cost of glass delivered to the feeder.

Batch handling

The methods used in mixing the batch and feeding it into the furnace range from the crudest hand-mixing and shoveling, which is still practiced in many plants, to largely automatic processes.

The raw materials are delivered in ground or powdered form, and crushing operations are not necessary. Sand, limestone, lime, soda ash, and salt-cake are usually delivered in paper-lined freight cars, from which they are unloaded by shoveling and wheelbarrows, by power shovels and coveyor belts, by suction or travelling elevator. Materials delivered in barrels or bags, such as borax and boric acid, lead oxide, and nitrates, are usually handled by trucks. Weighing procedures range from hand-barrow operations with platform or flush scales, to systems in which power-operated cars are run successively under the various hopper bins and the requisite weights of ingredient delivered. The batches are delivered to mixers, or the ingredients themselves are collected in portable mixers. The mixed bath is then transported to the feeding floor, in larger plants, by travelling containers. Cullet is usually added in the mixers. Special care must be taken to prevent segregation since uniformity of the batch is highly important for producing good quality glass. In pot operation the batch is fed with scoops which are either hand or machine-operated. Filling of pots is necessarily an intermittent operation. The batch is filled in, melted down, more batch added and the process repeated until the pot is full of glass. Tank operation is more nearly continuous, with filling taking place according to a schedule which depends on rate of pull and time of reversal of the furnaces. Filling takes place from a built-in structure connected with the tank, called a dog-house, which is usually filled by gravity by a chute from an overhead bin. A recent development is the triangular dog-house, from which the batch is delivered automatically, according to a predetermined time schedule, by a motor-driven pusher in each feeder. The quantity of batch is determined by the size of the batch feeders themselves, the length of the pusher stroke, the number of strokes per minute, and the length of the feeding time.

Annealing

A heat treatment of glass in the viscosity range 10^{13} to 10^{14} poises is important to secure stabilisation. It is in the same range of low fluidity that glass is annealed to prevent the development of permanent strain, or to remove strains introduced by too rapid or non-uniform cooling. When glass is cooled a temperature gradient is introduced. If the gradient is introduced when the glass behaves as an elastic substance, as it does for stresses of short duration at the strain point, the stress resulting from the temperature will remain until the gradient has disappeared and will disappear with the gradient. If the gradient is introduced at the softening point, at which the glass yields quickly under stress, no strain will result from introduction of the temperature gradient, the glass will cool free from strain until the temperature gradient is removed, and the removal of the temperature gradient will produce a permanent stress. The permanent stresses in poorly annealed glasses are developed by the removal of a temperature gradient at low temperatures, and the amount of this stress is the difference between that caused by the removal of the temperature gradient, which produces a compression on the surface, and the temporary strain carried down from the high temperature at which the gradient was introduced. In short, the permanent stress is equal and opposite in sign to the stress lost by flow in the first part of the cooling.

In a typical commercial annealing lehr the change in temperature with distance in a lehr is used for annealing bottles. The front part of the lehr is usually heated by gas. Where the bloom on the ware caused by sulphur in city gas is objectionable, either natural or propane gas is used, and electrically-heated lehrs are in development. The delivery end of the lehr is not heated and is usually lower than the front end in order to aid in controlling convection, which usually should be from the cold to the hot end. The ware is carried through the lehr on a woven wire belt.

Properties of Glass

The resistance which glass offers to the corroding action of water, atmospheric agencies (primarily water and carbon dioxide), and aqueous solutions of acids, bases and salts is a property of great practical significance, and is denoted by the term 'chemical durability'. In a large proportion of the uses to which glass is put its power of resisting such attack is the chief reason for its preference over competing materials. An example is the use of glass containers, of which enormous numbers are used for the distribution of commodities ranging from milk to medicine and acids. In this field the superiority of glass leaves it without a competitor.

Even in chemical manufacturing, where the requirements are more exacting, glass is being used to an increasing extent as an engineering material because of the resistance which it offers to surface attack under extreme conditions. In other uses of glass chemical durability is a secondary factor, but the requirement of a chemical durability sufficient for the service contemplated, places a limit on the compositions which may be employed. Many glasses possessing desirably optical or mechanical properties are unsuitable because of their susceptibility to corrosion; others may be suitable for optical purposes in protected lens systems and still be worthless for laboratory uses. The methods of testing glass thus become of fundamental importance.

Physical properties

The physical properties of glass are determined primarily by chemical composition, but when measured at ordinary temperatures they are affected by the thermal history of the glass. As the glass is cooled from a high temperature some molecular re-arrangement takes place with a general decrease in interatomic distances and possibly an increase in regularity of atomic distribution. At high temperatures and low viscosities this re-arrangement is practically instantaneous.

As the glass is cooled through the upper part of the annealing range it takes from a few minutes to hours for the glass to become stabilised, and below the annealing range impracticably long times are required for the glass to attain the equilibrium conditions. The stabilisation affects all the properties of glass, although it is usually a second-order effect. In a precise consideration of the relationship between composition and property it is necessary to define the thermal history. It is not sufficient to state that the glass was well annealed.

Measurement and specification of properties are necessary in desiging of equipment using glass and for control of process and product. Some properties can be calculated from composition by assuming an additive contributions from each component proportional to the amount of it present, but all such calculations are rough approximations. Most measurements are made at ordinary temperatures, but viscosity and surface tension can be measured only at high temperatures where the glass is in a fluid or molten condition.

Silica-free Glasses

Besides common silica-based glasses, many other inorganic and organic materials may also form glasses, including plastics (e.g. acrylic glass), carbon, metals, carbon dioxide (see below), phosphates, borates, chalcogenides, fluorides, germanates (glasses based on GeO_2), tellurites (glasses based on TeO_2), antimonates (glasses based on Sb_2O_3), arsenates (glasses based on As_2O_3), titanates (glasses based on TiO_2), tantalates (glasses based on Ta_2O_5), nitrates, carbonates and many other substances. Some glasses that do not include silica as a major constituent may have physico-chemical properties useful for their application in fibre optics and other specialised technical applications. These include fluoride glasses (fluorozirconates, fluoroaluminates), aluminosilicates, phosphate glasses and chalcogenide glasses.

Under extremes of pressure and temperature solids may exhibit large structural and physical changes which can lead to polyamorphic phase transitions. In 2006 Italian scientists created an amorphous phase of carbon dioxide using extreme pressure. The substance was named amorphous carbonia (α-CO_2) and exhibits an atomic structure resembling that of silica.

Fibreglass

Fibreglass is material made from extremely fine fibres of glass. It is used as a reinforcing agent for many polymer products; the resulting composite material, properly known as fibre-reinforced polymer (FRP) or glass-reinforced plastic (GRP), is called 'fibreglass' in popular usage.

Fibreglass is formed when thin strands of silica-based or other formulation glass is extruded into many fibres with small diameters suitable for textile processing. Glass, even as a fibre, has little crystalline structure. The properties of the structure of glass in its softened stage are very much like its properties when spun into fibre. One definition of glass is 'an inorganic substance in a condition which is continuous with, and analogous to the liquid state of that substance, but which, as a result of a reversible change in viscosity during cooling, has attained so high a degree of viscosity as to be, for all practical purposes, rigid'.

The technique of heating and drawing glass into fine fibres has been known for millennia; however, the use of these fibres for textile applications is more recent. Two types of fibreglass most commonly used are S-glass and E-glass. E-glass has good insulation properties and it will maintain its properties up to 1500°F (815°C). S-glass has a high tensile strength and is stiffer than E-glass.

Properties

Glass fibres are useful because of their high ratio of surface area to weight. However, the increased surface area makes them much more susceptible to chemical attack. By trapping air within them, blocks of glass fibre make good thermal insulation, with a thermal conductivity on the order of 0.05 W/(mK).

Glass strengths are usually tested and reported for virgin fibres: those which have just been manufactured. The freshest, thinnest fibres are the strongest because the thinner fibres are more ductile. The more the surface is scratched, the less the resulting tenacity. Because glass has an amorphous structure, its properties are the same along the fibre and across the fibre. Humidity is an important factor in the tensile strength. Moisture is easily adsorbed, and can worsen microscopic cracks and surface defects, and lessen tenacity.

In contrast to carbon fibre, glass can undergo more elongation before it breaks. There is a correlation between bending diameter of the filament and the filament diameter. The viscosity of the molten glass is very important for manufacturing success. During drawing (pulling of the glass to reduce fibre circumference), the viscosity should be relatively low. If it is too high, the fibre will break during drawing. However, if it is too low, the glass will form droplets rather than drawing out into fibre.

Manufacturing processes

Melting

There are two main types of glass fibre manufacture and two main types of glass fibre product. First, fibre is made either from a direct melt process or a marble remelt process. Both start with the raw materials in solid form. The materials are mixed and melted in a furnace. Then, for the marble process, the molten material is sheared and rolled into marbles which are cooled and packaged. The marbles are taken to the fibre manufacturing facility where they are inserted into a can and remelted. The molten glass is extruded to the bushing to be formed into fibre. In the direct melt process, the molten glass in the furnace goes right to the bushing for formation.

Formation

The bushing plate is the most important part of the machinery. This is a small, metal furnace containing nozzles for the fibre to be formed through. It is almost always made of platinum alloyed with rhodium for durability. Platinum is used because the glass melt has a natural affinity for wetting it. When bushings were first used they were 100 per cent platinum and the glass wetted the bushing so easily it ran under the plate after exiting the nozzle and accumulated on the underside. Also, due to its cost and the tendency to wear, the platinum was alloyed with rhodium. In the direct melt process, the bushing serves as a collector for the molten glass. It is heated slightly to keep the glass at the correct temperature for fibre formation. In the marble melt process, the bushing acts more like a furnace as it melts more of the material.

The bushings are what make the capital investment in fibreglass production expensive. The nozzle design is also critical. The number of nozzles ranges from 200 to 4000 in multiples of 200. The important part of the nozzle in continuous filament manufacture is the thickness of its walls in the exit region. It was found that inserting a counterbore here reduced wetting. Today, the nozzles are designed to have a minimum thickness at the exit. The reason for this is that as glass flows through the nozzle it forms a drop which is suspended from the end. As it falls, it leaves a thread attached by the meniscus to the nozzle as long as the viscosity is in the correct range for fibre formation. The smaller the annular ring of the nozzle or the thinner the wall at exit, the faster the drop will form and fall, and the lower its tendency to wet the vertical part of the nozzle. The surface tension of the glass is what influences the formation of the meniscus. For E-glass it should be around 400 mN per metre.

The attenuation (drawing) speed is important in the nozzle design. Although slowing this speed can make coarser fibre, it is not economical to run at speeds for which the nozzles were not designed.

Continuous filament process

In the continuous filament process, after the fibre is drawn, a size is applied. This size helps protect the fibre as it is wound onto a bobbin. The particular size applied relates to its use. While some sizes are processing aids, others cause the fibre to have an affinity for a certain resin, if the fibre is to be used in a composite. Size is usually added at 0.5–2.0 per cent by weight. Winding then takes place at around 1000 metre per minutes.

Staple fibre process

In staple fibre production, there are a number of ways to manufacture the fibre. The glass can be blown or blasted with heat or steam after exiting the formation machine. Usually these fibres are made into some sort of mat. The most common process used is the rotary process. Here, the glass enters a rotating spinner, and due to centrifugal force, is thrown out horizontally. The air jets push it down vertically and binder is applied. Then the mat is vacuumed to a screen and the binder is cured in the oven.

Glass-reinforced Plastic

Glass-reinforced plastic (GRP) is a composite material or fibre-reinforced plastic made of a plastic reinforced by fine glass fibres. Like graphite-reinforced plastic, the composite material is commonly referred to by the name of its reinforcing fibres (fibreglass). Chemosetting plastics are normally used for GRP production—most often polyester (using butanone as a catalyst), but vinylester or epoxy are also used. The glass can be in the form of a chopped strand mat (CSM) or a woven fabric.

As with many other composite materials (such as reinforced concrete), the two materials act together, each overcoming the deficits of the other. Whereas the plastic resins are strong in compressive loading and relatively weak in tensile strength, the glass fibres are very strong in tension but have no strength against compression. By combining the two materials, GRP becomes a material that resists both compressive and tensile forces well. The two materials may be used uniformly or the glass may be specifically placed in those portions of the structure that will experience tensile loads.

Uses for regular fibreglass include mats, thermal insulation, electrical insulation, reinforcement of various materials, tent poles, sound absorption, heat- and corrosion-resistant fabrics, high-strength fabrics, arrows, bows and crossbows, translucent roofing panels, automobile bodies, electrical insulation and boat hulls.

Glass Wool

Glass wool is a form of fibreglass where very thin strands of glass are arranged into a spongy texture similar to steel wool. Glass wool is used widely as an insulating material.

Manufacturing process

After the fusion of a mixture of natural sand and recycled glass at 1450°C, the glass that is produced is converted into fibres. The cohesion and mechanical strength of the product is obtained by the presence of a binder that 'cements' the fibres together. Ideally, a drop of bonder is placed at each fibre intersection. This fibre mat is then heated to around 200°C to polymerise the resin and is calendered to give it strength and stability. The final stage involves cutting the wool and packing it in rolls or panels under very high pressure before palletising the finished product in order to facilitate transport and storage.

Thanks to its intertwined flexible fibres, glass wool offers excellent fire-resistant properties as a thermal insulation material (in loft-of-wall cavity insulation, for example) and is also widely used as an absorbent material in acoustic treatments, such as sound-insulating absorbent ceiling tiles. Its light weight, flexibility and elasticity make it easy to install, which is another essential condition for effective insulation. Glass wool is an excellent heat insulator. It can be woven into a cloth which has the additional properties of being light, strong, waterproof and corrosion free.

Fibreglass Moulding

Fibreglass moulding is a process in which fibreglass reinforced resin plastics are formed into useful shapes.

Mould making

The fibreglass mould process begins with an object known as the plug or buck. This is an exact representation of the object to be made, and can be made from a variety of different materials. Certain types of foam are commonly used. After the plug has been formed, it is sprayed with a mould release agent. The release agent will allow the mould to be separated from the plug once it is finished. The mould release agent is a special wax, and/or polyvinyl alcohol (PVA). Polyvinyl alcohol, however, is said to have negative effects on the final mould's surface finish. Once the plug has its release agent applied, gelcoat is applied with a roller, brush or specially-designed spray gun. The gelcoat is pigmented resin, and gives the mould surface a harder, more durable finish.

Once the release agent and gelcoat are applied, layers of fibreglass and resin are laid-up onto the surface. The fibreglass used will typically be identical to that which will be used in the final product.

In the laying-up process, a layer of fibreglass mat is applied, and resin is applied over it. A special roller is then used to remove air bubbles. If left in the curing resin, air bubbles would significantly reduce the strength of the finished mould. The fibreglass spray lay-up process is also used to produce moulds, and can provide good filling of corners and cavities where a glass mat or weave may prove to be too stiff. Once the final layers of fibreglass are applied to the mould, the resin is allowed to set-up and cure. Wedges are then driven between the plug and the mould in order to separate the two. Advanced techniques such as resin transfer moulding are also used.

Making a component

The component-making process involves building up a component on the fibreglass mould. The mould is a negative image of the component to be made, so the fibreglass will be applied inside the mould, rather than around it. As in the mould-making process, release agent is first applied to the mould. Coloured gelcoat is then applied. Layers of fibreglass are then applied, using the same procedure as before. Once completed and cured, the component is separated from the mould using wedges, compressed air or both.

Glass Microsphere

Glass microspheres are microscopic spheres of glass manufactured for a wide variety of uses in research, medicine, consumer goods and various industries. Glass microspheres are usually between 1 to 1000 micrometers in diameter. The term is also used for glass spheres between 100 nanometers to 5 millimeters in diameter. Hollow glass microspheres, sometimes termed microballoons, have diameters ranging from 10 to 300 micrometers. Hollow spheres are used as a lightweight filler in composite materials such as syntactic foam and light weight concrete. Microballoons give syntactic foam its light weight, low thermal conductivity, and a resistance to compressive stress that far exceeds that of other foams. These properties are exploited in the hulls of submersibles and deep-sea oil drilling equipment, where other types of foam would implode. Hollow spheres of other materials create syntactic foams with different properties, for example ceramic balloons can make a light syntactic aluminium foam.

Hollow spheres also have uses ranging from storage and slow release of pharmaceuticals and radioactive tracers to research in controlled storage and release of hydrogen. Microspheres are also used in composites to fill polymer resins for specific characteristics such as weight, sandability and sealing surfaces. When making surfboards for example, shapers seal the EPS foam blanks with epoxy and microballoons to create an impermeable and easily sanded surface upon which fibreglass laminates are applied. Glass microspheres can be made by heating tiny droplets of dissolved water glass in a process known as ultrasonic spray pyrolysis, and properties can be improved somewhat by using an acid treatment to remove some of the sodium.

Optical Fibre

An optical fibre (or fibre) is a glass or plastic fibre that carries light along its length. Fibre optics is the overlap of applied science and engineering concerned with the design and application of optical fibres. Optical fibres are widely used in fibre-optic communications, which permits transmission over longer distances and at higher bandwidths (data rates) than other forms of communications. Fibres are used instead of metal wires because signals travel along them with less loss, and they are also immune to electromagnetic interference. Fibres are also used for illumination, and are wrapped in bundles so they can be used to carry images, thus allowing viewing in tight spaces. Specially designed fibres are used for a variety of other applications, including sensors and fibre lasers.

Light is kept in the core of the optical fibre by total internal reflection. This causes the fibre to act as a waveguide. Fibres which support many propagation paths or transverse modes are called multi-mode fibres (MMF), while those which can only support a single mode are called single-mode fibres (SMF). Multi-mode fibres generally have a larger core diameter, and are used for short-distance communication links and for applications where high power must be transmitted. Single-mode fibres are used for most communication links longer than 550 meters (600 yards).

Joining lengths of optical fibre is more complex than joining electrical wire or cable. The ends of the fibres must be carefully cleaved, and then spliced together either mechanically or by fusing them together with an electric arc. Special connectors are used to make removable connections.

Glass Recycling

Glass recycling is the process of turning waste glass into usable products. Depending on the end use, this commonly includes separating it into different colours. Glass normally comes in a number of colours.

Glass makes up a large component of household and industrial waste due to its weight and density. The glass component in municipal waste is usually made up of bottles, broken glassware, light bulbs and other items. Glass recycling uses less energy than manufacturing glass from sand, lime and soda. Every ton of waste glass recycled into new items saves 315 kg of carbon dioxide. Glass that is crushed and ready to be remelted is called cullet. Reuse of glass containers is preferable to recycling according to the waste hierarchy. Refillable bottles are used extensively in many European countries, Canada and until relatively recently, in the United States. In Denmark 98 per cent of bottles are refillable and 98 per cent of those are returned by consumers. A similarly high number is reported for beer bottles in Canada. These systems are typically supported by container deposit laws and other regulations. In some developing nations like India and Brazil, the cost of new bottles often forces manufacturers to collect and refill old glass bottles for selling carbonated and other drinks.

Soda-lime Glass

Soda–lime glass, also called soda–lime–silica glass, is the most prevalent type of glass, used for windowpanes, and glass containers (bottles and jars) for beverages, food, and some commodity items. Also some of the Pyrex brand kitchen glassware is made of soda-lime glass.

Soda-lime glass is prepared by melting the raw materials, such as soda, lime, silica, alumina, and small quantities of fining agents (e.g. sodium sulphate, sodium chloride) in a glass furnace at temperatures locally up to 1675°C. The temperature is only limited by the quality of the furnace superstructure material and by the glass composition. Green and brown bottles are obtained from raw materials containing iron oxide. For lowering the price of the raw materials, pure chemicals are not used, but relatively inexpensive minerals such as trona, sand, and feldspar. The mix of raw materials is termed batch. Soda-lime glass is divided technically into glass used for windows, called float glass or flat glass, and glass for containers, called container glass. Both types differ in the application, production method (float process for windows, blowing and pressing for containers), and chemical composition. Float glass has a higher magnesium oxide and sodium oxide content as compared to container glass, and a lower silica, calcium oxide, and aluminium oxide content. From this follows a slightly higher quality of container glass concerning the chemical durability against water, which is required especially for storage of beverages and food.

Lead Glass

Glass consists of a network former, typically silica (SiO_2), and network modifiers, including alkali fluxes such as potassium oxide or sodium oxide, and a stabiliser, typically calcium oxide. Lead oxide acts as both a flux and a stabiliser. Lead glass forms part of the silica-potassium-lead system, where lead replaces the calcium content of typical potash glasses. Lead glass contains typically 18–35 mol% PbO, whilst modern lead crystal, historically also known as flint glass due to the original silica source, contains a minimum of 24 per cent lead oxide. Technically, the term crystal should never be applied to glass, as glass by definition lacks a crystalline structure, but the use of the term lead crystal remains popular due to historical and commercial reasons, originally stemming from the Venetian use of the word cristallo to describe the rock crystal imitated by Murano glassmakers. This is a naming convention which has been maintained to the present day to describe decorative hollowware.

Borosilicate Glass

Borosilicate glass is a type of glass with the main glass-forming constituents silica and boron oxide. Borosilicate glasses are most well known for having very low coefficient of thermal expansion ($\sim 5 \times 10^{-6}/°C$ at 20°C), making them resistant to thermal shock, more so than any other common glass.

Glass-Ceramic

Glass-ceramic materials share many properties with both glass and more traditional crystalline ceramics. It is formed as a glass, and then made to crystallise partly by heat treatment. Unlike sintered ceramics, glass-ceramics have no pores between crystals. While materials such as vaseline glass are also glass-ceramics, the term mainly refers to a mix of lithium, silicon, and aluminium oxides which yields an array of materials with interesting thermomechanical properties. The most commercially important of these have the distinction of being impervious to thermal shock. Originally developed for use in the mirrors and mirror mounts of astronomical telescopes, these materials have become known and entered the domestic market through its use in glass-ceramic cooktops, as well as cookware and bakeware.

The crystalline component of thermal glass-ceramics, beta spodumene, has a negative coefficient of thermal expansion, which contrasts with the positive coefficient of the glass. Adjusting the proportion of these two materials offers a wide range of possible coefficients in the finished composite.

When an interface between materials will be subject to thermal fatigue, glass-ceramics can be adjusted to match the coefficient of the material they will be bonded to. At a certain point, generally between 70 and 78 per cent crystallinity, the two coefficients balance such that the glass-ceramic as a whole has a thermal expansion coefficient that is very close to zero. Glass-ceramic is a mechanically strong material and can sustain repeated and quick temperature changes up to 800°–1000°C. At the same time, it has a very low heat conduction coefficient and can be made nearly transparent (15–20 per cent loss in a typical cooktop) for radiation in the infrared wavelengths. It is not, however, totally unbreakable. There have been instances where users reported damage to their cooktops when the surface was struck with a hard or blunt object (such as a can falling from above or other heavy items).

Crown Glass (Window)

Crown glass was an early type of window glass. In this process, glass was blown into a 'crown' or hollow globe. This was then flattened by reheating and spinning out the bowl-shaped piece of glass (bullion) into a flat disk by centrifugal force, up to 5 or 6 feet (1.5 to 1.8 metres) in diameter. The glass was then cut to the size required.

The thinnest glass was in a band at the edge of the disk, with the glass becoming thicker and more opaque toward the center.

Due to the distribution of the best glass, in order to fill large window spaces many small diamond shapes would be cut from the edge of the disk and these would be mounted into a lead lattice work and fitted in the window. Known as a bullseye, the centre area was used for less expensive windows.

Crown glass is one of many types of hand-blown glass. Other methods include: broad sheet, blown plate, polished plate and cylinder blown sheet. These methods of manufacture lasted at least until the end of the 19th century. The early 20th century marks the move away from hand-blown to machine manufactured glass such as rolled plate, machine drawn cylinder sheet, flat drawn sheet, single and twin ground polished plate and float glass.

Flat Glass

Flat glass, sheet glass or plate glass is a type of glass, initially produced in plane form, commonly used for windows, glass doors, transparent walls, and windshields. For modern architectural and automotive applications, the flat glass is sometimes bent after production of the plane sheet. Flat glass stands in contrast to container glass (used for bottles, jars, cups) and fibreglass (used for thermal insulation and optical communication). Most flat glass is soda-lime glass, produced by the float glass process.

Float Glass

Float glass is a sheet of glass made by floating molten glass on a bed of molten tin. This method gives the sheet uniform thickness and very flat surfaces. Modern windows are made from float glass. Most float glass is soda-lime glass, but relatively minor quantities of specialty borosilicate and flat panel display glass are also produced using the float glass process.

Manufacture

Float glass uses common glass making raw materials, typically consisting of sand, soda ash (sodium carbonate), dolomite, limestone, and salt cake (sodium sulphate). Other materials may be used as colourants, refining agents or to adjust the physical and chemical properties of the glass. The raw materials are mixed in a batch mixing process, then fed together with suitable cullet (waste glass), in a controlled ratio, into a furnace where it is heated to approximately 1500°C. Common flat glass furnaces are 9 metre wide, 45 metre long, and contain more than 1200 tons of glass. Once molten, the temperature of the glass is stablised to approximately 1200°C to ensure a homogeneous specific gravity.

The molten glass is fed into a tin bath, a bath of molten tin (about 3–4 metre wide, 50 metre long, 6 cm deep), through a delivery canal. The amount of glass allowed to pour onto the molten tin is controlled by a gate.

Tin is suitable for the float glass process because it has a high specific gravity, is immiscible, and is cohesive. Tin, however, is highly reactive with oxygen and oxidises in a natural atmosphere to form tin dioxide (SnO_2). Known in the production process as dross, the tin dioxide adheres to the glass. To prevent oxidation, the tin bath is provided with a positive pressure protective atmosphere consisting of a mixture of nitrogen and hydrogen.

The glass flows onto the tin surface forming a floating ribbon with perfectly smooth surfaces on both sides and an even thickness. As the glass flows along the tin bath, the temperature is gradually reduced from 1100°C until the sheet can be lifted from the tin onto rollers at approximately 600°C. The glass ribbon is pulled off the bath by rollers at a controlled speed. Variation in the flow speed and roller speed enables glass sheets of varying thickness to be formed.

Top rollers positioned above the molten tin may be used to control both the thickness and the width of the glass ribbon.

Once off the bath, the glass sheet passes through a lehr kiln for approximately 100 metre, where it is further cooled gradually so that it anneals without strain and does not crack from the change in temperature. On exiting the cold end of the kiln, the glass is cut by machines.

Incandescent Light Bulb

The incandescent light bulb, incandescent lamp or incandescent light globe is a source of electric light that works by incandescence, (a general term for heat-driven light emissions which includes the simple case of black body radiation). An electric current passes through a thin filament, heating it until it produces light. The enclosing glass bulb prevents the oxygen in air from reaching the hot filament, which otherwise would be destroyed rapidly by oxidation. Incandescent bulbs are also sometimes called electric lamps, a term also applied to the original arc lamps.

Incandescent bulbs are made in a wide range of sizes and voltages, from 1.5 volts to about 300 volts. They require no external regulating equipment and have a low manufacturing cost, and work well on either alternating current or direct current. As a result the incandescent lamp is widely used in household and commercial lighting, for portable lighting, such as table lamps, some car headlamps and electric flashlights, and for decorative and advertising lighting. Some applications of the incandescent bulb make use of the heat generated, such as incubators, brooding boxes for poultry, heat lights for reptile tanks, infrared heating for industrial heating and drying processes, and the easy-bake oven toy. In cold weather the heat shed by incandescent lamps contributes to building heating, but in hot climates lamp losses increase the energy used by air conditioning systems.

Incandescent light bulbs are gradually being replaced in many applications by (compact) fluorescent lamps, high-intensity discharge lamps, light-emitting diodes (LEDs), and other devices, which give more visible light for the same amount of electrical energy input. Some jurisdictions, such as the European Union are in the process of banning the use of incandescent light bulbs in favour of more energy-efficient lighting.

Cathode ray tube

The cathode ray tube (CRT) is a vacuum tube containing an electron gun (a source of electrons) and a fluorescent screen, with internal or external means to accelerate and deflect the electron beam, used to create images in the form of light emitted from the fluorescent screen. The image may represent electrical

264 Chemical Process Industries

waveforms (oscilloscope), pictures (television, computer monitor), radar targets and others.

Colour CRTs have three separate electron guns (shadow mask) or electron guns that share some electrodes for all three beams. The CRT uses an evacuated glass envelope which is large, deep, heavy, and relatively fragile.

Colour CRTs

Colour tubes use three different phosphors which emit red, green, and blue light respectively. They are packed together in stripes (as in aperture grille designs) or clusters called triads. Colour CRTs have three electron guns, one for each primary colour, arranged either in a straight line or in a triangular configuration (the guns are usually constructed as a single unit).

Each gun's beam reaches the dots of exactly one colour; a grille or mask absorbs those electrons that would otherwise hit the wrong phosphor.

Since each beam starts at a slightly different location within the tube, and all three beams are perturbed in essentially the same way, a particular deflection charge will cause the beams to hit a slightly different location on the screen (called a sub pixel). Colour CRTs with the guns arranged in a triangular configuration are known as delta-gun CRTs, because the triangular formation resembles the triangular shape of the Greek letter.

Glass Tube

Glass tubes or glass tubing are hollow pieces of borosilicate glass used in laboratory glassware. They are commercially available in various thicknesses and lengths, according to known standards.

Glass tubes can be cut by scoring with a diamond cutter, and bending, giving a break with a clean edge. The ends are preferably flame polished before use to remove the edge. Hose barbs can be added to give a better grip and seal when used with rubber tubing. Glass tubes can be bent by heating to red heat in a non-luminous Bunsen flame. The glass tubes are fitted to rubber bungs by drilled holes.

In the past, scientists constructed their own laboratory apparatus prior to the ubiquity of interchangeable ground glass joints.

Today, commercially available parts connected by ground glass joints are preferred; where specialised glassware are required, they are made to measure using commercially available glass tubes by specialist glassblowers.

POLLUTION IN GLASS INDUSTRY

Glass is a vital component for day to day domestic use as well as of the industry. Glasswares provide a more hygienic way of life style. The greatest advantage of the glass items is that it is absolutely non- polluting as a waste. The whole glass waste materials are recycled and re-used.

Coal Fired Pot Furnaces

The direct coal fired pot furnaces existing in the small sector at Firozabad, generate air pollution in the form of particulate matter. The particulate matter mainly consists of unburnt coal particles and raw material fines. The higher level of emissions are due to improper firing and carry-over of the raw material. The limits for particulate matter emissions for pot furnaces melting soda lime glass is 1200 mg/nm^3. The emission for sulphur dioxide are generally low as it depends on the sulphur content of the fuel. The average capacity of such furnaces is 4 tons per day of melt glass.

Coal Fired Tank Furnaces

These furnaces also fall into the category of the small scale sector. The capacity of such furnaces may range from 5 tons glass melt to 30 tons glass melt per day. The crude type of static producers are used to produce gas which is burnt in the furnace for glass melting. In these furnaces, particulate matter is the main pollutant. The limits of particulate emission in this furnace is 2 kg/hr. Sulphur dioxide emission should meet the criteria of stack height.

Oil Fired Tank Furnaces

The oil fired tank furnaces are in use for glass melting in small as well as large scale industry for many years. In these furnaces, the emission limits for particulate matter is 2 kg/hr upto a product draw capacity of 60 MT/day. For the product drawn capacity more than 60 MT/day additional particulate emission of 0.5 kg/T of product is permitted.

Air Pollution from Melting Lead Glass

Lead glass

Furnace of any capacity melting lead glass, should not emit particulate matter more than 50 mg/nm^3 and the lead contribution should not be more than 20 mg/nm^3.

Chapter 11

Salt and Miscellaneous Sodium Compounds

INTRODUCTION

Many sodium salts are of definite industrial necessity. Most of them are derived directly or indirectly from ordinary salt so far as their sodium content is concerned. In one sense, the sodium may be viewed simply as a carrier for the more active anion to which the compound owes its industrial importance. For instance, in sodium sulphide, it is the sulphide part that is the more active. Similarly, this is the case with sodium thiosulphate and sodium silicate. The corresponding potassium salt could be used in most cases; however, sodium salts can be manufactured more cheaply and in sufficient purity to meet industrial demands.

Salt is also the basic raw material for a large number of inorganic chemical industries such as caustic soda and chlorine, soda ash, sodium sulphate, hydrochloric acid, etc. It also finds use in a large number of other industries such as oil hydrogenation, soap manufacture, dyes, leather, textile, food processing, etc. Salt also finds agricultural and medicinal uses.

SODIUM CHLORIDE

Sodium chloride, also known as common salt, table salt, or halite, is a chemical compound with the formula NaCl. Sodium chloride is the salt most responsible for the salinity of the ocean and of the extracellular fluid of many multicellular organisms. As the major ingredient in edible salt, it is commonly used as a condiment and food preservative.

Manufacture of Sodium Chloride

Salt is currently produced by evaporation of sea water or brine from other sources, such as brine wells and salt lakes and by mining rock salt called halite. The important methods are:
1. Solar evaporation.
2. Artificial evaporation.

Solar Evaporation

In most countries of the world salt is manufactured by this method. The initial density of sea brine is 3°–3.5°Be. This is conveyed through channels to a reservoir of sufficiently large size usually built at a height so that brine can subsequently flow by gravity from the same. The purpose of the reservoir is to store brine, remove suspended impurities by sedimentation and concentrate it to 10°Be by solar energy. The brine of 10°Be is transfered by gravity to condensers through the channels. Here it is concentrated to 25°Be by solar evaporation. The wind velocity, humidity and the surface of exposure determine the rate of evaporation. As the brine gradually concentrates due to evaporation of water, at 7.4°Be, most of the ferrous iron present separates out as ferric oxide. On further concentration to 10°Be the calcium carbonate starts precipitating till 12°Be when calcium sulphate also starts precipitating and a mixture of calcium carbonate and sulphate separates out. At 16.4°Be about half the calcium sulphate initially present is removed, the balance being removed till the brine reaches a concentration of 25°Be.

The brine from the condensers is now conveyed through channels to the crystallisers. The channels are all made of impermeable clay to avoid leakage through seepage. Salt separates out in the crystallisers from 25.4° to 30°Be. When other impurities also start separating. The salt is scraped out and the mother liquor known as 'bittern' is used for the recovery of other by-products. For every ton of salt made, one ton of bittern is produced.

The same procedure for evaporations is followed in case of other sources of brine. However, the degree of separation of salts differ since these brines contain different salts.

Artificial Evaporation

In France and Germany where brine is weak, both solar and artificial evaporation methods are used. In UK, Germany, USA and other cold countries where solar evaporation cannot be used, artificial evaporation method is used.

Open pan method is now obsolete. In this method, purified brine was evaporated in direct fired open pans. The fuel efficiency was rather low.

VACUUM EVAPORATION METHOD

Vacuum pan salt is made by boiling brine at less than atmospheric pressure. The grains are cubical. Purified brine is pumped from the storage tanks to the vacuum pan system. Calcium sulphate is removed by counter flow hydraulic washing with brine. The lighter calcium sulphate washes out leaving high purity sodium chloride for further processing.

Generally the triple effect evaporators are used, the bodies being of cast iron with steel sheets and copper tubes. Vapour lines are of steel. Formation of salt scale on the tubes limits the operation. They must be cleaned by periodic boiling with water. Operating cycle between cleanings varies from 48 hrs to a week. Salt slurry is drawn continuously from each evaporator through the salt leg. The brine is fed into the bottom of the leg so that the salt slurry is washed, by incoming brine. Impurities are thus washed back into the pans where they are allowed to accumulate until the boiling out period.

Salt slurry goes by gravity to a cone-shaped tank. From this tank the slurry is pumped to a feed tank feeding rotary vacuum pump for dewatering and drying. The filtered and partially dried salt goes to a rotary drier for final drying. The dryer is usually of monel metal. The dry crystals pass through a scalping screen to remove the lumps. The salt is carried to storage bins, screened, sized and packaged.

Salt Evaporator

Salt evaporator shown in Fig. 11.1. There may be only one evaporator body (single effect) or up to four evaporators can be connected in series so that the steam produced in the preceding evaporator body is used in the following evaporator. Thus, a quadruple effect evaporator can theoretically evaporate four kilograms water from brine for every kilogram of steam fed to the first effect. However, thermal losses decrease the steam economy to somewhat over three kilograms of water evaporated by each kilogram of steam. The bodies of the various effects are exactly alike.

Vacuum pans vary in size from 5.5–6.5 metres in diameter and 10.5 to 14 metres in height. They are usually made in three sections; the top and bottom are conical while the central section is cylindrical being larger in diameter at the top than bottom. These sections may be of cast iron and bolted together at flanges or may be of all steel and welded construction. The top section has a vapour outlet pipe leading to the next effect.

Extending across the central section are two tube sheets. Between these tube sheets are expanded a number of copper tubes. In the centre is very much larger tube called the downtake. Before starting, the evaporator is filled with brine to a level slightly above the top tube sheet and steam is turned on through 'steam inlet'. The steam surrounds the heating element and gives up its heat to the brine. The condensate is removed through pipe by a pump. Air or other noncondensible gases present in steam are vented off through pipe known as condensible gases outlet pipe. Brine is fed through 'brine feeding pipe' in the lower section. The vapour from first effect leaving at 'pipe leading to second effect' enters the steam chest of second effect serving as the source of heat. In the end the vapour leaving the last effect passes to a condenser through cold water supply connection. The condenser is provided with a number of baffles and as the vapour passes up, it is condensed. The condensed steam along with water leaves 'steam condensate outlet pipe'. The condenser is set high enough (about 10.5 metres above the hot well) so that the water is discharged by gravity. A vacuum pump is connected to the top of the condenser and serves to remove any air that may reach the condenser through 'noncondensible gases outlet pipe' and that liberated from the condenser water.

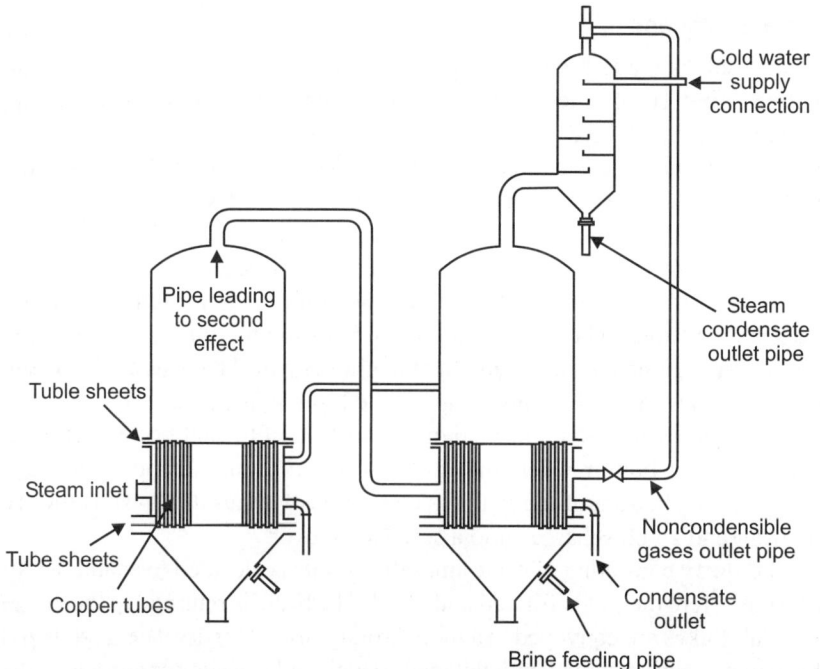

Fig. 11.1. Salt evaporator.

The steam surrounding the tubes being hotter than the brine in the tubes condenses and gives up its heat to the brine. The condensation of vapour coming through 'pipe leading to second effect' produces partial vacuum in the first effect and lowers the boiling point of brine. As the brine boils inside the 'tube sheets' the lifting effect of the steam bubbles causes the brine to circulate through downtake 'copper tubes'. The propeller assists in the circulation. The condenser maintains the vacuum in the last effect.

Many salt evaporators use parallel feed. If there are appreciable quantities of impurities like calcium and magnesium compounds, it may be desirable to concentrate these impurities in one effect. In such cases either a forward or backward feed may be used.

The salt nuclei produced in the evaporated brine are uniformly surrounded with saturated solution and the crystal grows in its normal cubic shape. The size of the crystals is controlled by the rate of evaporation and amount of agitation which keeps the crystal suspended in the solution where growth can occur. These crystals along with the brine are pumped to a setting tank by means of a centrifugal pump attached to the cone of the evaporator. The excess brine overflowing from the top of the tank is returned to the evaporator. The salt from the bottom of the setting tank is centrifuged, dried and packaged.

Grainer Process

Grainer salt is made by evaporating brine in long, shallow pans below the boiling point. The grainer consists of a steel vessel 30 to 45 metres long, 3 to 6 metres wide and 5.5 decimetres deep. The inside lining of these grainers is protected by brine resistant coating as neoprene. The internal mechanism is made of stainless steel, monel or ordinary steel covered with phenolic resin to prevent metallic contamination of the product. A heating elements is provided in the form of number of passes of 10 cm pipe, parallel to the long axis of the grainer.

Purified brine is fed into the grainer and continuously circulated through the heat exchanger where the temperature is raised to 100°C, a filter (where the calcium sulphate is removed) and the grainer pan. The pans are kept filled at all the times. Average temperature of the brine in the pans is 102°C.

Salt crystals are slowly formed on the surface of the hot brine in grainers. They fall to the bottom where they grow in size. A monel scrapping conveyor system on the bottom of the grainers continuously rakes the crystals towards the front end and discharges into a screw conveyor.

The slurry consisting of brine and salt crystals is elevated to a tank feeding salt filter. Here the salt is filtered and dried. The brine is returned to the grainer pan. Salt flakes are conveyed to a monel rotary drier. Hot dry salt crystals pass through a scalping screen and elevated to a glazed tile sile for storage.

Flake salt is screened in stainless steel screens and divided into many specialised grades for the food industry. The various size products are conveyed from storage bins to packaging machines.

Grainer salt crystallised mostly on the brine surface where evaporation takes place. The crystals grow, as additional cubes of salt become attached to the original nuclei. As the small cube nucleus grows, it immerses slightly, additional growth is, therefore, on the edge nearest to the surface. The nucleus tends to take a rectangular form at the surface so the crystal has flat sides formed by many cubes. These crystals are supported by the surface tension of the brine until their weight or some disturbance causes them to sink.

Different Grades of Salt

Different grades of salt can be produced by adding different compounding materials. Rock salt is used in cattle feed and in such manufacturing operations as synthetic rubber and ice-cream. Flake salt from the grainer process becomes cheese salt or butter salt or anti-oxidant salt (with a chemical to retard rancidity in certain foodstuff). Granulated or cube salt from the vacuum evaporators becomes free-running (by adding one per cent of finely ground inert material such as calcium phosphate or magnesia) or shaker or table salt. By addition of 0.025 per cent of potassium iodide to the above shaker or table salt, it makes iodised salt. In very fine size, it is known as 'pop corn salt.' Special products include sulphurised block for stock feed, calcium sulphate salt tablets for tomato canners, dextrose salt tablets for heat relief, pure tablets for general canning.

Stainless steel is largely used to prevent corrosion and contamination. Ni-resist is used for hot brine pumps. Slurry pumps have rubber stators. Most of the bolts and nuts are of stainless steel. All screws conveyors may be covered with a protective coaating of phenolic resin. Neoprene coating may also be used. Silos may be made of glazed tile. Hard wood may be used for temporary storage.

By-Products

The important by-products of salt industry are: (i) calcium carbonate, (ii) calcium sulphate, (iii) bromine, (iv) magnesium sulphate, (v) magnesium chloride, (vi) potassium chloride, and (vii) sodium sulphate.

The operation are carried out in tower packed with rasching rings in nine sections. Steam at $4.5 \, kg/cm^2$ is passed at the bottom section and chlorine at the second section from bottom. Figure 11.2 shows the flow chart of the bromine manufacturing unit using bitterns as raw material.

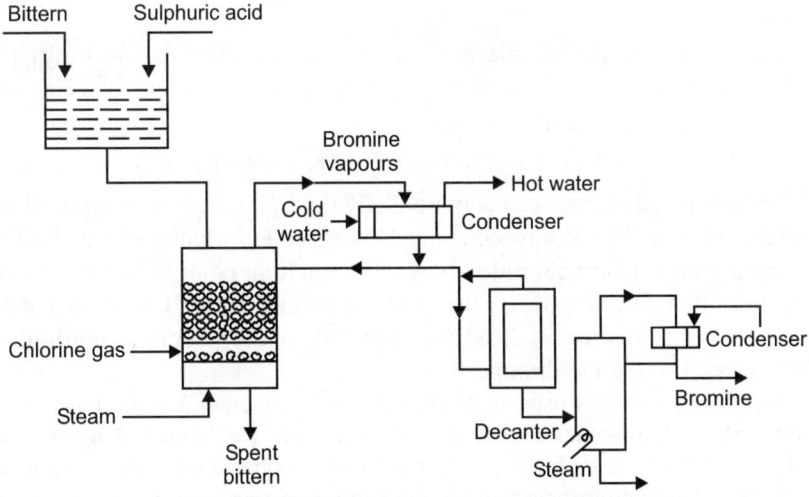

Fig. 11.2. Manufacture of bromine from bittern.

The bittern solution is sprayed countercurrent to the flow of chlorine and steam. Bromine vapours escape at 78°C from the eight section of the tower. The bromine vapour is condensed by cooling with water in tantalum condenser. The impure bromine contains chlorine and water. It is passed through U-tube another end of which is connected to the ninth section of the tower. From here it goes to the gravity separator. Here most of the water is separated and the water layer is recycled to the fourth section of the tower to recover bromine present in it. Liquid bromine is purified by distillation with 2.8 kg/cm² closed steam.

Uses

Salt is currently mass-produced by evaporation of seawater or brine from other sources, such as brine wells and salt lakes, and by mining rock salt, called halite. Sodium chloride is used in cooking, salt is used in many applications, from manufacturing pulp and paper, to setting dyes in textiles and fabric, to producing soaps, detergents, and other bath products.

It is the major source of industrial chlorine and sodium hydroxide, and used in almost every industry. Sodium chloride is sometimes used as a cheap and safe desiccant due to its hygroscopic properties, making salting an effective method of food preservation historically. Even though more effective desiccants are available, few are safe for humans to ingest.

Synthetic uses

Sodium chloride is also the raw material used to produce chlorine which itself is required for the production of many modern materials including PVC and pesticides.

Industrially, elemental chlorine is usually produced by the electrolysis of sodium chloride dissolved in water. Along with chlorine, this chloralkali process yields hydrogen gas and sodium hydroxide, according to the chemical equation:

$$2NaCl + 2H_2O \rightarrow Cl_2 + H_2 + 2NaOH$$

Sodium metal is produced commercially through the electrolysis of liquid sodium chloride. This is now done in a Down's cell in which sodium chloride is mixed with calcium chloride to lower the melting point below 700°C. As calcium is more electropositive than sodium, no calcium will be formed at the cathode. This method is less expensive than the previous method of electrolysing sodium hydroxide.

Sodium chloride is used in other chemical processes for the large-scale production of compounds containing sodium or chlorine. In the Solvay process, sodium chloride is used for producing sodium carbonate and calcium chloride.

In the Mannheim process and in the Hargreaves process, it is used for the production of sodium sulphate and hydrochloric acid.

Biological uses

Many micro-organisms cannot live in an overly salty environment: water is drawn out of their cells by osmosis. For this reason salt is used to preserve some foods, such as smoked bacon or fish. It can also be used to detach leeches that have attached themselves to feed. It has also been used to disinfect wounds.

Optical uses

Pure NaCl crystal is an optical compound with a wide transmission range from 200 nm to 20 μm. It was often used in the infrared spectrum range and it is still used sometimes.

NaCl crystal is soft, hygroscopic and cheap. This limits its application to protected environment or for short term uses (prototyping). Exposed to free air NaCl optics will rot. Today tougher crystals like ZnSe are used instead of NaCl (for the IR spectral range).

Household uses

Since at least medieval times, people have used salt as a cleansing agent rubbed on household surfaces. It is also used in many brands of shampoo.

Biological Functions

In humans, a high-salt intake has long been known to generally raise blood pressure, especially in certain individuals. More recently, it was demonstrated to attenuate nitric oxide production. Nitric oxide (NO) contributes to vessel homeostasis by inhibiting vascular smooth muscle contraction and growth, platelet aggregation, and leukocyte adhesion to the endothelium.

Road Salt

Calcium chloride is preferred over sodium chloride, since $CaCl_2$ releases energy upon forming a solution with water, heating any ice or snow it is in contact with. It also lowers the freezing point, depending on the concentration. NaCl does not release heat upon solution; however, it does lower the freezing point. It is also more readily available and does not have any special handling or storage requirements, unlike calcium chloride.

Additives

Table salt sold for consumption today is not pure sodium chloride. In 1911 magnesium carbonate was first added to salt to make it flow more freely. In 1924 trace amounts of iodine in form of sodium iodide, potassium iodide or potassium iodate were first added, to reduce the incidence of simple goiter.

Salt for de-icing in the UK typically contains sodium hexacyanoferrate(II) at less than 100 ppm as an anti-caking agent. In recent years this additive has also been used in table salt.

Common Chemicals

Chemicals used in de-icing salts are mostly found to be sodium chloride (NaCl) or calcium chloride ($CaCl_2$). Both are similar and are effective in de-icing roads. When these chemicals are produced, they are mined/made, crushed to fine granules, then treated with an anti-caking agent. Adding salt lowers the freezing point of the water, which allows the liquid to be stable at lower temperatures and allows the ice to melt. Alternative de-icing chemicals have also been used. Chemicals such as calcium magnesium acetate and potassium formate are being produced. These chemicals have few of the negative chemical effects on the environment commonly associated with NaCl and $CaCl_2$.

SODIUM SULPHATE (SALT CAKE AND GLAUBER'S SALT)

Sodium sulphate is the sodium salt of sulphuric acid. Anhydrous, it is a white crystalline solid of formula Na_2SO_4 known as the mineral thenardite; the decahydrate $Na_2SO_4 \cdot 10H_2O$ has been known as Glauber's salt. Other solid is the heptahydrate, which transforms to mirabilite when cooled. It is one of the world's major commodity chemicals and one of the most damaging salts in structure conservation: when it grows in the pores of stones it can achieve high levels of pressure, causing structures to crack. Sodium sulphate can be manufactured from natural brines and from salt and sulphuric acid.

From Natural Brines

Natural brines are now the principal sources of sodium sulphate. From the Searles Lake brines, sodium sulphate may be considered either a coproduct or a by-product of the production of sodium carbonate, and the process is considered under that heading (Fig. 11.3).

$$Na_2SO_4 + 10H_2O \longrightarrow Na_2SO_4 \cdot 10H_2O$$

$$Na_2SO_4 \cdot 10H_2O \longrightarrow Na_2SO_4 + 10H_2O$$

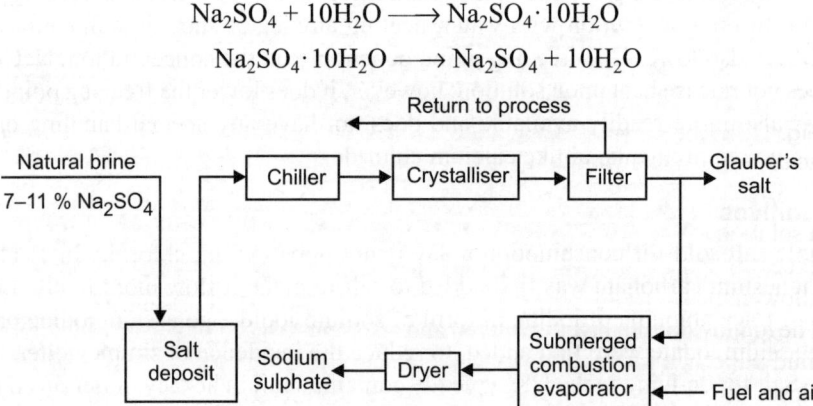

Fig. 11.3. Flow diagram for manufacture of sodium sulphate from natural brines.

The specific process used for recovery of natural sodium sulphate depends of course on the composition of the brine, and to some extent on its location. The natural brines available contain 7 to 11 per cent sodium sulphate, plus some sodium chloride and magnesium sulphate. To lower the sodium sulphate solubility of the brine, it is saturated with sodium chloride by pumping it into a salt deposit. The salt-enriched brine leaving the well is chilled to $-10°$ to $-6°C$ in ammonia-cooled coils and sent to a crystalliser. The resulting Glauber's salt crystals are separated from the mother liquor by filtration. The mother liquor is returned to the process. The Glauber's salt crystals are charged to a submerged combustion evaporator where they are melted, and most of the resulting water is removed by evaporation. The wet salt product is then dried in a rotary kiln.

From Salt and Sulphuric Acid

Still a major source of salt cake, but of diminishing importance, is the salt-sulphuric acid process for the production of hydrochloric acid. Salt cake is a by-product. Salt and sulphuric acid, sp. gr. 1.72 (slight excess) are charged to a furnace equipped with a rake agitator (Mannheim furnace), where the reacting mass is slowly heated to a temperature just below fusion (843°C). Hydrogen chloride is evolved and led through a cooling and condensing system to the absorbers. Salt cake (crude sodium sulphate) is continuously discharged from the periphery of the furnace (Fig. 11.4).

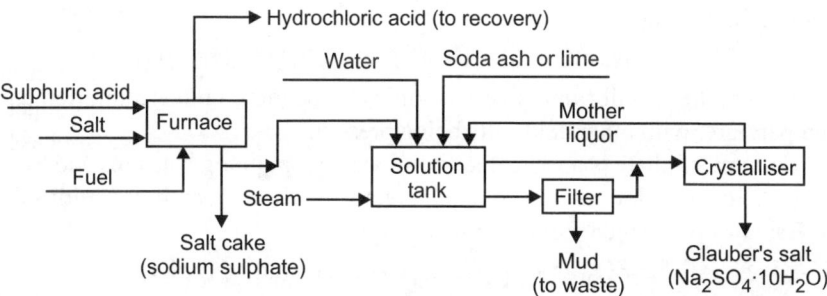

Fig. 11.4. Flow diagram for manufacture of sodium sulphate from salt and sulphuric acid.

When Glauber's salt is desired, the salt cake is dissolved in hot water to form a solution of sp. gr. 1.29. Either soda ash or lime is then added to neutralise excess sulphuric acid and to precipitate iron and alumina. The precipitate is allowed to settle, and the clear supernatant liquor is pumped to the crystalliser. The muddy bottom layer is filtered and also sent to the crystalliser. The filtered mud cake is discarded. After crystallisation the Glauber's salt is stored in closed bins to prevent desiccation. The mother liquor is returned to the solution tank for use in subsequent batches. In order for the crystals to be colourless, the crystalliser liquor must be maintained on the acid side of neutral. Free acid in the finished product will be of the order of 0.01 per cent. The following reactions takes place.

$$2NaCl + H_2SO_4 \longrightarrow 2HCl + Na_2SO_4$$
98% yield
$$Na_2SO_4 + 10H_2O \longrightarrow Na_2SO_4 \cdot 10H_2O$$
95% yield

Niter cake (sodium bisulphate) may be used in the Mannheim process in place of sulphuric acid; it is mixed with the salt and fed to the furnace.

Uses

Sodium sulphate is mainly used for the manufacture of detergents and in the kraft process of paper pulping. About two-thirds of the world's production is from mirabilite, the natural mineral form of the decahydrate, and the remainder from by-products of chemical processes such as hydrochloric acid production.

Sodium sulphate is chemically very stable, being unreactive toward most oxidising or reducing agents at normal temperatures. At high temperatures, it can be reduced to sodium sulphide. It is a neutral salt, which forms aqueous solutions with pH of 7. The neutrality of such solutions reflects the fact that Na_2SO_4 is derived, formally speaking, from the strong acid sulphuric acid and a strong base sodium hydroxide. Sodium sulphate reacts with an equivalent amount of sulphuric acid to give an equilibrium concentration of the acid salt sodium bisulphate:

$$Na_2SO_4(aq) + H_2SO_4(aq) \rightarrow 2NaHSO_4(aq)$$

In fact, the equilibrium is very complex, depending on concentration and temperature, with other acid salts being present.

Sodium sulphate is a typical ionic sulphate, containing Na^+ ions and SO_4^{2-} ions. Aqueous solutions can produce precipitates when combined with salts of Ba^{2+} or Pb^{2+}, which form insoluble sulphates.

$$Na_2SO_4(aq) + BaCl_2(aq) \rightarrow 2NaCl(aq) + BaSO_4(s)$$

Sodium sulphate has unusual solubility characteristics in water. Its solubility rises more than ten-fold between 0° to 32.4°C, where it reaches a maximum of 49.7 g Na_2SO_4 per 100 g water. At this point the solubility curve changes slope, and the solubility becomes almost independent of temperature. In the presence of NaCl, the solubility of sodium sulphate is markedly diminished. Such changes provide the basis for the use of sodium sulphate in passive solar heating systems, as well is in the preparation and purification of sodium sulphate. This nonconformity can be explained in terms of hydration, since 32.4°C corresponds with the temperature at which the crystalline decahydrate (Glauber's salt) changes to give a sulphate liquid phase and an anhydrous solid phase.

Sodium sulphate decahydrate is also unusual among hydrated salts in having a measureable residual entropy (entropy at absolute zero) of $6.32 \text{ J} \cdot \text{K}^{-1} \cdot \text{mol}^{-1}$. This is ascribed to its ability to distribute water much more rapidly compared to most hydrates.

Sodium sulphate displays a moderate tendency to form double salts. The only alums formed with common trivalent metals are $NaAl(SO_4)_2$ (unstable above 39°C) and $NaCr(SO_4)_2$, in contrast to potassium sulphate and ammonium sulphate which form many stable alums. Double salts with some other alkali metal sulphates are known, including $Na_2SO_4 \cdot 3K_2SO_4$ which occurs naturally as the mineral glaserite. Formation of glaserite by reaction of sodium sulphate with potassium chloride has been used as the basis of a method for producing potassium sulphate, a fertiliser.

Other double salts include $3Na_2SO_4 \cdot CaSO_4$, $3Na_2SO_4 \cdot MgSO_4$ (vanthoffite) and $NaF \cdot Na_2SO_4$.

Chemical Industry

About one third of the world's sodium sulphate is produced as by-product of other processes in chemical industry. Most of this production is chemically inherent to the primary process, and only marginally economical. By effort of the industry, therefore, sodium sulphate production as by-product is declining.

The most important chemical sodium sulphate production is during hydrochloric acid production, either from sodium chloride (salt) and sulphuric acid, in the Mannheim process, or from sulphur dioxide in the Hargreaves process. The resulting sodium sulphate from these processes are known as salt cake.

Mannheim: $\qquad 2NaCl + H_2SO_4 \rightarrow 2HCl + Na_2SO_4$

Hargreaves: $\quad 4NaCl + 2SO_2 + O_2 + 2H_2O \rightarrow 4HCl + 2Na_2SO_4$

The second major production of sodium sulphate are the processes where surplus sulphuric acid is neutralised by sodium hydroxide, as applied on a large scale in the production of rayon. This method is also a regularly applied and convenient laboratory preparation.

$$2NaOH(aq) + H_2SO_4(aq) \rightarrow Na_2SO_4(aq) + 2H_2O$$

Formerly, sodium sulphate was also a by-product of the manufacture of sodium dichromate, where sulphuric acid is added to sodium chromate solution forming sodium dichromate, or subsequently chromic acid. Alternatively, sodium sulphate is or was formed in the production of lithium carbonate, chelating agents, resorcinol, ascorbic acid, silica pigments, nitric acid, and phenol. Bulk sodium sulphate is usually purified via the decahydrate form, since the anhydrous form tends to attract iron compounds and organic compounds. The anhydrous form is easily produced from the hydrated form by gentle warming.

Applications

Drying a wet organic phase using sodium sulphate, which clumps, indicating that more sodium sulphate is needed. Drying a fairly dry organic phase using sodium sulphate, which does not clump, indicating that the solution is dry.

Thermal storage

The high heat storage capacity in the phase change from solid to liquid, and the advantageous phase change temperature of 32°C (90°F) makes this material especially appropriate for storing low grade solar heat for later release in space heating applications. In some application the material is incorporated into thermal tiles that are placed in an attic space while in other applications the salt is incorporated into cells surrounded by solar-heated water. The phase change allows a substantial reduction in the mass of the material required for effective heat storage (83 calories per gram stored across the phase change, versus one calorie per gram per degree Celsius using only water), with the further advantage of a consistency of temperature as long as sufficient material in the appropriate phase is available.

Small-scale applications

In the laboratory, anhydrous sodium sulphate is widely used as an inert drying agent, for removing traces of water from organic solutions. It is more efficient, but slower-acting, than the similar agent magnesium sulphate. It is only effective below about 30°C, but it can used with a variety of materials since it is chemically fairly inert. Sodium sulphate is added to the solution until the crystals no longer clump together.

Glauber's salt, the decahydrate, was historically used as a laxative. It is effective for the removal of certain drugs such as acetaminophen from the body, for example, after an overdose.

Other uses for sodium sulphate include de-frosting windows, in carpet fresheners, starch manufacture, and as an additive to cattle feed.

Lately, sodium sulphate has been found effective in dissolving very finely electroplated micrometre gold that is found in gold electroplated hardware on electronic products such as pins, and other connectors and switches. It is safer and cheaper than other reagents used for gold recovery, with little concern for adverse reactions or health effects.

Safety

Although sodium sulphate is generally regarded as non-toxic, it should be handled with care. The dust can cause temporary asthma or eye irritation; this risk can be prevented by using eye protection and a paper mask.

SODIUM BISULPHATE

It is also known as sodium sulphate and niter cake. It is in the form of colourless crystals or white, fused lumps; aqueous solution is strongly acid. Soluble in water, sp. gr. 2.435 (13°C); m.p. > 315°C, sp. gr. 2.103 (13°C); m.p. 58.5°C, noncombustible. It is a by-product in the manufacture of hydrochloric and nitric acids. It can be purified by recrystallisation. It is strong irritant to tissue.

SODIUM BISULPHITE

Sodium hydrogen sulphite or sodium bisulphite is a chemical compound with the chemical formula $NaHSO_3$. Sodium bisulphite is a food additive with E number E222. Sodium bisulphite can be prepared by bubbling sulphur dioxide in a solution of sodium carbonate in water. Sodium bisulphite in contact with chlorine bleach (aqueous solution of sodium hypochlorite) will release harmful fumes.

Uses

In organic chemistry sodium bisulphite has several uses. It forms a bisulphite adduct with aldehyde groups and with certain cyclic ketones to a sulphonic acid. This reaction has limited synthetic value(s) but it is used in purification procedures. Contaminated aldehydes in a solution precipitate as the bisulphate adduct which can be isolated by filtration. The reverse reaction takes place in presence of a base such as sodium bicarbonate or sodium hydroxide and the bisulphite is liberated as sulphur dioxide.

The other main use of sodium bisulphite is as a mild reducing agent in organic synthesis in particular in purification procedures. It can efficiently remove traces or excess amounts of chlorine, bromine, iodine, hypochlorite salts, osmate esters, chromium trioxide and potassium permanganate.

A third use of sodium bisulphite is as a decolouration agent in purification procedures obviously because it can reduce strongly colourised oxidising agents and conjugated alkenes and carbonyl compounds. Sodium bisulphite is also the key ingredient in the Bucherer reaction.

Sodium bisulphite is used in almost all commercial wines, to prevent oxidation and preserve flavour. In fruit canning, sodium bisulphite is used to prevent browning (caused by oxidation) and to kill microbes.

In the case of wine making, sodium bisulphite releases sulphur dioxide gas when added to water or products containing water. The sulphur dioxide kills yeasts, fungi, and bacteria in the grape juice before fermentation. When the sulphur dioxide levels have subsided (about 24 hrs), fresh yeast is added for fermentation.

It is later added to bottled wine to prevent oxidation (which makes vinegar), and to protect the colour of the wine from oxidation, which causes browning.

The sulphur dioxide displaces oxygen in the bottle and dissolved in the wine. Oxidised wine can turn orange or brown, and taste like raisins or cough syrup.

Sodium bisulphite is also added to leafy green vegetables in salad bars and elsewhere, to preserve apparent freshness, under names like LeafGreen. The concentration is sometimes high enough to cause serious allergic reactions.

SODIUM SULPHITE

Sodium sulphite is a soluble compound of sodium. It is a product of SO_2 scrubbing, a part of the flue gas desulphurisation process. It is also used as a preservative to prevent dried fruit from discolouring, and for preserving meats, and is used in the same way as sodium thiosulphate to convert elemental halides to their respective acids, in photography and for reducing chlorine levels in pools. It is in the form of white crystals or powder; saline, sulphurous taste, soluble in water; sparingly soluble in alcohol; sp. gr. 2.633.

It is prepared when sulphur dioxide is reacted with soda ash and water, and a solution of the resulting sodium bisulphite thus produced is treated with additional soda ash. It is obtained as a by-product of the caustic fusion process for phenol. Its use is prohibited in meats and other sources of vitamin B_1.

Uses

Sodium sulphite is primarily used in the pulp and paper industry. It is used in water treatment as an oxygen scavenger agent, in the photographic industry to protect developer solutions from oxidation and (as hypo clear solution) to wash fixer (sodium thiosulphate) from film and photo-paper emulsions, in the textile industry as a bleaching, desulphurising and dechlorinating agent and in the leather trade for the sulphitisation of tanning extracts. It is used in the purification of TNT for military use. It is used in chemical manufacturing as a sulphonation and sulphomethylation agent. It is used in the production of sodium thiosulphate. It is used in other applications, including ore flotation, oil recovery, food preservatives, making dyes, and detergent. It forms a bisulphite adduct with aldehydes, and with ketones forms a sulphonic acid. It is used to purify or isolate aldehydes and ketones.

Sodium sulphite is decomposed by even weak acids, giving up sulphur dioxide gas.

$$Na_2SO_3 + 2H^+ \rightarrow 2Na^+ + H_2O + SO_2$$

A saturated aqueous solution has pH of 9. Solutions exposed to air are eventually oxidised to sodium sulphate. If sodium sulphite is allowed to crystallise from aqueous solution at room temperature or below, it does so as a heptahydrate. The heptahydrate crystals effloresce in warm dry air. Heptahydrate crystals also oxidise in air to form the sulphate. The anhydrous form is much more stable against oxidation by air. Parenteral injection containing sodium sulphite may cause life-threatening allergic reactions.

SODIUM THIOSULPHATE

Sodium thiosulphate ($Na_2S_2O_3$), is a colourless crystalline compound that is more familiar as the pentahydrate, $Na_2S_2O_3 \cdot 5H_2O$, an efflorescent, monoclinic crystalline substance also called sodium hyposulphite or 'hypo'. It has specific gravity 1.729 (17°C), m.p. 48°C. It is soluble in water and oil of turpentine, insoluble in alcohol.

The thiosulphate anion is tetrahedral in shape and is notionally derived by replacing one of the oxygen atoms by a sulphur atom in a sulphate anion. The S-S distance indicates a single bond, implying that the sulphur bears significant negative charge and the S-O interactions have more double bond character. The first protonation of thiosulphate occurs at sulphur.

On an industrial scale, sodium thiosulphate is produced chiefly from liquid waste products of sodium sulphide or sulphur dye manufacture. In the laboratory, this salt can be prepared by heating an aqueous solution of sodium sulphite with sulphur. Thiosulphate anion characteristically reacts with dilute acids to produce sulphur, sulphur dioxide and water:

$$Na_2S_2O_3 + 2HCl \rightarrow 2NaCl + S + SO_2 + H_2O$$

This reaction has been employed to generate colloidal sulphur.

Applications

Medical

It is used as an antidote to cyanide poisoning. Thiosulphate acts as a sulphur donor for the conversion for cyanide to thiocyanate (which can then be safely excreted in the urine), catalysed by the enzyme rhodanase. It has also been used as treatment of calciphylaxis in hemodialysis patients with end-stage renal disease.

Other uses

Sodium thiosulphate is also used:

1. As a component in hand warmers and other chemical heating pads that produce heat by exothermic crystallisation of a supercooled solution.

2. In pH testing of bleach substances. The universal indicator and any other liquid pH indicator are destroyed by bleach, rendering them useless for testing the pH. If one first adds sodium thiosulphate to such solutions, it will neutralise the colour-removing effects of bleach and allow one to test the pH of bleach solutions with liquid indicators.

The relevant reaction is akin to the iodine reaction: thiosulphate reduces the hypochlorite (active ingredient in bleach) and in so doing becomes oxidised to sulphate. The complete reaction is:

$$4NaClO + Na_2S_2O_3 + 2NaOH \rightarrow 4NaCl + 2Na_2SO_4 + H_2O$$

3. To lower chlorine levels in swimming pools and to remove iodine stains, e.g. after the explosion of nitrogen triiodide.
4. In bacteriological water assessment.
5. In the tanning of leather.
6. Often used in pharmaceutical preparations as an anionic surfactant to aid in dispersion.

SODIUM NITRITE

Sodium nitrite, with chemical formula $NaNO_2$, is used as a colour fixative and preservative in meats and fish. When pure, it is a white to slight yellowish crystalline powder. It is very much soluble in water and is hygroscopic. It is also slowly oxidised by oxygen in the air to sodium nitrate, $NaNO_3$. The compound is a strong oxidising agent. It has specific gravity 2.157, m.p. 271°C, explodes at 1000°F (537°C), decomposes at 320°C.

It is also used in manufacturing diazo dyes, nitroso compounds, and other organic compounds; in dyeing and printing textile fabrics and bleaching fibres; in photography; as a laboratory reagent and a corrosion inhibitor; in metal coatings for phosphatising and detinning; and in the manufacture of rubber chemicals.

Sodium nitrite also has been used in human and veterinary medicine as a vasodilator, a bronchodilator, and an antidote for cyanide poisoning.

Uses

Food additive

As a food additive, it serves a dual purpose in the food industry since it both alters the colour of preserved fish and meats and also prevents growth of *Clostridium botulinum*, the bacteria which causes botulism.

Disease treatment

Recently, sodium nitrite has been found to be an effective means to increase blood flow by dilating blood vessels, acting as a vasodilator. Research is ongoing to investigate its applicability towards treatments for sickle cell anemia, cyanide poisoning, heart attacks, brain aneurysms, and pulmonary hypertension in infants.

Synthetic reagent

Sodium nitrite is used to convert amines into diazo compounds. The synthetic utility of such a reaction is to render the amino group labile for nucleophilic substitution, as the N_2 group is a better leaving group.

SODIUM SILICATE

Sodium silicate is the common name for a compound sodium metasilicate, Na_2SiO_3, also known as water glass or liquid glass.

It is available in aqueous solution and in solid form and is used in cements, passive fire protection, refractories, textile and lumber processing, and art. Sodium carbonate and silicon dioxide react when molten to form sodium silicate and carbon dioxide:

$$Na_2CO_3 + SiO_2 \rightarrow Na_2SiO_3 + CO_2$$

Anhydrous sodium silicate contains a chain polymeric anion composed of corner shared (SiO_4) tetrahedra, and not a discrete SiO_3^{2-} ion. In addition to the anhydrous form there are a number of hydrates with the formulae $Na_2SiO_3 \cdot nH_2O$ (where, $n = 5, 6, 8, 9$) which contain the discrete approximately tetrahedral anion $SiO_2(OH)_2^{2-}$ with water of hydration e.g. the commercially available sodium silicate pentahydrate, $Na_2SiO_3 \cdot 5H_2O$ is formulated $Na_2SiO_2(OH)_2 \cdot 4H_2O$ and the nonahydrate, $Na_2SiO_3 \cdot 9H_2O$ is formulated $Na_2SiO_2(OH)_2 \cdot 8H_2O$.

Sodium silicate is a white solid that is readiliy soluble in water, producing an alkaline solution. It is one of a number of related compounds which include sodium orthosilicate, Na_4SiO_4; sodium pyrosilicate, $Na_6Si_2O_7$, and others.

All are glassy, colourless and dissolve in water. Sodium silicate is stable in neutral and alkaline solutions. In acidic solutions, the silicate ion reacts with hydrogen ions to form silicic acid, which when heated and roasted forms silica gel, a hard, glassy substance.

Uses

Metal repair

Sodium silicate is used, along with magnesium silicate, in muffler repair and fitting paste. When dissolved in water, both sodium silicate, and magnesium silicate form a thick paste that is easy to apply. When the exhaust system of an internal combustion engine heats up to its operating temperature, the heat drives out all of the excess water from the paste. The silicate compounds that are left over have glass-like properties, making a somewhat permanent, brittle repair.

Automotive repair

Sodium silicate can be used to seal leaks at the head gasket. A common use is when an alloy cylinder head motor is left sitting for extended periods or the coolant is not changed at proper intervals, electrolysis can eat out sections of the head causing the gasket to fail.

Aquaculture

Sodium silicate gel is also used for creating algae in aquaculture hatcheries.

Cement uses

Sodium silicate has been widely used as a general purpose cement, but especially for applications involving cementing objects exposed to heat or fire.

Food preservation

Sodium silicate was also used as an egg preservation agent in the early 20th century with large success. When fresh eggs are immersed in it, bacteria which cause the eggs to spoil are kept out and water is kept in. Eggs can be kept fresh using this method for up to nine months. When boiling eggs preserved this way, it is well advised to pin-prick the egg to allow steam to escape because the shell is no longer porous.

Concrete and general Masonry treatment

Concrete treated with a sodium silicate solution helps to significantly reduce porosity in most masonry products such as concrete, stucco, plasters. A chemical reaction occurs with the excess $Ca(OH)_2$ in the concrete that permanently binds the silicates with the surface making them far more wearable and water repellent.

Passive fire protection (PFP)

Sodium silicates are inherently intumescent. They come in prill (solid beads) form, as well as the liquid, water glass. The solid sheet form (Palusol) must be waterproofed to ensure longterm passive fire protection.

Refractory use

Water glass is a useful binder of solids, such as vermiculite and perlite. When blended with the aforementioned lightweight aggregates, water glass can be used to make hard, high-temperature insulation boards used for refractories, passive fire protection and high temperature insulations, such as moulded pipe insulation applications.

Water treatment

Water glass is used as a water treatment in waste-water treatment plants. Water glass will bind to heavier molecules and drag them out of the water.

SODIUM PEROXIDE

Sodium peroxide, Na_2O_2, is the normal product when sodium is burned. It is a strong oxidiser. It has specific gravity 2.805, m.p. 460°C, b.p. 657°C. It is soluble in cold water.

Chemical Properties

Sodium peroxide is hydrolysed by water to form sodium hydroxide plus hydrogen peroxide according to the reaction:

$$Na_2O_2 + 2H_2O \rightarrow 2NaOH + H_2O_2$$

The hydrogen peroxide thus formed decomposes rapidly in the ensuing basic solution, producing water and oxygen. The reaction is substantially exothermic and can set fire to combustible materials.

Sodium peroxide will also set fire to many organic liquids on contact (particularly alcohols and glycols), and reacts violently with powdered metals and numerous other compounds after minimal initiation.

Preparation

Sodium peroxide can be synthesised by direct reaction with sodium and oxygen at 130°–200°C. Lower temperature (0°–20°C) synthesis can be achieved by passing O_2 over a dilute (0.1–5.0 mole per cent) sodium amalgam, thus oxidising the sodium. It may also be produced by passing ozone gas over solid sodium iodide inside a platinum or palladium tube. The ozone oxidises the sodium to form sodium peroxide. The iodine is freed into iodine crystals, which can be sublimed by mild heating. The platinum or palladium catalyses the reaction and is not attacked by the sodium peroxide.

Uses

Given its strong oxidation properties, sodium peroxide is used to bleach wood pulp for the production of paper. It has also been used for the extraction of minerals from various ores. Sodium peroxide may go by the commercial names of *Solozone* and *Flocool.* In chemistry preparations, sodium peroxide is used as an oxidising agent. It is also used as an oxygen source by reacting it with carbon dioxide to produce oxygen and sodium carbonate; it is thus particularly useful in scuba gear, submarines, etc.

SODIUM PERBORATE

Sodium perborate (SPB) is a white, odourless, water soluble chemical compound with chemical formula $NaBO_3$. It crystallises as the monohydrate, $NaBO_3 \cdot H_2O$, tetrahydrate, $NaBO_3 \cdot 4H_2O$ and trihydrate, $NaBO_3 \cdot 3H_2O$. The monohydrate and tetrahydrate are the commercially important forms. It melts at 63°C, loses H_2O at 130°–150°C, pH of aqueous solution 10–10.3.

Sodium perborate is manufactured by reaction of disodium tetraborate pentahydrate, hydrogen peroxide, and sodium hydroxide.

The monohydrate form dissolves better than the tetrahydrate and has higher heat stability; it is prepared by heating the tetrahydrate. Sodium perborate undergoes hydrolysis in contact with water, producing hydrogen peroxide and borate.

Uses

It serves as a source of active oxygen in many detergents, laundry detergents, cleaning products, and laundry bleaches. It is also present in some tooth bleaching formulas. It is used as a bleaching agent for internal bleaching of a tooth that has had root canal treatment. The sodium perborate is placed inside the tooth and left in place for an extended period of time to allow it to diffuse into the tooth and bleach stains from the inside out. It has antiseptic properties and can act as a disinfectant. It is also used as a disappearing preservative in some brands of eye drops. Sodium perborate is a less aggressive bleach than sodium hypochlorite, causing less degradation to dyes and textiles. Borates also have some non-oxidative bleaching properties. Sodium perborate releases oxygen rapidly at temperatures over 60°C. To make it active at lower temperatures (40°–60°C), it has to be mixed with a suitable activator, typically tetraacetylethylenediamine (TAED). It is a skin irritant.

SODIUM AMIDE

Sodium amide, commonly called sodamide, is the chemical compound with the formula $NaNH_2$. This solid, which is dangerously reactive toward water, is white when pure, but commercial samples are typically gray due to the presence of small quantities of metallic iron from the manufacturing process. Such impurities do not usually affect the utility of the reagent. $NaNH_2$ has been widely employed as a strong base in organic synthesis. It decomposes in water and hot alcohol, m.p. 210°C, b.p. 400°C.

Sodium amide can be prepared by the reaction of sodium with ammonia gas, but it is usually prepared by the reaction in liquid ammonia using iron(III) nitrate as a catalyst. The reaction is fastest at the boiling point of the ammonia, ca. –33°C.

$$2Na + 2NH_3 \rightarrow 2NaNH_2 + H_2$$

$NaNH_2$ is a salt-like material and as such, crystallises as an infinite polymer. The geometry about sodium is tetrahedral. In ammonia, $NaNH_2$ forms conductive solutions, consistent with the presence of $Na(NH_3)_6^+$ and NH_2^- anions.

Uses

Sodium amide is used in the industrial production of indigo, hydrazine, and sodium cyanide. It is the reagent of choice for the drying of ammonia (liquid or gaseous) and is also widely used as a strong base in organic chemistry, often in liquid ammonia solution. One of the main advantages to the use of sodamide is that it is an excellent base and rarely serves as a nucleophile. It is, however, poorly soluble and its use has been superseded by the related reagents such as sodium hydride, sodium bis(trimethylsilyl)amide (NaHMDS), and lithium diisopropylamide (LDA).

Safety

Sodium amide reacts violently with water to produce ammonia and sodium hydroxide and will burn in air to give oxides of sodium and nitrogen.

$$NaNH_2 + H_2O \rightarrow NH_3 + NaOH$$

$$2NaNH_2 + 4O_2 \rightarrow Na_2O_2 + 2NO_2 + 2H_2O$$

In the presence of limited quantities of air and moisture, such as in a poorly closed container, explosive mixtures of oxidation products can form. This is accompanied by a yellowing or browning of the solid. As such, sodium amide should always be stored in a tightly closed container, if possible under an atmosphere of nitrogen gas. Sodium amide samples which are yellow or brown in colour should be destroyed immediately: one method for destruction is the careful addition of ethanol to a suspension of sodium amide in a hydrocarbon solvent. Sodium amide may be expected to be corrosive to the skin, eyes and mucous membranes. Care should be taken to avoid dispersal of the dust.

SODIUM CYANIDE

Sodium cyanide is the inorganic compound with the formula NaCN. This highly toxic colourless salt, is used mainly in gold mining but has other niche applications. It is the conjugate base of the weak acid hydrogen cyanide. It melts at 563°C, b.p. 149°C.

Sodium cyanide is produced by treating hydrogen cyanide with sodium hydroxide:

$$HCN + NaOH \rightarrow NaCN + H_2O$$

It is prepared by the Castner-Kellner process involving the reaction of sodium amide with carbon at elevated temperatures.

$$NaNH_2 + C \rightarrow NaCN + H_2$$

The structure of solid NaCN is related to that of sodium chloride. The anions and cations are each six-coordinate (KCN has a similar structure). Each Na^+ forms pi-bonds to two CN^- groups as well as two bent K—CN and two bent K—NC links.

Because the salt is derived from a weak acid, NaCN readily reverts back to HCN by hydrolysis: the moist solid emits small amounts of hydrogen cyanide, which smells like bitter almonds (not everyone can smell it—the ability thereof is due to a genetic trait). Sodium cyanide reacts rapidly with strong acids to release hydrogen cyanide. This dangerous process represents a significant risk associated with cyanide salts. It is detoxified most efficiently with hydrogen peroxide:

$$NaCN + H_2O_2 \rightarrow NaOCN + H_2O$$

Applications

Cyanide mining

Sodium cyanide is mainly used to extract gold and other precious metals in mining. This application exploits the high affinity of gold(I) for cyanide, which induces gold metal to oxidise and dissolve in the presence of air and water.

$$4Au + 8NaCN + O_2 + 2H_2O \rightarrow 4Na[Au(CN)_2] + 4NaOH$$

Few alternative methods exist for this extraction process.

Chemical feedstock

Several commercially significant chemical compounds are derived from cyanide, including cyanuric chloride, cyanogen chloride, and many nitriles. In organic synthesis, cyanide, which is classified as a strong nucleophile, is used to prepare nitriles, which occur widely in many specialty chemicals, including pharmaceuticals.

Niche uses

Being highly toxic, sodium cyanide is used to kill or stun rapidly such as in illegal cyanide fishing and in collecting jars used by entomologists.

Toxicity

Cyanide salts are among the most rapidly acting of all known poisons. Cyanide is a potent inhibitor of respiration, acting on mitochondrial cytochrome oxidase and hence blocking electron transport.

This results in decreased oxidative metabolism and oxygen utilisation. Lactic acidosis then occurs as a consequence of anaerobic metabolism.

SODIUM FERROCYANIDE

Sodium ferrocyanide, also known as tetrasodium hexacyanoferrate or sodium hexacyanoferrate (II), is a coordination compound of formula $Na_4Fe(CN)_6$ which forms semi-transparent yellow crystals at room temperature, and which decomposes at its boiling point. It is soluble in water and insoluble in alcohol, and the solution can react with acid or photodecompose to release hydrogen cyanide gas. In its hydrous form, $Na_4Fe(CN)_6 \cdot 10H_2O$ (sodium ferrocyanide decahydrate), it is generally known as yellow prussiate of soda. It has specific gravity 1.458.

Uses

As yellow prussiate of soda, it is added to road and food grade salt as an anticaking agent. When combined with iron, it converts to a deep blue pigment which is the main component of Prussian blue. In photography it is used for bleaching, toning and fixing. It is used as a stabiliser for the coating on welding rods. In the petroleum industry it is used for removal of mercaptans.

Chapter 12

Chlor-alkali Industries: Soda Ash, Caustic Soda and Chlorine

INTRODUCTION

The global chlor-alkali industry is a very important economic contributor to the global economy. The chlor-alkali industry comprises the following products: caustic soda (also known as sodium hydroxide), chlorine and soda ash (sodium carbonate). The industry is highly energy intensive, with electricity and other utilities accounting for 40–50 per cent of production costs.

The global chlor-alkali industry has generally been facing maturing demand, a problem made even more challenging by any weakness in the economy. The distinguishing feature of the industry, as compared to other segments of the chemical industry, is the unique coproduct relationship between chlorine and caustic soda.

Chlorine markets follow the economy closely. Some of the most significant end-products for chlorine, such as PVC, used in the housing and automotive industries, and ethylene dichloride (EDC)—the raw material for PVC—tend to see weak demand in an economic downturn. As a result, chlorine prices come under additional pressure from falling PVC and EDC prices, impacting the margins of chlor-alkali producers.

Caustic soda, on the other hand, does not respond as readily to economic changes because of the diverse nature of its markets, such as pulp and paper and chemical processing. Another advantage for caustic soda is that it can be more easily stored, which helps flatten out variable demand.

The imbalance between supply and demand for chlorine and caustic soda, generally leads to production cut backs and increased prices for either chlorine or caustic soda.

MANUFACTURING PROCESS

The chlor-alkali industry is one of the largest electrochemical technologies in the world. It is the second largest consumer of electricity (2400 bn kWh) among electrolytic industries.

289

Chlorine is produced electrolytically using three types of electrolytic cells. In 2003, about 63 per cent of the total world chlorine capacity of about 43.4 MT was produced electrolytically using diaphragm and membrane cells, while about 35 per cent was made using mercury cells (Fig. 12.1).

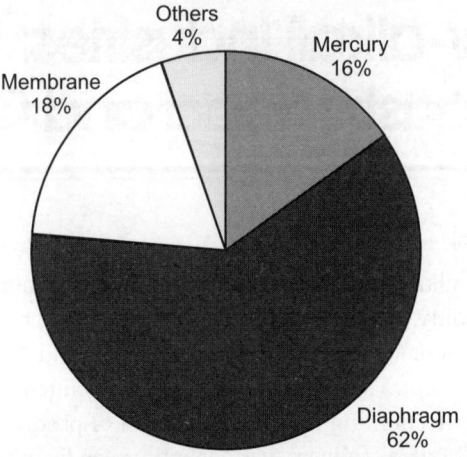

Fig. 12.1. Chlorine manufacturing technologies.

Chlorine is produced by the electrolysis of sodium chloride solution, called 'brine'. When sodium chloride is dissolved in water, it dissociates into sodium cations and chloride anions. The chloride ions are oxidised at the anode to form chlorine gas and water molecules are reduced at the cathode to form hydroxyl anions and hydrogen gas. The sodium ions in the solution and the hydroxyl ions produced at the cathode constitute the components of sodium hydroxide.

The main difference in three chlorine process technologies lies in the manner by which the chlorine gas and the sodium hydroxide are prevented from mixing with each other to ensure generation of pure products. Thus, in diaphragm cells, brine from the anode compartment flows through the separator to the cathode compartment, the separator material being either asbestos or polymer-modified asbestos composite deposited on a perforated cathode.

In membrane cells, on the other hand, an ion-exchange membrane is used as a separator. Anolyte-catholyte separation is achieved in the diaphragm and membrane cells using separators and ion-exchange membranes, respectively, whereas mercury cells contain no diaphragm or membrane and the mercury itself acts as a separator. The anode in all technologies is titanium metal coated with an electrocatalytic layer of mixed oxides.

All modern cells use these so-called 'dimensionally stable anodes' (DSA). Earlier cells used carbon based anodes. The cathode is typically in steel in diaphragm cells, nickel in membrane cells, and mercury in mercury cells.

These cell technologies are schematically depicted in Figs 12.2 to 12.4 are described below.

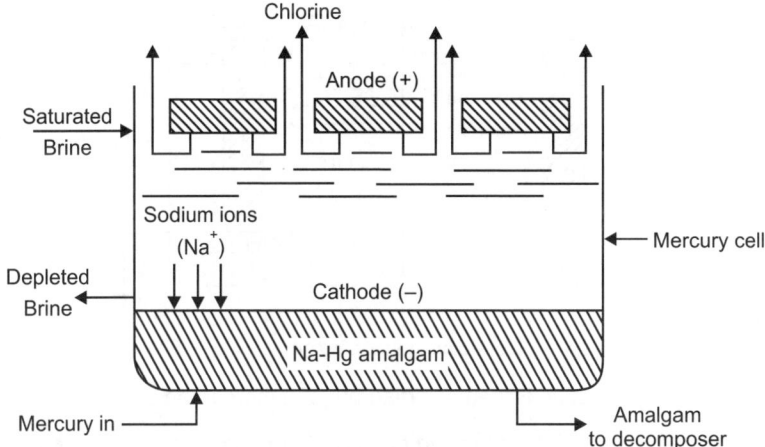

Fig. 12.2. Schematic diagram of a mercury cell.

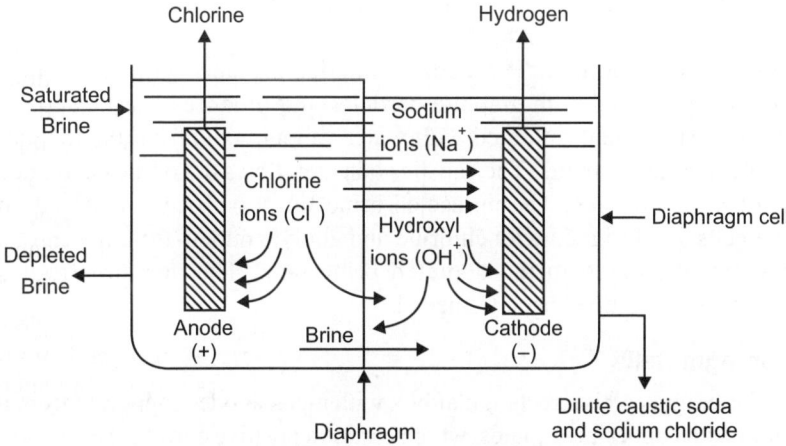

Fig. 12.3. Schematic diagram of a diaphragm cell.

Mercury Cells

The mercury cell has steel bottoms with rubber-coated steel sides, as well as end boxes for brine and mercury feed and exit streams with a flexible rubber or rubber-coated steel cover. Adjustable metal anodes hang from the top, and mercury (which forms the cathode of the cell) flows on the inclined bottom. The current flows from the steel bottom to the flowing mercury.

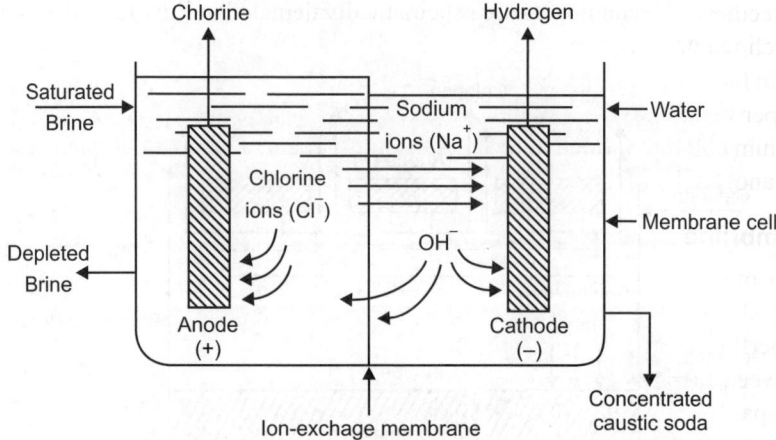

Fig. 12.4. Schematic diagram of a membrane cell.

Saturated brine fed from the end box is electrolysed at the anode to produce the chlorine gas, which flows from the top portion of the trough and then exits. The sodium ion generated reacts with the mercury to form sodium amalgam (an alloy of mercury and sodium), which flows out of the end box to a vertical cylindrical tank. About 0.25 per cent to 0.5 per cent sodium amalgam is produced in the cell. The sodium amalgam reacts with water in the decomposer, packed with graphite particles and produces caustic soda and hydrogen. Hydrogen, saturated with water vapour, exits from the top along with the mercury vapours. The caustic soda then flows out of the decomposer as 50 per cent caustic. The unreacted brine flows out of the exit end box. Some cells are designed with chlorine and anolyte outlets from the end box, which are separated in the depleted brine tank. The mercury from the decomposer is pumped back to the cell.

Diaphragm Cells

The diaphragm cell is a rectangular box with metal anodes supported from the bottom with copper-base plates, which carries a positive current. The cathodes are metal screens or punch plates connected from one end to the other end of the rectangular tank. Asbestos, dispersed as a slurry in a bath, is vacuum deposited onto the cathodes, forming a diaphragm. Saturated brine enters the anode compartment and the chlorine gas liberated at the anode during electrolysis, exits from the anode compartment. It is saturated with water vapour at a partial pressure of water over the anolyte. The sodium ions are transported from the anode compartment to the cathode compartment, by the flow of the solution and by electromigration, where they combine with the hydroxyl ions generated at the cathode during the formation of the hydrogen from the water molecules. The diaphragm resists the back migration of the hydroxyl ions, which would otherwise react with the chlorine in the anode compartment.

In the cathode compartment, the concentration of the sodium hydroxide is 12 per cent, and the salt concentration is ~14 per cent. There is also some sodium chlorate formed in the anode compartment, dependent upon the pH of the anolyte.

Membrane Cells

In a membrane cell, an ion-exchange membrane separates the anode and cathode compartments. The separator is generally a bilayer membrane made of perfluorocarboxylic and perfluorosulphonic acid-based films, sandwiched between the anode and the cathode. The saturated brine is fed to the anode compartment where chlorine is liberated at the anode, and the sodium ion migrates to the cathode compartment.

Unlike in the diaphragm cells, only the sodium ions and some water migrate through the membrane. The unreacted sodium chloride and other inert ions remain in the anolyte. About 30–32 per cent caustic soda is fed to the cathode compartment, where sodium ions react with hydroxyl ions produced during the course of the hydrogen gas evolution from the water molecules. This forms caustic, which increases the concentration of caustic solution to 35 per cent. The hydrogen gas, saturated with water, exits from the catholyte compartment.

Only part of the caustic soda product is withdrawn from the cathode compartment. The remaining caustic is diluted to 32 per cent and returned to the cathode compartment.

Thus, all three basic cell technologies generate chlorine at the anode, and hydrogen along with caustic soda in the cathode compartment. The distinguishing difference between the technologies lies in the manner by which the anolyte and the catholyte streams are prevented from mixing with each other. Separation is achieved in a diaphragm cell by a separator, and in a membrane cell by an ion-exchange membrane. In mercury cells, the cathode itself acts as a separator by forming an alloy of sodium and mercury (sodium amalgam) which is subsequently reacted with water to form sodium hydroxide and hydrogen in a separate reactor.

Chlorine Processing

The chlorine gas from the anode compartment contains moisture, by-product oxygen, and some back-migrated hydrogen. In addition, if the brine is alkaline, it will contain carbon dioxide and some oxygen and nitrogen from the air leakage via the process or pipelines. Chlorine is first cooled to 60°F (16°C) and passed through demisters to remove the water droplets and the particulates of salt and sodium sulphate. The cooled gas goes to sulphuric acid circulating towers, which are operated in series. Commonly, three towers are used for the removal of moisture. The dried chlorine then goes through demisters before it is compressed and liquefied at low temperatures. The non-condensed gas, called snift gas, is used for producing hypochlorite or hydrochloric acid. If there is no market for hydrochloric acid, the snift gas is neutralised with caustic soda or lime (calcium hydroxide) to form hypochlorite. The hypochlorite is either sold as bleach or decomposed to form salt and oxygen.

Health Effects

Chlorine is a toxic gas that irritates the respiratory system. Because it is heavier than air, it tends to accumulate at the bottom of poorly ventilated spaces. Chlorine gas is a strong oxidiser, which may react with flammable materials. Chlorine is detectable in concentrations of as low as 1 ppm. Coughing and vomiting may occur at 30 ppm and lung damage at 60 ppm. About 1000 ppm can be fatal after a few deep breaths of the gas. Breathing lower concentrations can aggravate the respiratory system, and exposure to the gas can irritate the eyes. Chlorine's toxicity comes from its oxidising power. When chlorine is inhaled at concentrations above 30-ppm it begins to react with water and cells which change it into hydrochloric acid (HCl) and hypochlorous acid (HClO).

Caustic Soda Processing

Caustic soda is marketed as 50 per cent, 73 per cent or anhydrous (dry) beads or flakes. The mercury cell can produce 50 per cent and 73 per cent caustic directly. The caustic from the decomposer is cooled and passed once or twice through an activated carbon filter to reduce the mercury levels in the caustic. After filtration, the mercury concentration is lowered to the parts-per-million (ppm) levels. Even these low levels of mercury may be unacceptable to some customers, who then have to switch to using membrane grade caustic soda. The mercury cell caustic soda has a few ppm salt and <5 ppm sodium chlorate. The mercury cell caustic is the highest purity caustic that can be made electrolytically if trace concentrations of mercury are tolerable in the end-use of caustic.

The membrane cell caustic is concentrated in a multiple effect falling film evaporator, which increases the caustic soda concentration to 50 per cent with a high steam economy. Caustic soda from membrane cells generally has 30 ppm sodium chloride and 5–10 ppm sodium chlorate.

The catholyte from the diaphragm cells contains 12 per cent sodium hydroxide, 14 per cent sodium chloride, 0.25–0.3 per cent sodium sulphate, and 100–500 ppm sodium chlorate. The catholyte is evaporated in a multi-effect evaporator. Most of the salt from the catholyte will precipitate during the concentration of the caustic soda to 50 per cent sodium hydroxide. The

50 per cent caustic soda product will contain about 1 per cent sodium chloride. The 50 per cent caustic also has a high chlorate concentration (0.1 per cent) compared to the caustic from membrane or mercury cells (~10 ppm). The salt, separated from the caustic during evaporation, is used to resaturate the brine fed to the cell. An additional single-effect evaporator is needed to produce 73 per cent caustic soda. Anhydrous (dry) caustic soda is produced in a rising film evaporator, operating at 725°F (385°C) and at a few inches of water vacuum.

Applications of Chlorine

The major application of chlorine is in the manufacture of EDC, which in turn is used to make vinyl chloride and subsequently PVC (Fig. 12.5). PVC is a very versatile thermoplastic, used in a wide variety of daily products.

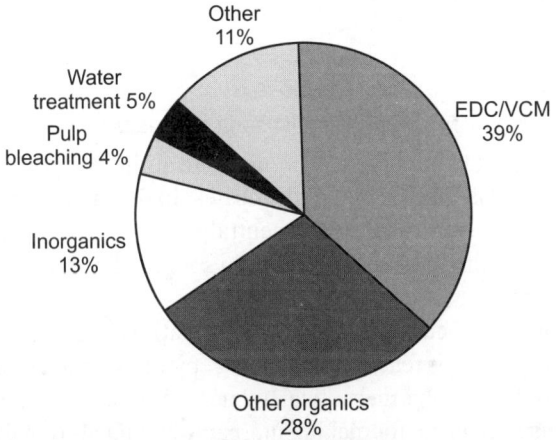

Fig. 12.5. Chlorine end-uses.

Chlorine is used in pulp and paper manufacturing operations for bleaching to produce a high quality whitened material and in water treatment operations as a disinfectant.

The major use of chlorine in the production of inorganic chemicals is for titanium dioxide (a widely used pigment), manufactured from naturally occurring ores (ilmenite or rutile).

Applications of Caustic Soda

The end-uses of caustic soda are diverse compared to the uses of chlorine (Fig. 12.6). More than half of caustic soda produced is used in the chemical industry and the rest used in the production of consumer goods such as soaps, detergents and textiles as well as in water treatment, aluminium production and oil refining.

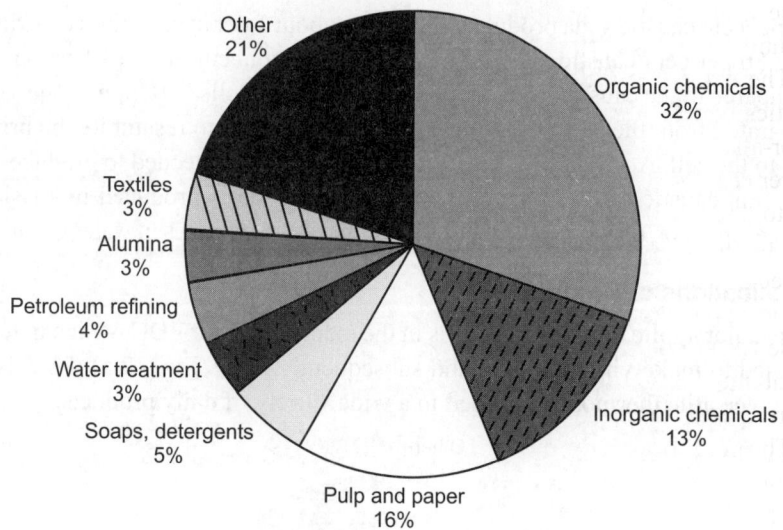

Fig. 12.6. Caustic soda end-uses.

Its primary applications are in the neutralisation reactions and forming anionic species such as aluminates and zincates. In the manufacture of organic chemicals, caustic is employed for the neutralisation of acids, pH control, off-gas scrubbing, dehydrochlorination, and as a source of sodium during various chemical reactions.

For example, it is used in the dehydrochlorination stage of the epoxy resin production and hydrolysis reactions involving epichlorohydrin in the formation of glycerine, used in the pharmaceutical, tobacco and food/beverage industries.

The major use of caustic for making inorganic chemicals is in the production of hypochlorite for household and industrial bleaching purposes. Also, its use in the pulp and paper industry is in the production of sodium sulphide and sodium hydrosulphide for mechanical pulping. It is also used in the pH food processing applications, which include skin removal of potatoes, tomatoes, etc. for further processing.

Energy Efficiency

There are good reasons for this. Operation with membrane cells is more efficient in terms of energy consumed per ton of caustic soda produced than the mercury cell process. Chlor-alkali units that draw their power needs from the state electricity grids are saddled with power costs that are high and any improvements in efficiency of energy usage are welcome. Many units have now resorted to using captive power (which is almost always cheaper), in a bid to rein in operating costs and to ensure continuous supply, and a switchover to membrane cells will certainly add to the savings. All these switches also come with a small expansion of capacity, which implies that the power cost per unit of production is even lower.

The chlor-alkali industry is energy intensive, with electricity and other utilities typically accounting for 40–50 per cent of production costs. As the chlor-alkali (and other industries) have been harping for more than a decade, power costs in India are among the highest in the world. To some extent this has to do with hydrocarbon (oil, gas and coal) pricing, but to a larger extent due to the cross-subsidy mechanism that makes industrial consumers (especially large ones) pay more in order to serve seemingly social needs of providing deserving sections with low (and sometimes, free) power. Yet another cause for the escalation in power are the inefficiencies in transmission and distribution (T and D losses, in power industry jargon, but only half-jokingly referred to as theft and dacoity, by cynics).

The industry has coped with this state of affairs by setting up captive power stations, using whatever fuel it can lay its hands on. From the exorbitant diesel generating sets, to plants based on coal, fuel oil or natural gas (if the industry is lucky enough to get a linkage to a gas source), the chlor-alkali industry now operates a large portion of its capacity on captively generated power. The cost of the power generated varies, depending obviously on the fuel used, the size of the power plants etc. More often than not, however, this power is cheaper than grid electricity, and comes with a reliability the latter has never been able to ensure.

Opportunities for Chlorine Utilisation

Caustic soda and chlorine are basic inorganic chemicals produced together from salt and electricity by an electrochemical process that permanently links the fortunes of the two products together. With 1.1 ton of chlorine produced per ton of caustic soda, the key to successful operation of the industry lies in ensuring adequate offtake of the two materials in the ratio in which they are produced. This is easier said than done — more so in India. The end-use sectors the products serve and, importantly, their market dynamics, differ. The problem is compounded by the fact that unlike caustic soda, which is a liquid in its most widely traded form, chlorine is a toxic gas that does not lend itself to easy transportation, particularly over long distances by road.

The balanced utilisation of chlorine and caustic soda is a challenge for operators across the world, with one important difference from India. In most countries, be it developed ones in Europe and North America, or developing China, chlor-alkali plants are essentially built for the chlorine they produce. This is usually consumed in or around the plant to produce a variety of chemicals of which the most important are polyvinyl chloride (PVC), its monomer vinyl chloride (VCM) or its traded intermediate, ethylene dichloride

(EDC). In fact, more than a third of all chlorine produced ends up in this vinyl chain. Other significant chlorine consumers include producers of isocyanates (for polyurethane), titanium dioxide (used in coatings and plastics as pigment), propylene oxide (an intermediate for polyether polyols used in polyurethanes), epichlorohydrin (for epoxy resins), chloromethanes and chloroethanes (both of which find use as solvent and as intermediate for fluorinated products). Together, these account for a sizeable portion of all chlorine consumed.

SODA ASH

Soda ash (sodium carbonate) forms an important part of the chlor-alkali segment of the Indian chemical industry. It is a solid at normal temperature and pressure and is broadly classified into light soda ash (LSA) and dense soda ash (DSA) based on density. It is a high volume, low value product and finds applications in production of detergents, glass, chemicals, sodium silicate, pulp and paper and water treatment. Soda ash is moderately soluble in water and the solution remains strongly alkaline. Although it is low in toxicity, intake can be harmful to the human body. Product dust may produce irritation of eyes, nose, throat and lungs (Table 12.1).

Table 12.1. Properties of soda ash.

Commerical name	Soda ash	
Chemical name	Disodium carbonate (sodium carbonate)	
Chemical formula	Na_2CO_3	
State	Granular solid	
Melting point	851°C	
Odour	None	
Colour	White	
Types	Dense	Light
Granular size	0.1–0.2 mm	Up to 0.3 mm
Bulk density	0.96–1.04 g/cc	0.40–0.70 g/cc

Properties by Applications

Soda ash used in different industries requires different properties depending on its function in the manufacturing process. Table 12.2 shows the type of density and functions of soda ash in different applications.

Technology

There are three processes currently used to process soda ash:
1. Standard Solvay process.
2. Dual Solvay process.
3. Dry liming process.

Table 12.2. Importance of soda ash properties by application.

Applications	Type of soda ash	Important variable	Comment
Container glass	Mainly dense, rarely light	Assay	Poor soda ash quality results in cloudy containers and high rejects. Quality container manufacturing requires high purity soda ash. Processes such as high speed container formation or light weight need the additional clarity, viscosity and chemical control
Float glass	Dense only	Assay, particle size, level of chloride, colour	Float glass manufacturers require soda ash with: assays of 99%, proper particle size and distribution for viscosity control and low chloride levels to preserve furnace life
Detergents	Light only	Colour, chloride level	Speed of dissolving is important in the detergent industry, as well as colour and chloride levels. Manufacturers will add optical brighteners to increase whiteness of the final product so colour is a critical variable
Sodium tripoly-phosphate (STPP)	Mainly light, rarely dense	Chloride level, colour	When using soda ash as a source of alkali STPP manufacturers prefer low chloride content. High chloride soda ash may corrode the steel in scrubbers
Sodium silicate	Mainly light, rarely dense	Colour particle size	The colour of sodium silicate is important in many of its end-uses. One of the critical factors in determining colour is the purity of soda ash used in the manufacturing process. White colour and low impurity content is preferred for high quality sodium silicate manufacturing

Standard Solvay process

The Solvay process is often-called the ammonia process. Brine (saltwater) solution is reacted with ammonia to form ammoniated brine, which is then passed over carbonators to react with carbon dioxide to form sodium bicarbonate slurry.

This slurry is then passed to a centrifuge section where crystals of soda bicarbonate are collected and passed through steam tube drier to form soda ash. Mixing limestone and coke in vertical shaft kilns gives carbon dioxide.

The major drawbacks of this process are:

1. Low utilisation of raw salt and the requirement of good quality limestone and coke.
2. Large effluent facilities for the treatment of by-products.

Modified Solvay process

The main raw material in this process is common salt. This process does away with the use of limestone in the previous process. Ammonia and carbon dioxide are supplied from outside. In this process, the lime kiln and the ammonia recovery sections are replaced with ammonium chloride section where ammonium chloride is crystallised and recovered. The end-products are ammonium chloride and soda ash. This process has major advantages of better salt utilisation and low effluent generation.

These plants are located proximal to the fertiliser plants. The major drawbacks of this process are the availability of refined salts and the production of ammonium chloride, which finds very little usage today.

Dry liming process

The basic advantage is the perfect steam balance and reduction in the raw material inputs, resulting in substantial savings in energy. The consumption of steam and lime is very less as compared to other processes. The mother liquor (the remains after sodium bicarbonate have been removed) is fed to the prelimer. The lime from the vertical shaft kiln is crushed and passed to the prelimer unit directly instead of making slurry. The heat of hydration of the lime and the chemical reaction raise the liquor to boiling temperature.

Natural Soda Ash

Natural soda ash is manufactured from the natural deposits of trona, which is generally available near seashore. USA is the major source of trona. The trona is processed through crushing, calcining, dissolving and filtering to manufacture soda ash. Natural soda ash is produced with very low cost (almost half) as compared to the synthetic soda ash. Thus, more than half of the international trade originates from USA.

Conversion of light soda ash to dense soda ash

A densification plant is installed to increase the density of the soda ash, which ultimately converts LSA into DSA. Thus, DSA is manufactured from LSA only.

Applications

Soda ash finds its application in various industries. This makes the industry unique as the market dynamics of various industries are required to be understood in detail.

Soaps and detergents industry

This sector consumes the majority of soda ash production. Soda ash is used as a filler in the production of soaps and detergents, however, it can also be used as an active ingredient.

Glass industry

The raw material for manufacture of glass is soda ash, sand and cullet, dolomite, feldspar, power and fuel. Soda ash constitutes nearly 60–70 per cent of the total raw material costs and nearly 25 per cent of the weight of glass. It is used for bringing down the melting temperature of glass, which is important for reducing the manufacturing cost of glass. Quality of soda ash is very important for maintaining the quality of float glass.

Sheet or float glass accounts for 40 per cent of the soda ash used in the glass industry, followed by container glass (30 per cent) and cullet (30 per cent).

Chemical industry

The chemical sector uses soda ash as a prime input, especially for the manufacturing of sodium carbonate monohydrate, sodium aluminate, sodium bichromate and sodium tripolyphosphate. Dyestuffs like vinyl sulphone dyes and H acid also require soda ash during the manufacturing process.

Soda ash is also used to produce sodium silicates. Other major end-uses are silica-based catalysts, silica gels and laundry soaps. They are also used in the manufacture of adhesives, metal cleaning and pigments.

Water treatment

Soda ash finds application in the water treatment segment for adjusting the pH of water when it is in acidic condition. As soda ash is safer to handle than other alkalies and its price is more stable over time, companies are increasingly choosing soda ash for water treatment.

Pulp and paper

Soda ash has a variety of uses in paper industry ranging from pulp digestion to water treatment.

SODIUM BICARBONATE

Sodium bicarbonate or sodium hydrogen carbonate is the chemical compound with the formula $NaHCO_3$. Sodium bicarbonate is a white solid that is crystalline but often appears as a fine powder. It has a slight alkaline taste resembling that of washing soda (sodium carbonate). It is a component of the mineral natron and is found dissolved in many mineral springs. The natural mineral form is known as nahcolite. It is also produced artificially. Since it has long been known and is widely used, the salt has many related names such as baking soda, bread soda, cooking soda, bicarbonate of soda.

Manufacture

$NaHCO_3$ is mainly prepared by the Solvay process, which is the reaction of calcium carbonate, sodium chloride, ammonia, and carbon dioxide in water.

$NaHCO_3$ may be obtained by the reaction of carbon dioxide with an aqueous solution of sodium hydroxide. The initial reaction produces sodium carbonate:

$$CO_2 + 2NaOH \rightarrow Na_2CO_3 + H_2O$$

Further addition of carbon dioxide produces sodium bicarbonate, which at sufficiently high concentration will precipitate out of solution:

$$Na_2CO_3 + CO_2 + H_2O \rightarrow 2NaHCO_3$$

Commercial quantities of baking soda are also produced by a similar method: soda ash, mined in the form of the ore trona, is dissolved in water and treated with carbon dioxide. Sodium bicarbonate precipitates as a solid from this method:

$$Na_2CO_3 + CO_2 + H_2O \rightarrow 2NaHCO_3$$

Chemistry

Sodium bicarbonate is an amphoteric compound. Aqueous solutions are mildly alkaline:

$$HCO_3^- + H_2O \rightarrow H_2CO_3 + OH^-$$

Sodium bicarbonate can be used as a wash to remove any acidic impurities from a crude liquid, producing a purer sample. Reaction of sodium bicarbonate and an acid to give a salt and carbonic acid, which readily decomposes to carbon dioxide and water:

$$NaHCO_3 + HCl \rightarrow NaCl + H_2CO_3$$
$$H_2CO_3 \rightarrow H_2O + CO_2 \text{ (gas)}$$

Reaction of sodium bicarbonate and acetic acid:

$$NaHCO_3 + CH_3COOH \rightarrow CH_3COONa + H_2O + CO_2 \text{ (gas)}$$

Sodium bicarbonate reacts with bases:

$$NaHCO_3 + NaOH \rightarrow Na_2CO_3 + H_2O$$

Thermal decomposition

Above 70°C, it gradually decomposes into sodium carbonate, water and carbon dioxide. The conversion is fast at 250°C:

$$2NaHCO_3 \rightarrow Na_2CO_3 + H_2O + CO_2$$

Most bicarbonates undergo this dehydration reaction. Further heating converts the carbonate into the oxide (at around 1000°C):

$$Na_2CO_3 \rightarrow Na_2O + CO_2$$

These conversions are relevant to the use of $NaHCO_3$ as a fire-suppression agent (BC powder) in some dry powder fire extinguishers.

Applications

Cooking

Sodium bicarbonate is primarily used in cooking (baking) where it reacts with other components to release carbon dioxide, that helps dough rise. The acidic compounds that induce this reaction include phosphates, cream of tartar, lemon juice, yogurt, buttermilk, cocoa, vinegar, etc. Sodium bicarbonate can be substituted for baking powder provided sufficient acid reagent is also added to the recipe. Many forms of baking powder contain sodium bicarbonate combined with one or more acidic phosphates or cream of tartar.

Neutralisation of acids and bases

Many laboratories keep a bottle of sodium bicarbonate powder within easy reach, because sodium bicarbonate is amphoteric, reacting with acids and bases. Furthermore, as it is relatively innocuous in most situations, there is no harm in using excess sodium bicarbonate. A wide variety of applications follows from its neutralisation properties. Sodium bicarbonate can be added as a simple solution for raising the pH balance of water that has a high level of chlorine, such as in swimming pools and aquariums.

Medical uses

Sodium bicarbonate is used as an antacid to treat acid indigestion and heartburn. An aqueous solution is administered intravenously for cases of acidosis, or when there are insufficient sodium or bicarbonate ions in the blood. This compound has also been used for patients who have had a ureterosigmoidostomy. It is used as well for treatment of hyperkalemia and cardiorespiratory arrest.

Sodium bicarbonate may also be used as an anti-fungal for dandruff caused by fungus. Sodium bicarbonate is also being used as an alternative to shampoo; one of the many reasons for this is widespread awareness of the impact of plastic shampoo bottles on the environment. Sodium bicarbonate is also a key material used in hemodialysis, one of the two types of kidney dialysis.

As a cleaning agent

1. A paste from baking soda can be very effective when used in cleaning and scrubbing.
2. A solution in warm water will remove the tarnish from silver when the silver is in contact with a piece of aluminium foil.
3. It has been used for many years informally as a tooth whitening agent.
4. It is more effective than vinegar, salt, or hot water alone in removing pesticides from vegetables.

BLEACHING POWDER (CALCIUM HYPOCHLORITE)

Calcium hypochlorite is a chemical compound with formula $Ca(ClO)_2$. It is widely used for water treatment and as a bleaching agent (bleaching powder). This chemical is considered to be relatively stable and has greater available chlorine than sodium hypochlorite (liquid bleach). It has specific gravity 2.35, decomposes at 100°C. It decomposes in water and alcohol both. It is not hygroscopic.

Preparation

It is manufactured using the calcium process or the sodium process.
  Calcium process:
$$2Ca(OH)_2 + 2Cl_2 \rightarrow Ca(ClO)_2 + CaCl_2 + 2H_2O$$
Sodium process:
$$2Ca(OH)_2 + 3Cl_2 + 2NaOH \rightarrow Ca(ClO)_2 + CaCl_2 + 2H_2O + 2NaCl$$
  Bleaching powder is actually a mixture of calcium hypochlorite $Ca(ClO)_2$ and the basic chloride $CaCl_2$, $Ca(OH)_2$, H_2O with some slaked lime, $Ca(OH)_2$.

Properties

It is a yellow white solid which has a strong smell of chlorine. Calcium hypochlorite is not highly soluble in water. For that reason it should preferably be used in soft to middle hard water. There are two types of calcium hypochlorite — a dry form and a hydrated form. The hydrated form is safer to handle.

  Calcium hypochlorite reacts with carbon dioxide to form calcium carbonate and release chlorine:
$$2Ca(ClO)_2 + 2CO_2 \rightarrow 2CaCO_3 + 2Cl_2 + O_2$$
  Calcium hypochlorite reacts with hydrochloric acid to form calcium chloride:
$$Ca(ClO)_2 + 4HCl \rightarrow CaCl_2 + 2 H_2O + 2 Cl_2$$

Extreme care should be used in handling this product. Always keep in a cool dry place away from any organic material. When mixing it with water, it is safest to add the calcium hypochlorite to water. This material has been known to undergo self-heating and rapid decomposition accompanied by the release of toxic chlorine gas.

Uses

Calcium hypochlorite is used for the disinfection of drinking water or swimming pool water. For use in outdoor swimming pools, calcium hypochlorite can be used as a sanitiser in combination with a cyanuric acid stabiliser. The stabiliser will reduce the loss of chlorine because of UV radiation. Calcium does make the water hard and tends to clog up some filters. However, some types of calcium hypochlorite do contain anti-scaling agents in order to prevent clogging up of pipes/filters. This grade of calcium hypochlorite can also be used in hard waters. The main advantage of calcium hypochlorite is that it is unstabilised unlike chlorinated isocyanurates such as sodium dichloroiso-cyanurate or trichloroisocyanuric acid. Latter products do contain cyanuric acid. If the level of cyanuric acid becomes too high, it will influence the performance of the chlorine. Pools running on chlorinated isocyanurates should maintain a free chlorine level between 2 and 5 ppm (mg/l), whereas pools running on calcium hypochlorite should have a chlorine level of 1–2 ppm (mg/l). Calcium hypochlorite (known as bleaching powder) is also used for bleaching cotton and linen and is used in the manufacture of chloroform.

SODIUM HYPOCHLORITE

Sodium hypochlorite is a chemical compound with the formula $NaOCl$. Sodium hypochlorite solution, commonly known as bleach, is frequently used as a disinfectant or a bleaching agent. It is unstable in air unless mixed with sodium hydroxide. It has disagreeable sweetish odour and pale-greenish colour. It is soluble in cold water and decomposes in hot water, m.p. 18°C.

Sodium hypochlorite ($NaOCl$) and sodium chloride ($NaCl$) is formed when chlorine is passed into cold and dilute sodium hydroxide solution. It is prepared industrially by electrolysis minimal separation between the anode and the cathode. The solution must be kept below 40°C (by cooling coils) to prevent the undesired formation of sodium chlorate.

$$3Cl_2 + 6NaOH \rightarrow 5NaCl + NaClO_3 + 3H_2O$$

Sodium hydroxide and chlorine are commercially produced by the chlor-alkali process, and there is no need to isolate them to prepare sodium hypochlorite. Hence, chlorine is simultaneously reduced and oxidised. The commercial solutions always contain significant amounts of sodium chloride (common salt) as the main by-product, as seen in the equation above.

Uses

Bleaching

In household bleach form, sodium hypochlorite is used for removal of stains from laundry. It is particularly effective on cotton fibre, which stains easily but bleaches well.

Disinfection

A weak solution of 1 per cent household bleach in warm water is used to sanitise smooth surfaces prior to brewing of beer or wine. Surfaces must be rinsed to avoid imparting flavours to the brew; these chlorinated by-products of sanitising surfaces are also harmful.

Water treatment

For shock chlorination of wells or water systems, a 2 per cent solution of household bleach is used. The alkalinity of the sodium hypochlorite solution also causes the precipitation of minerals such as calcium carbonate, so that the shock chlorination is often accompanied by a clogging effect. The precipitate also preserves bacteria, making this practice somewhat less effective. Sodium hypochlorite has been used for the disinfection of drinking water.

Endodontics

Sodium hypochlorite is now used in endodontics during root canal treatments. It is the medicament of choice due to its efficacy against pathogenic organisms and pulp digestion.

Waste-water treatment

An alkaline solution (pH 11.0) of sodium hypochlorite is used to treat dilute (<1 g/l) cyanide waste-water, e.g. rinsewater from an electroplating shop. In batch treatment operations, sodium hypochlorite has been used to treat more concentrated cyanide wastes, such as silver cyanide plating solutions. A well-mixed solution is fully treated when an excess of chlorine is detected. Sodium hypochlorite in the form of household bleach is often used to oxidise foul-smelling thiol wastes generated in a chemistry laboratory.

Oxidation

Household bleach, with a phase-transfer catalyst, has been reported to oxidise alcohols to the corresponding carbonyl compound.

Safety

Sodium hypochlorite is a strong oxidiser. Products of the oxidation reactions are corrosive. Solutions burn skin and cause eye damage, particularly when used in concentrated forms.

SODIUM CHLORITE

It is in the form of white crystals or crystalline powder, slightly hygroscopic. It is soluble in water, m.p. 180°–200°C.

Manufacture

The free acid, chlorous acid, $HClO_2$, is only stable at low concentrations. Since it cannot be concentrated, it is not a commercial product. However, the corresponding sodium salt, sodium chlorite, $NaClO_2$ is stable and inexpensive enough to be commercially available. The corresponding salts of heavy metals (Ag^+, Hg^+, Tl^+, Pb^{2+}, and also Cu^{2+} and NH_4^+) decompose explosively with heat or shock.

Sodium chlorite is derived indirectly from sodium chlorate, $NaClO_3$. First, the explosively unstable gas chlorine dioxide. ClO_2 is produced by reducing sodium chlorate in a strong acid solution with a suitable reducing agent (for example, sodium chloride, sulphur dioxide, or hydrochloric acid). The chlorine dioxide is then absorbed into an alkaline solution and reduced with hydrogen peroxide, H_2O_2 yielding sodium chlorite.

Usage

The main application of sodium chlorite is the generation of chlorine dioxide for bleaching and stripping of textiles, pulp, and paper. It is also used for disinfection of a few municipal water treatment plants after conversion to chlorine dioxide. An advantage in this application, as compared to the more commonly used chlorine, is that trihalomethanes are not produced from organic contaminants. Sodium chlorite, $NaClO_2$ also finds application as a component in therapeutic rinses, mouthwashes, toothpastes and gels, mouth sprays, chewing gums and lozenges, and also in contact lens cleaning solution under the trade name purite. In organic synthesis, sodium chlorite is frequently used for the oxidation of aldehydes to carboxylic acids. The reaction is usually performed in the presence of a chlorine scavenger. Sodium chlorite, like many oxidising agents, should be protected from inadvertent contamination by organic materials to avoid the formation of an explosive mixture. Recently, sodium chlorite has been used as a oxidising agent to covert alkyl furans to the corresponding 4-oxo-2-alkenoic acids in a simple one pot synthesis. It is flammable, strong oxidising agent. The solution is strong irritant to skin and tissue.

IMPROVE OPERATIONAL EFFICIENCY OF CAUSTIC SODA PLANTS

Industrial grade raw salt, water and power are the three main inputs in caustic soda manufacture. The industrial grade raw salt having sodium chloride 98 per cent; calcium 0.2 per cent maximum; magnesium 0.14 per cent max; sulphates 0.7 per cent; iron <2 ppm and insoluble 1 per cent maximum; moisture

5 per cent is dissolved in a resaturator to obtain the raw brine of sodium chloride concentration 300–310 gm/l. The salts of magnesium, calcium and sodium sulphate impurities present in the raw brine are treated with sodium hydroxide, sodium carbonate and barium carbonate solutions and converted into magnesium hydroxide, calcium carbonate and barium sulphates which settle as sludge in a clarifier. The clarified brine is filtered in anthracite beds and further passed through candle filters and ion exchange resin beds to reduce the calcium and magnesium to less than 20 ppb. This ultra pure brine is heated to about 60°C and sent as feed to cells/electrolysers.

The mono-polar membrane cells have titanium coated metal anodes and activated carbon coated cathodes suitable for the medium handled and high current densities of 3–4.5 ka/m^2. Direct current is passed through the brine and electrolysis of sodium chloride takes place. Chlorine gas evolved is removed along with brine as overflow from the anode chambers and the 32 per cent caustic solution along with the evolved hydrogen gas is removed from the cathode chambers.

The depleted brine at a concentration of about 200–210 gm/l and about 90°C is vacuum flushed to remove all the free chlorine and acidified. It is then treated with sodium bisulphite solution to remove all the free chlorine which may pose odour problems and affect ion exchange resin. This dechlorinated brine known as lean brine is sent back to the resaturator for dissolving fresh raw salt and recycled back to the cells after chemical treatment and purification. The wet gaseous chlorine is water cooled to remove the moisture, brine is removed in a mist filter and divided into two parts using a blower discharge. On one side the chlorine is cooled further with chilled water to remove maximum moisture and dried in a drying tower using circulating concentrated sulphuric acid (98 per cent sulphuric acid is added as makeup). This dried chlorine gas is then compressed to 3–3.5 KScg and liquefied and stored in chlorine storage tanks for further distribution through one ton cylinders. Another portion of wet chlorine gas is burned with it slight excess of hydrogen gas (hydrogen gas produced from the cells is moisture removed using cooling water coolers) in the HCl reactors to form hydrogen chloride gas. This hydrogen chloride gas will be absorbed in DM water to form 33 per cent, mercury free, commercial grade HCl acid.

The excess hydrogen gas produced could be used as fuel. The 32 per cent caustic lye produced from the cells is further concentrated to 48–50 per cent caustic lye in multiple effect steam evaporators and sold as commercial lye. Also, the caustic lye is further concentrated and flaked in the caustic fusion plant and sold as flakes in bags.

Reactions involved:

$$NaCl + H_2O \longrightarrow NaOH + \tfrac{1}{2}H_2 + \tfrac{1}{2}Cl$$

58.5 kg of NaCl will give 40 kg of caustic soda and 1 kg of hydrogen and 35.5 kg of chlorine. Based on the same, we see that a 100 tpd caustic soda plant will require 146.25 tons of salt as 100 per cent, and will produce 2.5 tons of hydrogen gas (28000 nm^3/day) and 88.75 tons of chlorine gas as by-products.

As the concentration of the brine solution gets depleted across the cell by 100 kg/m^3 only, this 146.25 tons of salt required will be obtained by circulating 60 m^3/hr of 300 gpl brine solution. The per day consumption of the treatment chemicals will be calculated based on the following equations:

$$SO_4^- + BaCO_3 \longrightarrow BaSO_4 + CO_2^-$$

$$CaCl_2 + Na_2CO_3 \longrightarrow CaCO_3 + 2NaCl$$

$$Mg^{++} + 2NaOH \longrightarrow 2Na^+ + Mg(OH)_2$$

Though the theoretical decomposition potential of sodium is 2.3 V/DC only, in practice the minimum achievable voltage works out to 3.225 V for a current density of 3–4.5 ka/m^2. This is due to the drops across various connecting bus bars, resistance of anodes, cathodes and membrane apart from the temperature of brine and other operating conditions.

As per Faraday's law of electrolysis, 1 coloumb of electricity or 96450 amp. sec. of electricity will produce 1 g equivalent of caustic soda which is 40 g. Therefore, 1 kg × 1 hr × 1 cell will produce:

$$\frac{1000 \times 3600 \times 1}{96450 \times 100} \times 40 = 1.493 \text{ kg/hr of NaOH}$$

Current efficiency is defined as the ratio of the amount of product actually produced to the amount of the product that should have been produced theoretically. Because of the back migration of hydroxyl ions in the electrolyser a small amount of sodium chlorate is always formed in the process.

'This along with other electrical losses will bring down the current efficiency in the range of 95 to 96.5 per cent and thereby reduce the actual production of caustic soda. Hence, power required per ton of caustic will be 2240 units with fresh cells and membranes. With ageing, this power increases to 2375 units in 4 years time.

Operational Issues

Control of power cost, marketing of chlorine/hydrochloric acid/caustic soda due to the cyclic demand for the products are the major operational issues.

Caustic soda manufacturing through electrolysis is a power intensive process. In fact, power cost forms 80–90 per cent of variable cost of caustic soda. Hence, cutting the power cost is the key operational issue. As the technology improved, the power consumption has come down from about

3500 units per ton to 2300 units per ton. Generally, the power consumption varies with the following:

1. As the feed temperature goes up the power consumption comes down, limited by brine boiling.
2. The power consumption comes down as the average molar concentrations of the feed brine and concentration of caustic lye formed comes down.
3. Power consumption goes up due to fouling of membranes with the impurities in the brine with ageing, and degradation.
4. Power consumption goes up with the degradation of titanium oxide coatings on the anodes.
5. Power consumption increases with the aged, deteriorated nickel coating of the cathodes.

Even though the decomposition potential for sodium is only 2.3 volt, due to various reasons including the cell geometry the minimum practical decomposition voltage achieved is in the range of 3.2 V plus for the DC 3–4.5 ka/m². When the system is very new with new membranes and electrodes, under operating conditions, the voltage achieved is about 3.225 volts. With ageing, in 5–6 years, assuming that standard operating procedures are followed, the voltage gradually increases by 200–250 milli-volt, there by increasing the power consumption per ton of production. The specific power consumption of caustic which is about 2240 DC units per ton during initial conditions (3.225 volts) will increase gradually to 2375 units per ton over a period of over 4–5 years, necessitating the change of membrane and checking the coatings of anode and cathodes.

Periodical thorough cleaning of the interconnecting bus bars of the cells, recoating of the titanium wire mesh anodes, activated carbon coated cathodes and maintaining 60° to 90°C brine temperature across the cell will reduce the increase in milli-volts. Again proper maintaining of the feed brine quality and replacement of membranes once in 5 to 6 yrs as per life/cell voltage will check the excess voltage and hence the power.

It has been observed that use of membranes over a period of 5 to 6 years lead to excess voltage in a range of 200–250 mV. It could be economical to replace them considering the escalating power cost.

The non-cell house power can be controlled effectively by the use of hydrogen as fuel, and installation of waste heat recovery systems like waste heat recovery boilers, vapour absorption chillers, etc.

Installing self power generation (coal/gas/oil based) units and effective monitoring standard operating parameters and maintenance practices could bring down the over all power cost.

Caustic soda and chlorine have cyclic market demands. Being coproducts, disposal of one is an issue when the demand is very high for the other. At least caustic lye can be converted into flakes and stored, but disposal of chlorine could be a real issue. Normally, caustic soda plants in India dispose chlorine in the form of gaseous chlorine for polymer applications or convert them into either liquid chlorine or hydrogen chloride gas/HCl acid/ammonium chloride.

Large quantities of HCl is used by metal processing industries and for water treatment purposes. Liquid chlorine is used in municipal water treatment, paper and pulp industries, petrochemicals like epichlorohydrin, ethylene/propylene glycol, PVC, etc. The main uses of caustic soda are in aluminium manufacturing, water treatment, paper and pulp, sodium based chemicals manufacturing, etc.

Thus, based on the above points projected, it is evident that for any new caustic soda project to be viable on a sustainable basis, the following factors need to be taken care of:

1. As power forms 80–90 per cent of the variable cost of production of caustic soda, availability of low cost power at the operating site is essential. It can be a part of a bigger complex instead of a stand-alone project where surplus power and utilities are available.

2. Towards maintaining the higher capacity utilisation and on stream factor, it is imperative that such projects should have captive or assured markets for both caustic soda lye and chlorine or at least for chlorine positively.

With these, caustic soda plant operations can be maintained effectively with the best returns on investment.

Chapter 13

Electrolytic Industries

INTRODUCTION

Electric energy is extensively consumed by the chemical process industries, not only to furnish power through electric motors, but to give rise to elevated temperatures and directly to cause chemical change. Energy in the form of electricity causes chemical reactions to take place in the electrolytic industries described in this chapter. The heat produced thereby is the basis for the high temperature required in the electrothermal industries. The materials manufactured with the aid of electricity vary from chemicals that are also produced by other methods, such as caustic soda, hydrogen, and magnesium, to chemicals that at present cannot be made economically in any other way, such as aluminium and calcium carbide. The cost of electric power is usually the deciding factor in the electrochemical industries. Thus these industries have tended to become established in regions of cheap electric power based on falling water.

ALUMINIUM

Aluminium is a silvery white and ductile member of the boron group of chemical elements. It has the symbol Al and atomic number 13. It is not soluble in water under normal circumstances. Aluminium is the most abundant metal in the earth's crust, and the third most abundant element therein, after oxygen and silicon. It makes up about 8 per cent by weight of the earth's solid surface. Aluminium is too reactive chemically to occur in nature as the free metal. Instead, it is found combined in over 270 different minerals. The chief source of aluminium is bauxite ore.

Aluminium is remarkable for its ability to resist corrosion (due to the phenomenon of passivation) and its low density. Structural components made from aluminium and its alloys are vital to the aerospace industry and very important in other areas of transportation and building. Its reactive nature makes it useful as a catalyst or additive in chemical mixtures, including being used in ammonium nitrate explosives to enhance blast power.

312

Aluminium is a soft, durable, lightweight, malleable metal with appearance ranging from silvery to dull grey, depending on the surface roughness. Aluminium is nonmagnetic and nonsparking. It is also insoluble in alcohol, though it can be soluble in water in certain forms. The yield strength of pure aluminium is 7–11 MPa, while aluminium alloys have yield strengths ranging from 200 MPa to 600 MPa. Aluminium has about one-third the density and stiffness of steel. It is ductile, and easily machined, cast, and extruded. Aluminium is a good thermal and electrical conductor, by weight better than copper. Aluminium is capable of being a superconductor, with a superconducting critical temperature of 1.2 kelvin and a critical magnetic field of about 100 gauss.

However, because of its strong affinity to oxygen, it is almost never found in the elemental state; instead it is found in oxides or silicates. Feldspars, the most common group of minerals in the earth's crust, are aluminosilicates. Native aluminium metal can be found as a minor phase in low oxygen fugacity environments, such as the interiors of certain volcanoes.

Although aluminium is an extremely common and widespread element, the common aluminium minerals are not economic sources of the metal. Almost all metallic aluminium is produced from the ore bauxite.

Production and Refinement

Although aluminium is the most abundant metallic element in the earth's crust (believed to be 7.5 to 8.1 per cent), it is rare in its free form, occurring in oxygen-deficient environments such as volcanic mud, and it was once considered a precious metal more valuable than gold.

Aluminium is a strongly reactive metal that forms a high-energy chemical bond with oxygen. Compared to most other metals, it is difficult to extract from ore, such as bauxite, due to the energy required to reduce aluminium oxide (Al_2O_3). For example, direct reduction with carbon, as is used to produce iron, is not chemically possible, since aluminium is a stronger reducing agent than carbon. Aluminium oxide has a melting point of about 2000°C. Therefore, it must be extracted by electrolysis. The cross-sectional diagram of an aluminium reduction pot is shown in Fig. 13.1. In this process, the aluminium oxide is dissolved in molten cryolite and then reduced to the pure metal. The operational temperature of the reduction cells is around 950° to 980°C. Cryolite is found as a mineral in Greenland, but in industrial use it has been replaced by a synthetic substance. Cryolite is a chemical compound of aluminium, sodium, and calcium fluorides (Na_3AlF_6).

The aluminium oxide (a white powder) is obtained by refining bauxite in the Bayer process of Karl Bayer. (Previously, the Deville process was the predominant refining technology.)

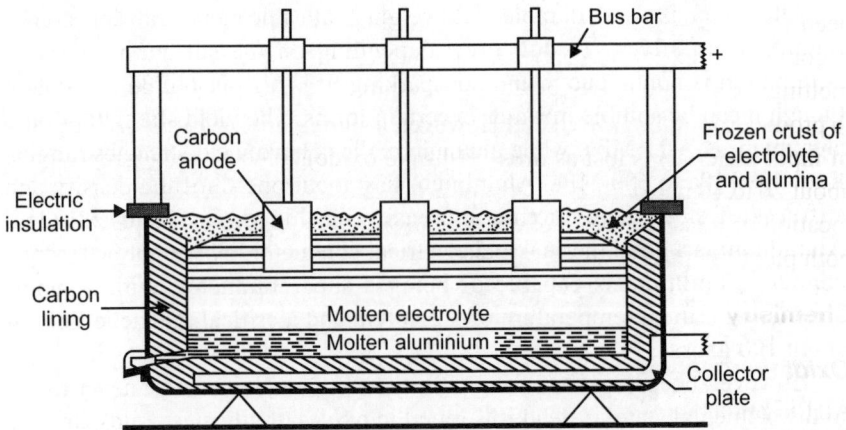

Fig. 13.1. Cross-section diagram of an aluminium reduction pot.

The electrolytic process replaced the Wöhler process, which involved the reduction of anhydrous aluminium chloride with potassium.

Both of the electrodes used in the electrolysis of aluminium oxide are carbon. Once the refined alumina is dissolved in the electrolyte, its ions are free to move around. The reaction at the cathode (negative electrode) is:

$$Al^{3+} + 3e^- \rightarrow Al$$

Here the aluminium ion is being reduced (electrons are added). The aluminium metal then sinks to the bottom and is tapped off. At the anode (positive electrode), oxygen is formed:

$$2O^{2-} \rightarrow O_2 + 4\,e^-$$

This carbon anode is then oxidised by the oxygen, releasing carbon dioxide.

$$O_2 + C \rightarrow CO_2$$

The anodes in a reduction cell must therefore be replaced regularly, since they are consumed in the process. Unlike the anodes, the cathodes are not oxidised because there is no oxygen present, as the carbon cathodes are protected by the liquid aluminium inside the cells. Nevertheless, cathodes do erode, mainly due to electrochemical processes and metal movement.

After five to ten years, depending on the current used in the electrolysis, a cell has to be rebuilt because of cathode wear. Aluminium electrolysis with the Hall-Héroult process consumes a lot of energy, but alternative processes were always found to be less viable economically and/or ecologically. The most modern smelters achieve approximately 12.8 kW·h/kg (46.1 MJ/kg). (Compare this to the heat of reaction, 31 MJ/kg, and the Gibbs free energy of reaction, 29 MJ/kg.) Reduction line currents for older technologies are typically 100 to 200 kA; state-of-the-art smelters operate at about 350 kA. Trials have been reported with 500 kA cells. Recovery of the metal via recycling has become an important facet of the aluminium industry. Recycling involves melting the scrap, a process that requires only five per cent of the energy used to produce aluminium from ore. However, a significant part (up to 15 per cent of input material) is lost as dross (ash-like oxide). Electric power represents about 20 to 40 per cent of the cost of producing aluminium, depending on the location of the smelter. Smelters tend to be situated where electric power is both plentiful and inexpensive.

Chemistry

Oxidation state one

AlH is produced when aluminium is heated in an atmosphere of hydrogen. Al_2O is made by heating the normal oxide, Al_2O_3, with silicon at 1800°C in a vacuum.

Al_2S can be made by heating Al_2S_3 with aluminium shavings at 1300°C in a vacuum. It quickly disproportionates to the starting materials. The selenide is made in a parallel manner.

AlF, AlCl and AlBr exist in the gaseous phase when the tri-halide is heated with aluminium. Aluminium halides usually exist in the form AlX_3, e.g. AlF_3, $AlCl_3$, $AlBr_3$, AlI_3, etc.

Oxidation state two

Aluminium monoxide, AlO, is present when aluminium powder burns in oxygen.

Oxidation state three

Fajans' rules show that the simple trivalent cation Al^{3+} is not expected to be found in anhydrous salts or binary compounds such as Al_2O_3. The hydroxide is a weak base and aluminium salts of weak acids, such as carbonate, can not be prepared. The salts of strong acids, such as nitrate, are stable and soluble in water, forming hydrates with at least six molecules of water of crystallisation. Aluminium hydride, $(AlH_3)_n$, can be produced from trimethylaluminium and an excess of hydrogen. It burns explosively in air. It can also be prepared by the action of aluminium chloride on lithium hydride in ether solution, but cannot be isolated free from the solvent. Alumino-hydrides of the most electropositive elements are known, the most useful being lithium aluminium hydride, $Li[AlH_4]$. It decomposes into lithium hydride, aluminium and hydrogen when heated, and is hydrolysed by water. It has many uses in organic chemistry, particularly as a reducing agent. The aluminohalides have a similar structure. Aluminium hydroxide may be prepared as a gelatinous precipitate by adding ammonia to an aqueous solution of an aluminium salt.

It is amphoteric, being both a very weak acid, and forming aluminates with alkalies. It exists in various crystalline forms.

Aluminium carbide, Al_4C_3 is made by heating a mixture of the elements above 1000°C. The pale yellow crystals have a complex lattice structure, and react with water or dilute acids to give methane. The acetylide, $Al_2(C_2)_3$, is made by passing acetylene over heated aluminium. Aluminium nitride, AlN, can be made from the elements at 800°C. It is hydrolysed by water to form ammonia and aluminium hydroxide. Aluminium phosphide, AlP, is made similarly, and hydrolyses to give phosphine.

Aluminium oxide, Al_2O_3, occurs naturally as corundum, and can be made by burning aluminium in oxygen or by heating the hydroxide, nitrate or sulphate. As a gemstone, its hardness is only exceeded by diamond, boron nitride, and carborundum. It is almost insoluble in water. Aluminium sulphide, Al_2S_3, may be prepared by passing hydrogen sulphide over aluminium powder. It is polymorphic. Aluminium iodide, AlI_3, is a dimer with applications in organic synthesis. Aluminium fluoride, AlF_3, is made by treating the hydroxide with HF or can be made from the elements. It consists of a giant molecule which sublimes without melting at 1291°C. It is very inert. The other trihalides are dimeric, having a bridge-like structure. When aluminium and fluoride are together in aqueous solution, they readily form complex ions such as $AlF(H_2O)_5^{+2}$, $AlF_3(H_2O)_3^0$, AlF_6^{-3}. Of these, AlF_6^{-3} is the most stable. This is explained by the fact that aluminium and fluoride, which are both very compact ions, fit together just right to form the octahedral aluminium hexafluoride complex. When aluminium and fluoride are together in water in a 1:6 molar ratio, AlF_6^{-3} is the most common form, even in rather low concentrations.

Organo-metallic compounds of empirical formula AlR_3 exist and, if not also giant molecules, are at least dimers or trimers. They have some uses in organic synthesis, for instance trimethylaluminium.

Uses of Aluminium

Aluminium is the most widely used non-ferrous metal. Relatively pure aluminium is encountered only when corrosion resistance and/or workability is more important than strength or hardness. A thin layer of aluminium can be deposited onto a flat surface by physical vapour deposition or (very infrequently) chemical vapour deposition or other chemical means to form optical coatings and mirrors. When so deposited, a fresh, pure aluminium film serves as a good reflector (approximately 92 per cent) of visible light and an excellent reflector (as much as 98 per cent) of medium and far infrared.

Pure aluminium has a low tensile strength, but when combined with thermo-mechanical processing, aluminium alloys display a marked improvement in mechanical properties, especially when tempered. Aluminium alloys form vital components of aircraft and rockets as a result of their high strength-to-weight ratio. Aluminium readily forms alloys with many elements such as copper, zinc, magnesium, manganese and silicon (e.g. duralumin). Some of the many uses for aluminium metal are in:

1. Transportation (automobiles, aircraft, trucks, railway cars, marine vessels, bicycles, etc.).
2. Packaging (cans, foil, etc.).
3. Construction (windows, doors, siding, building wire, etc.)
4. Cooking utensils.
5. Street lighting.
6. Electrical transmission lines for power distribution.
7. Super purity aluminium (SPA, 99.980 per cent to 99.999 per cent Al), used in electronics and CDs.
8. Heat sinks for electronic appliances such as transistors and CPUs.
9. Substrate material of metal-core copper clad laminates used in high brightness.
10. Powdered aluminium is used in paint, and in pyrotechnics such as solid rocket.
11. Aluminium is widely used in watch production as it provides durability and resists tarnishing and corrosion.

Aluminium Compounds

Some of the important compounds of aluminium are given below:

1. Aluminium ammonium sulphate ($[Al(NH_4)](SO_4)_2$), ammonium alum is used as a mordant, in water purification and sewage treatment, in paper production, as a food additive, and in leather tanning.
2. Aluminium acetate is a salt used in solution as an astringent.
3. Aluminium borate ($Al_2O_3 \ B_2O_3$) is used in the production of glass and ceramic.
4. Aluminium borohydride [$Al(BH_4)_3$] is used as an additive to jet fuel.
5. Aluminium chloride ($AlCl_3$) is used in paint manufacturing, in petroleum refining and in the production of synthetic rubber.
6. Aluminium chlorohydride is used as an antiperspirant and in the treatment of hyperhydrosis.
7. Aluminium fluorosilicate [$Al_2(SiF_6)_3$] is used in the production of synthetic gemstones, glass and ceramic.
8. Aluminium hydroxide [$Al(OH)_3$] is used as an antacid, as a mordant, in water purification, in the manufacture of glass and ceramic and in the waterproofing of fabrics.

9. Aluminium oxide (Al_2O_3), alumina, is found naturally as corundum (rubies and sapphires), emery, and is used in glass making. Synthetic ruby and sapphire are used in lasers for the production of coherent light.

10. Aluminium phosphate ($AlPO_4$) is used in the manufacture of glass and ceramic, pulp and paper products, cosmetics, paints and varnishes and in making dental cement.

11. Aluminium sulphate $[Al_2(SO_4)_3]$ is used in the manufacture of paper, as a mordant, in a fire extinguisher, in water purification and sewage treatment, as a food additive, in fireproofing, and in leather tanning.

12. In many vaccines, certain aluminium salts serve as an immune adjuvant (immune response booster) to allow the protein in the vaccine to achieve sufficient potency as an immune stimulant.

MAGNESIUM

Magnesium is a chemical element with the symbol Mg, atomic number 12, atomic weight 24.3050 and common oxidation number +2.

Magnesium, an alkaline earth metal is the ninth most abundant element in the universe by mass. The commonness of magnesium is related to the fact that it is easily built up in supernova stars from a sequential addition of three helium nuclei to carbon (which in turn is made from a single reaction between three helium nuclei at once). Magnesium constitutes about 2 per cent of the earth's crust by mass, which makes it the eighth most abundant element in the crust. Magnesium ion's high solubility in water helps ensure that it is the third most abundant element dissolved in seawater.

Magnesium is the 11th most abundant element by mass in the human body; its ions are essential to all living cells, where they play a major role in manipulating important biological polyphosphate compounds like ATP, DNA, and RNA. Hundreds of enzymes thus require magnesium ion in order to function. Magnesium is also the metallic ion at the centre of chlorophyll, and is thus a common additive to fertilisers. Magnesium compounds are used medicinally as common laxatives, antacids (i.e. milk of magnesia), and in a number of situations where stabalisation of abnormal nerve excitation and blood vessel spasm is required (i.e. to treat eclampsia). Magnesium ions are sour to the taste, and in low concentrations help to impart a natural tartness to fresh mineral waters. The free element (metal) is not found naturally on earth, since it is highly reactive (though once produced, is coated in a thin layer of oxide which partly masks this reactivity). The free metal burns with a characteristic brilliant white light, making it a useful ingredient in flares. The metal is now mainly obtained by electrolysis of magnesium salts obtained from brine. Commercially, the chief use for the metal is as an alloying agent to make aluminium-magnesium alloys, sometimes called 'magnalium' or 'magnelium'. Since magnesium is less dense than aluminium, these alloys are prized for their relative lightness and strength.

Manufacture of Magnesium

The cheapest method of making magnesium is by the electrolytic process. During World War II magnesium was made by two other processes, the silicothermic or ferrosilicon, process and the carbon reduction process. The carbon reduction process never operated satisfactorily, and its used has long been discontinued. The silicothermic process is still used in areas where power is expensive and only a small magnesium capacity is required (Fig. 13.2).

Electrolysis of magnesium chloride

The magnesium chloride is obtained: (i) from salines, (ii) from brine wells, and (iii) from the reaction of magnesium hydroxide (from sea water or dolomite) with hydrochloric acid (Fig. 13.2). Dow Chemical Co., makes magnesium by electrolysing magnesium chloride from sea water, using oyster shells for the lime needed. The oyster shells, which are almost pure calcium carbonate, are burned to lime, slaked, and mixed with the sea water, thus precipitating magnesium hydroxide. This magnesium hydroxide is filtered off and treated with hydrochloric acid prepared at high temperature from water and the chlorine evolved from the cells. This forms a magnesium chloride solution which is evaporated to solid magnesium chloride in direct fired evaporators, followed by shelf drying. After dehydrating, the magnesium chloride is fed to a dow electrolytic cell, where it is decomposed into the metal and chlorine gas. These cells are large rectangular ceramic-covered steel pots, with a capacity of about 10 tons molten magnesium chloride and salts. The internal parts of the cell act as the cathode and 22 graphite anodes are suspended vertically at the top of the cell. Sodium chloride is added to the bath to lower the melting point and also to increase the conductivity. The salts are kept molten by the electric current used to extract the magnesium plus external heat supplied by gas-fired outside furnaces. The usual operating temperature is 710°C, which is sufficient to melt the magnesium (m.p. 651°C). Each cell operates at about 6 V and 80,000 to 1,00,000 A, with a current efficiency of above 80 per cent. The power requirements are 8 kWh/lb of magnesium produced. The molten magnesium is liberated at the cathode and rises to the bath surface, where toughs lead to the metal wells in front of the cell. The 99.9 per cent pure magnesium metal is dipped out several times during the day, each dipperful containing enough metal to fill a 42 lb self-pelletising mould.

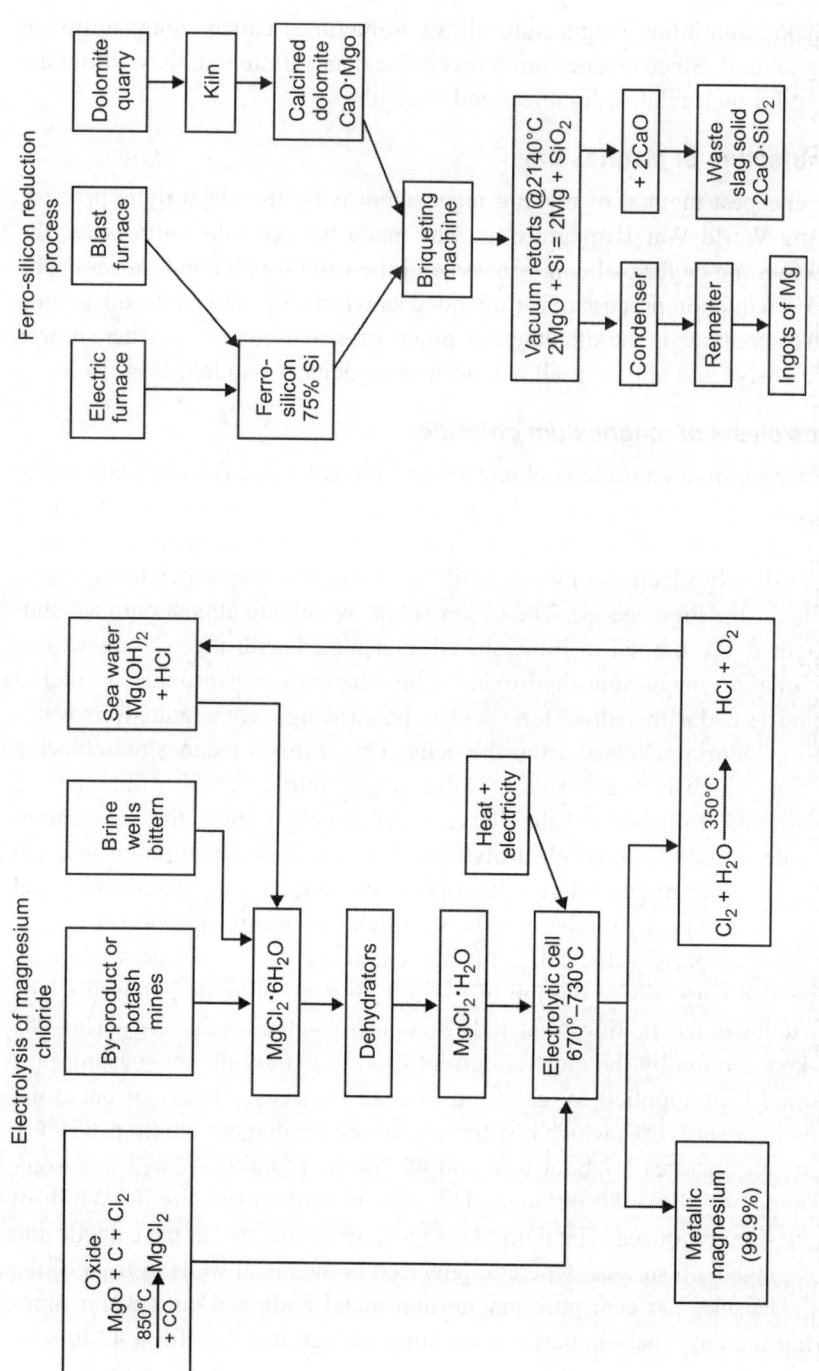

Fig. 13.2. Outline of processes for the production of metallic magnesium.

Silicothermic or Ferrosilicon Process

The process consists of mixing ground, dead-burned dolomite with ground, 70 to 80 per cent ferrosilicon and 1 per cent fluorspar (eutectic) and pelleting. The pellets are charged into the furnace. The essential reaction of this process is:

$$2(MgO \cdot CaO) + 1/6 FeSi_6 \longrightarrow 2Mg + (CaO)_2 SiO_2 + 1/6 Fe$$

High vacuum and heat (2140°F) are applied. The calcium oxide present in the burnt dolomite forms infusible dicalcium silicate that is easily removed from the retort at the end of the run. The chrome-alloy retort is equipped with a condenser tube with a removable lining. The reaction is run at very high vacuum (100 to 150 μ Hg), and the liberated magnesium is collected on the lining of the condenser. At the end of the run the furance is partly cooled, and the magnesium is removed from the condenser, by a procedure based on the difference in contraction of magnesium and steel, and taken to the remelt furnace.

Notable characteristics

Elemental magnesium is a fairly strong, silvery-white, light-weight metal (two-thirds the density of aluminium). It tarnishes slightly when exposed to air, although unlike the alkaline metals, storage in an oxygen-free environment is unnecessary because magnesium is protected by a thin layer of oxide which is fairly impermeable and hard to remove. Like its lower periodic table group neighbour calcium, magnesium reacts with water at room temperature, though it reacts much more slowly than calcium. When it is submerged in water, hydrogen bubbles will almost unnoticeably begin to form on the surface of the metal, though if powdered it will react much more rapidly. The reaction will occur faster with higher temperatures. Magnesium also reacts exothermically with most acids, such as hydrochloric acid (HCl). As with aluminium, zinc and many other metals, the reaction with hydrochloric acid produces the chloride of the metal and releases hydrogen gas.

Magnesium is a highly flammable metal, but while it is easy to ignite when powdered or shaved into thin strips, it is difficult to ignite in mass or bulk. Once ignited, it is difficult to extinguish, being able to burn in both nitrogen (forming magnesium nitride), and carbon dioxide (forming magnesium oxide and carbon). On burning in air, magnesium produces a brilliant white light. Thus magnesium powder (flash powder) was used as a source of illumination in the early days of photography. Later, magnesium ribbon was used in electrically ignited flash bulbs. Magnesium powder is used in the manufacture of fireworks and marine flares where a brilliant white light is required. Flame temperatures of magnesium and magnesium alloys can reach 1371°C (2500°F), although flame height above the burning metal is usually less than 300 mm.

Magnesium compounds are typically white crystals. Most are soluble in water, providing the sour-tasting magnesium ion Mg^{2+}. Small amounts of dissolved magnesium ion contributes to the tartness and taste of natural waters. Magnesium ion in large amounts is an ionic laxative, and magnesium sulphate (Epsom salts) is sometimes used for this purpose. So-called milk of magnesia is a water suspension of one of the few insoluble magnesium compounds, magnesium hydroxide. The undissolved particles give rise to its appearance and name. Milk of magnesia is a mild base commonly used as an antacid.

Applications

As the metal

Magnesium is the third most commonly used structural metal, following steel and aluminium. Magnesium compounds, primarily magnesium oxide (MgO), are used mainly as refractory material in furnace linings for producing iron, steel, nonferrous metals, glass and cement. Magnesium oxide and other compounds also are used in agricultural, chemical and construction industries. As a metal, this element's principal use is as an alloying additive to aluminium with these aluminium-magnesium alloys being used mainly for beverage cans.

Magnesium, in its purest form, can be compared with aluminium, and is strong and light, so it is used in several high volume part manufacturing applications, including automotive and truck components. Speciality, high-grade car wheels of magnesium alloy are called mag wheels.

The second application field of magnesium is electronic devices. Due to low weight, good mechanical and electrical properties, magnesium is widely used for manufacturing of mobile phones, laptop computers, cameras, and other electronic components.

Currently the use of magnesium alloys in aerospace is increasing, mostly driven by the increasing importance of fuel economy and the need to reduce weight.

Magnesium is also used:
1. To remove sulphur from iron and steel.
2. To refine titanium in the Kroll process.
3. To photoengrave plates in the printing industry.
4. To combine in alloys, where this metal is essential for airplane and missile construction.
5. In the form of turnings or ribbons, to prepare Grignard reagents, which are useful in organic synthesis.
6. As an alloying agent, improving the mechanical, fabrication and welding characteristics of aluminium.
7. As an additive agent in conventional propellants and the production of nodular graphite in cast iron.

8. As a reducing agent for the production of uranium and other metals from their salts.
9. As a desiccant, since it easily reacts with water.
10. As a sacrificial (galvanic) anode to protect underground tanks, pipelines, buried structures, and water heaters.

In magnesium compounds

1. The magnesium ion is necessary for all life, so magnesium salts are an additive for foods, fertilisers (Mg is a component of chlorophyll), and culture media.
2. Magnesium hydroxide is used in milk of magnesia, its chloride, oxide, gluconate, malate, orotate and citrate used as oral magnesium supplements, and its sulphate (Epsom salts) for various purposes in medicine. Oral magnesium supplements have been claimed to be therapeutic for some individuals who suffer from restless leg syndrome (RLS).
3. Magnesium borate, magnesium salicylate and magnesium sulphate are used as antiseptics.
4. Magnesium bromide is used as a mild sedative (this action is due to the bromide, not the magnesium).
5. Dead-burned magnesite is used for refractory purposes such as brick and liners in furnaces and converters.
6. Magnesium carbonate ($MgCO_3$) powder is also used by athletes, such as gymnasts and weightlifters, to improve the grip on objects—the apparatus or lifting bar.
7. Magnesium stearate is a slightly flammable white powder with lubricative properties. In pharmaceutical technology it is used in the manufacturing of tablets, to prevent the tablets from sticking to the equipment during the tablet compression process (i.e. when the tablet's substance is pressed into tablet form).
8. Magnesium sulphite is used in the manufacture of paper (sulphite process).
9. Magnesium phosphate is used to fireproof wood for construction.
10. Magnesium hexafluorosilicate is used in mothproofing of textiles.

Magnesium from sea water

The Mg^{2+} cation is the second most abundant cation in seawater (occurring at about 12 per cent of the mass of sodium there), which makes seawater and sea-salt an attractive commercial source of Mg. To extract the magnesium, calcium hydroxide is added to sea water to form magnesium hydroxide precipitate.

$$MgCl_2 + Ca(OH)_2 \rightarrow Mg(OH)_2 + CaCl_2$$

Magnesium hydroxide is insoluble in water so it can be filtered out, and reacted with hydrochloric acid to obtain concentrated magnesium chloride.

$$Mg(OH)_2 + 2HCl \rightarrow MgCl_2 + 2H_2O$$

From magnesium chloride, electrolysis produces magnesium.

Precautions

Magnesium metal and alloys are highly flammable in their pure form when molten, as a powder, or in ribbon form. Burning or molten magnesium metal reacts violently with water. Magnesium powder is an explosive hazard. One should wear safety glasses while working with magnesium, and if burning it, these should include a heavy UV filter, similar to welding eye protection. The bright white light (including ultraviolet) produced by burning magnesium can permanently damage the retinas of the eyes, similar to welding arc burns.

Water should not be used to extinguish magnesium fires, because it can produce hydrogen which will feed the fire, according to the reaction:

$$Mg_{(s)} + 2H_2O_{(g)} \rightarrow Mg(OH)_{2(s)} + H_{2(g)}$$

or in words:

Magnesium $_{(solid)}$ + steam $\rightarrow$ Magnesium hydroxide $_{(solid)}$ + Hydrogen $_{(gas)}$

Carbon dioxide fire extinguishers should not be used, because magnesium can burn in carbon dioxide (forming magnesium oxide, MgO, and carbon). An easy way to extinguish small metal fires is to place a polyethylene bag filled with dry sand atop the fire. The heat of the fire will melt the bag, releasing the sand onto the fire.

MAGNESIUM CHLORIDE

Magnesium chloride is the name for the chemical compounds with the formulas $MgCl_2$ and its various hydrates $MgCl_2(H_2O)_x$. These salts are typical ionic halides, being highly soluble in water. The hydrated magnesium chloride can be extracted from brine or sea water. Magnesium chloride as the natural mineral Bischofite is also extracted (solution mining) out of ancient seabeds, for example the Zechstein seabed in NW Europe. Anhydrous magnesium chloride is the principal precursor to magnesium metal, which is produced on a large scale.

Applications

Magnesium chloride serves as precursor to other magnesium compounds, for example by precipitation:

$$MgCl_2(aq) + Ca(OH)_2(aq) \rightarrow Mg(OH)_2(s) + CaCl_2(aq)$$

It can be electrolysed to give magnesium metal:

$$MgCl_2(l) \rightarrow Mg(l) + Cl_2(g)$$

The thermal dehydration of the hydrates $MgCl_2(H_2O)_x$ ($x = 6, 12$) does not occur straightforwardly. Magnesium chloride is used for a variety of other applications besides the production of magnesium: the manufacture of textiles, paper, fireproofing agents, cements and refrigeration brine, and dust and erosion control. Mixed with hydrated magnesium oxide, magnesium chloride forms a hard material called Sorel cement. Magnesium chloride is also used as an important component in the polymerase chain reaction, a procedure used to amplify DNA fragments. Magnesium chloride is also used in several medical and topical (skin related) applications.

Culinary use

Magnesium chloride is an important coagulant used in the preparation of tofu from soya milk. In Japan it is sold as nigari (derived from the Japanese word for bitter), a white powder produced from seawater after the sodium chloride has been removed, and the water evaporated. In China it is called lushui (in Chinese). It is also an ingredient in baby formula milk.

Use as an anti-icer

Magnesium chloride is much less toxic to plant life surrounding highways and airports, and is less corrosive to concrete and steel (and other iron alloys) than sodium chloride. The liquid magnesium chloride is sprayed on dry pavement (tarmac) prior to precipitation or wet pavement prior to freezing temperatures in the winter months to prevent snow and ice from adhering and bonding to the roadway. The application of anti-icers is utilised in an effort to improve highway safety. Magnesium chloride is also sold in crystal form for household and business use to de-ice sidewalks and driveways. In these applications, the compound is applied after precipitation has fallen or ice has formed.

Use in dust and erosion control

Road departments and private industry may apply liquid or powdered magnesium chloride to control dust and erosion on unimproved (dirt or gravel) roads and dusty job sites such as quarries. Its hygroscopy makes it absorb moisture from the air, controlling the number of small particles which become airborne. Similarly, owners of indoor arenas (e.g. for horse riding) may apply magnesium chloride to sand or other floor materials to control dust.

Use in hydrogen storage

Magnesium chloride has shown promise as a storage material for hydrogen. Ammonia, which is rich in hydrogen atoms, is used as an intermediate storage material. Ammonia can be effectively adsorbed onto solid magnesium chloride, forming $Mg(NH_3)_6Cl_2$. Ammonia is released by mild heat, and is then passed through a catalyst to give hydrogen gas.

SODIUM

Sodium is an element which has the symbol Na, atomic number 11, atomic mass 23 amu, and a common oxidation number +1. Sodium is a soft, silvery white, highly reactive element and is a member of the alkali metals within group 1 (formerly known as group IA). It has only one stable isotope, ^{23}Na. Sodium was first isolated by Sir Humphry Davy in 1807 by passing an electric current through molten sodium hydroxide. Sodium quickly oxidises in air and is violently reactive with water, so it must be stored in an inert medium, such as kerosene or mineral oil. Sodium is present in great quantities in the earth's oceans as sodium chloride (common salt). It is also a component of many minerals, and it is an essential element for animal life. As such, it is classified as a dietary inorganic macro-mineral.

At room temperature, sodium metal is so soft that it can be easily cut with a knife. The density of alkali metals generally increases with increasing atomic number, but sodium is denser than potassium.

Chemical Properties

Compared with other alkali metals, sodium is generally less reactive than potassium and more reactive than lithium, in accordance with periodic law: for example, their reaction in water, chlorine gas, etc.

Sodium reacts exothermically with water. Small pea-sized pieces will bounce across the surface of the water until they are consumed by it, whereas large pieces will explode. While sodium reacts with water at room temperature, the sodium piece melts with the heat of the reaction to form a sphere, if the reacting sodium piece is large enough. The reaction with water produces very caustic sodium hydroxide (lye) and highly flammable hydrogen gas. These are extreme hazards. When burned in air, sodium forms sodium peroxide Na_2O_2, or with limited oxygen, the oxide Na_2O (unlike lithium, the nitride is not formed). If burned in oxygen under pressure, sodium superoxide NaO_2 will be produced. In chemistry, most sodium compounds are considered soluble but nature provides examples of many insoluble sodium compounds such as the feldspars.

There are other insoluble sodium salts such as sodium bismuthate $NaBiO_3$, sodium octamolybdate $Na_2Mo_8O_{25}\cdot4H_2O$, sodium thioplatinate $Na_4Pt_3S_6$, sodium uranate Na_2UO_4. Sodium meta-antimonate's $2NaSbO_3\cdot7H_2O$ solubility is 0.3 g/l as is the pyro form $Na_2H_2Sb_2O_7\cdot H_2O$ of this salt. Sodium meta-phosphate $NaPO_3$ has a soluble and an insoluble form.

Compounds

Sodium compounds are important to the chemical, glass, metal, paper, petroleum, soap, and textile industries. Hard soaps are generally sodium salt of certain fatty acids (potassium produces softer or liquid soaps). The sodium compounds that are the most important to industries are common salt (NaCl), soda ash (Na_2CO_3), baking soda ($NaHCO_3$), caustic soda (NaOH), sodium nitrate ($NaNO_3$), di- and tri-sodium phosphates, sodium thiosulphate (hypo, $Na_2S_2O_3 \cdot 5H_2O$), and borax ($Na_2B_4O_7 \cdot 10H_2O$).

Manufacture of Sodium

Sodium was first produced commercially in 1855 by thermal reduction of sodium carbonate with carbon at 1100°C, in what is known as the Deville process.

$$Na_2CO_3 \text{ (liquid)} + 2C \text{ (solid)} \rightarrow 2Na \text{ (vapour)} + 3CO \text{ (gas)}.$$

Sodium is now produced commercially through the electrolysis of liquid sodium chloride. This is done in a downs cell in which the NaCl is mixed with calcium chloride to lower the melting point below 700°C. The cell for this electrolysis consist of a closed rectangular, refractory lined steel box (Fig. 13.3).

As calcium is less electropositive than sodium, no calcium will be formed at the anode. The anode and cathode are arranged in separate compartments to facilitate the recovery of sodium and chlorine.

Precautions

Extreme care is required in handling elemental/metallic sodium. Sodium is potentially explosive in water (depending on quantity) and is a corrosive substance, since it is rapidly converted to sodium hydroxide on contact with moisture. The powdered form may combust spontaneously in air or oxygen. Sodium must be stored either in an inert (oxygen and moisture free) atmosphere (such as nitrogen or argon), or under a liquid hydrocarbon such as mineral oil or kerosene.

The reaction of sodium and water is a familiar one in chemistry labs, and is reasonably safe if amounts of sodium smaller than a pencil eraser are used and the reaction is done behind a plastic shield by people wearing eye protection. However, the sodium-water reaction does not scale up well, and is treacherous when larger amounts of sodium are used. Larger pieces of sodium melt under the heat of the reaction, and the molten ball of metal is buoyed up by hydrogen and may appear to be stably reacting with water, until splashing covers more of the reaction mass, causing thermal runaway and an explosion which scatters molten sodium, lye solution, and sometimes flame (18.5 grams explosion).

Sodium is much more reactive than magnesium; a reactivity which can be further enhanced due to sodium's much lower melting point. When sodium catches fire in air (as opposed to just the hydrogen gas generated from water by means of its reaction with sodium) it more easily produces temperatures high enough to melt the sodium, exposing more of its surface to the air and spreading the fire.

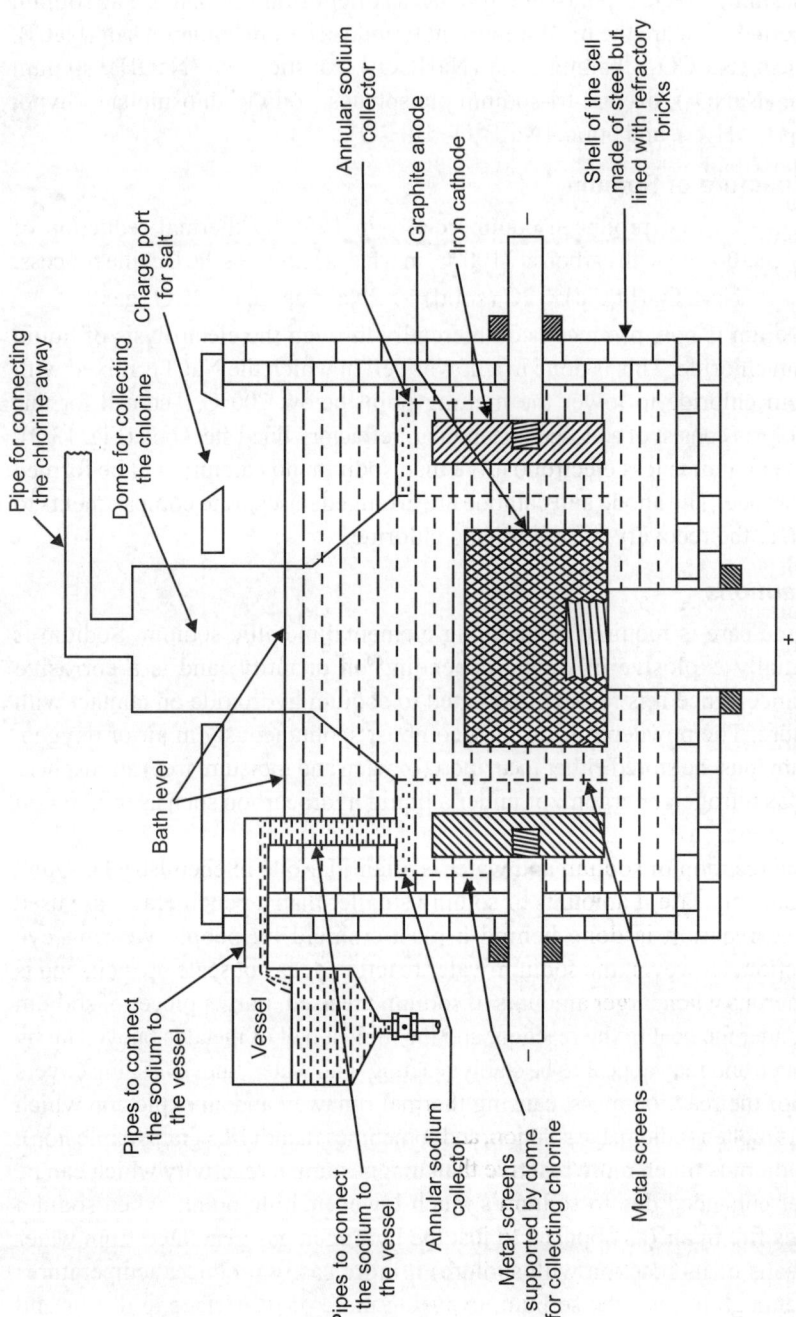

Fig. 13.3. Downs sodium cell for electrolysis of fused sodium chloride.

Because of the reaction scale problems discussed above, disposing of large quantities of sodium (more than 10 to 100 grams) must be done through a licensed hazardous materials disposer. Smaller quantities may be broken up and neutralised carefully with ethanol (which has a much slower reaction than water), or even methanol (where the reaction is more rapid than ethanol's but still less than in water), but care should nevertheless be taken, as the caustic products from the ethanol or methanol reaction are just as hazardous to eyes and skin as those from water. After the alcohol reaction appears complete, and all pieces of reaction debris have been broken up or dissolved, a mixture of alcohol and water, then pure water, may then be carefully used for a final cleaning. This should be allowed to stand a few minutes until the reaction products are diluted more thoroughly and flushed down the drain. The purpose of the final water soak and wash of any reaction mass which may contain sodium is to ensure that alcohol does not carry unreacted sodium into the sink trap, where a water reaction may generate hydrogen in the trap space which can then be potentially ignited, causing a confined sink trap explosion.

SODIUM CHLORATE

Sodium chlorate is a chemical compound with the chemical formula ($NaClO_3$). When pure, it is a white crystalline powder that is readily soluble in water. It is hygroscopic. It decomposes above 250°C to release oxygen and leave sodium chloride. It has specific gravity 2.49. It is colourless, odourless, saline taste. It should not be trituted with any combustible substance.

Synthesis

Industrially, sodium chlorate is synthesised from the electrolysis of hot sodium chloride solution in a mixed electrode tank:

$$NaCl + 3H_2O \rightarrow NaClO_3 + 3H_2$$

It can also be synthesised by passing chlorine gas to a hot sodium hydroxide solution. It is then purified by crystallisation.

Uses

Herbicides

Sodium chlorate is used as a non-selective herbicide. It is considered phytotoxic to all green plant parts. It can also kill through root absorption.

The herbicide is mainly used on non-crop land for spot treatment and for total vegetation control on roadsides, fenceways, ditches and suchlike. Sodium chlorate is also used as a defoliant and desiccant for: cotton, safflower, corn, flax, peppers, soyabeans, grain, sorghum, southern peas, dry beans, rice and sunflowers.

If used in combination with atrazine, it increases the persistence of the effect. If used in combination with 2,4-D, it improves performance of the material. Sodium chlorate has a soil-sterilant effect. Mixing with other herbicides in aqueous solution is possible to some extent, so long as they are not susceptible to oxidation.

Chemical oxygen generator

Chemical oxygen generator for example in commercial aircraft provide emergency oxygen to passengers to protect them from drops in cabin pressure by catalytic decomposition of sodium chlorate. The catalyst is normally some iron powder and barium peroxide (BaO_2) is used to absorb the chlorine which is a minor product in the decomposition.

Sodium chlorate is used in some aircraft as a source of supplemental oxygen. Iron powder is mixed with sodium chlorate and ignited by a charge activated by pulling on the emergency mask. The reaction produces more oxygen than is required for combustion.

Toxicity in humans

Due to its oxidative nature, sodium chlorate can be very toxic if ingested. The oxidative effect on haemoglobin leads to methaemoglobin formation, which is followed by denaturation of the globin protein and a cross-linking of erythrocyte membrane proteins with resultant damage to the membrane enzymes.

Formulations

Sodium chlorate comes in dust, spray and granule formulations. There is a risk of fire and explosion in dry mixtures with other substances, especially organic materials, that is other herbicides, sulphur, phosphorus, powdered metals, strong acids. Particularly when mixed with sugar it has explosive properties. If accidentally mixed with one of these substances do not store inside your home, garage etc.

Marketed formulations contain a fire depressant, but this has little effect if deliberately ignited. Most commercially available chlorate weedkillers contain approximately 53 per cent sodium chlorate with the balance being a fire depressant such as sodium metaborate or ammonium phosphates.

SODIUM PERCHLORATE

Sodium perchlorate is a perchlorate of sodium and has the formula $NaClO_4$. Sodium perchlorate melts with decomposition at 480°C. Its heat of formation is -382.75 kJ mol^{-1}. It is a white crystalline solid. It is hygroscopic. It is soluble in water and in alcohol. It usually comes as the monohydrate, which has a rhombic crystal structure. It has specific gravity 2.02, m.p. 482°C, dangerous fire and explosion in contact with organic materials and sulphuric acid.

Synthesis

$$ClO_3^- + H_2O \rightarrow ClO_4^- + 2H^+ + 2e^-$$

Sodium perchlorate is manufactured by anodic oxidation of sodium chloride or sodium chlorate at high current density, with platinum (or in some cases, lead dioxide, manganese dioxide, and possibly magnetite and cobalt oxide) anodes and graphite, steel, nickel, or titanium cathodes.

Uses

The present major use of perchlorate salts is as oxidisers in solid propellants such as NASA's solid rocket boosters. The potassium salt was first used and quickly followed by what is now the more important salt, ammonium perchlorate.

Lithium perchlorate, which has the highest weight percentage of oxygen of all compounds, has been tested as an oxidiser in solid propellants, but has not found favour with propellant manufacturers due to its hygroscopicity.

Sodium perchlorate itself finds only minimal use in pyrotechnics because it is hygroscopic; ammonium and potassium perchlorates are preferred. They are made by double decomposition from a solution of sodium perchlorate and potassium or ammonium chlorides.

It is also used in standard DNA extraction and hybridisation reactions in molecular biology.

PRIMARY CELL

A primary cell is any kind of electrochemical cell in which the electrochemical reaction of interest is not reversible, so used in disposable batteries. The most common primary cells today are found in alkaline batteries; earlier carbon-zinc cells, with a carbon post as cathode and a zinc shell as anode were prevalent. Unlike a secondary cell, attempting to reverse the reaction in a primary cell via recharging is dangerous and can lead to a battery explosion.

A related difference is that primary batteries use up the materials in one or both of their electrodes, while, ideally, the reversibility of the reactions in a secondary cell allows them to be restored to almost the same fully charged condition on each recharging.

Comparison with Rechargeables

Even though rechargeable batteries are more expensive than disposable batteries with equivalent voltages and shapes, the rechargeable batteries would be much cheaper if the main price is divided with the full number of recharge cycles, even including a battery charger, compared to the total cost of number of primary cells equivalent to recharge cycles of NiMH, NiCd and Li-ion batteries.

However, there are some battery uses that require long dormancy periods and few replacements, so major issue is charge retention. In these circumstances, certain rechargeable battery technologies may not be appropriate, as they may have a high self-discharge rate compared to equivalent non-rechargeable batteries.

For example, a flashlight used for emergencies must work when needed, even if it has sat on a shelf for an extended period of time. Primary cells are also more cost-efficient in this case, as rechargeable batteries would use only a small fraction of available recharge cycles.

Primary Cells

Some of the important primary cells are given below:
1. Laclanché cell.
2. Daniell cell.
3. Bunsen cell.
4. Chromic acid cell.
5. Clark cell.
6. Weston cell.

Laclanché cell

Georges Laclanché invented and patented in 1866 his battery, the Laclanché cell. It contained a conducting solution (electrolyte) of ammonium chloride, a cathode (positive terminal) of carbon, a depolariser of manganese dioxide, and an anode (negative terminal) of zinc.

The chemical process which produces electricity in a Laclanché cell begins when zinc atoms on the surface of the anode oxidise, i.e. they give up both their electrons to become positively-charged ions. As the zinc ions move away from the anode, leaving their electrons on its surface, the anode becomes more negatively charged than the cathode. When the cell is connected in an external electrical circuit, the excess electrons on the zinc anode flow through the circuit to the carbon rod, the movement of electrons forming an electrical current.

When the electrons enter the rod, they combine with molecules of manganese dioxide and molecules of water, which react with each other to produce manganese oxide and negatively charged hydroxide ions. This is accompanied by a secondary reaction in which the negative hydroxide ions react with positive ammonium ions in the ammonium chloride electrolyte to produce molecules of ammonia and water.

$$Zn(s) + 2MnO_2(s) + 2NH_4Cl(aq) \rightarrow ZnCl_2 + Mn_2O_3(s) + 2NH_3(aq) + H_2O$$

The electromotive force (emf) produced by a Laclanché cell is typically around 1.5 volts with a resistance of several ohms where a porous pot is used. It saw extensive usage in telegraphy, signalling, electric bells and similar applications where intermittent current was required and it was desirable that a battery should require little maintenance.

The Laclanché battery (or wet cell as it was referred to) was the forerunner of the modern dry cell zinc-carbon battery.

Daniell cell

The Daniell cell, also called the gravity cell or crowfoot cell was invented in 1836 by John Frederic Daniell, who was a British chemist and meteorologist. The Daniell cell was a great improvement over the voltaic pile used in the early days of battery development. The Daniell cell's theoretical voltage is 1.1 volts and the chemical reaction is:

$$Zn(s) + Cu^{2+}(aq) \rightarrow Zn^{2+}(aq) + Cu(s).$$

In the Daniell cell, copper and zinc electrodes are immersed in a solution of copper (II) sulphate and zinc sulphate respectively. At the anode, zinc is oxidised per the following half reaction:

$$Zn_{(s)} \rightarrow Zn^{2+}_{(aq)} + 2e^-$$

At the cathode, copper is reduced per the following reaction:

$$Cu^{2+}_{(aq)} + 2e^- \rightarrow Cu_{(s)}$$

In the Daniell cell which, due to its simplicity, is often used in classroom demonstrations, a wire and light bulb may connect the two electrodes. Electrons that are pulled from the zinc travel through the wire, providing an electrical current that illuminates the bulb. In such a cell, the sulphate ions play an important role. Having a negative charge, these anions build up around the anode to maintain a neutral charge. Conversely, at the cathode the copper (II) cations accumulate to maintain this neutral charge. These two processes cause copper solid to accumulate at the cathode and the zinc electrode to dissolve into the solution. Since neither half reaction will occur independently of the other, the two half cells must be connected in a way that will allow ions to move freely between them. A porous barrier or ceramic disk may be used to separate the two solutions while allowing ion flow. When the half cells are placed in two entirely different and separate containers, a salt bridge is often used to connect the two cells. In the above wet-cell, sulphate ions move from the cathode to the anode via the salt bridge and the Zn^{2+} cations move in the opposite direction to maintain neutrality.

Bunsen cell

The Bunsen cell is a zinc-carbon primary cell (colloquially called a battery) composed of a zinc anode in dilute sulphuric acid separated by a porous pot from a carbon cathode in nitric or chromic acid.

The Bunsen cell voltage is about 1.9 volts and arises from the following reaction:

$$Zn + H_2SO_4 + 2HNO_3 \rightarrow ZnSO_4 + 2H_2O + 2NO_2$$

The cell is named after its inventor, German chemist Robert Wilhelm Bunsen. Bunsen used this cell to extract metals from their salts by electrolysis, enabling him to isolate metallic magnesium for the first time.

Chromic acid cell

The chromic acid cell was a type of primary cell which used chromic acid as a depolariser. The chromic acid was usually made by acidifying (with sulphuric acid) a solution of potassium dichromate. The old name for potassium dichromate was potassium bichromate and the cell was often called a bichromate cell. This type of cell is now only of historical interest. The cell was made in two forms—the single-fluid type, attributed to Poggendorff and the two-fluid type, attributed to Fuller. In both cases, cell voltage was about 2 volts.

Clark cell

The Clark cell, invented by English engineer Josiah Latimer Clark in 1873, is a wet-chemical cell (colloquially: battery) that produces a highly stable voltage usable as a laboratory standard.

Clark cells use a zinc or zinc amalgam, anode and a mercury cathode in a saturated aqueous solution of zinc sulphate, with a paste of mercurous sulphate as depolariser.

Clark's original cell was set up in a glass jar in a similar way to a Daniell cell. The copper cathode was replaced by a pool of mercury at the bottom of the jar. Above this was the mercurous sulphate paste and, above that, the zinc sulphate solution. A short zinc rod dipped into the zinc sulphate solution. The zinc rod was supported by a cork with two holes—one for the zinc rod and the other for a glass tube reaching to the bottom of the cell. A platinum wire, fused into the glass tube, made contact with the mercury pool. When complete, the cell was sealed with a layer of marine glue. Clark cells were later made obsolete by the more temperature-independent Weston cell design.

Weston cell

The Weston cell, invented by Edward Weston in 1893, is a wet-chemical cell that produces a highly stable voltage suitable as a laboratory standard for calibration of voltmeters. It was adopted as the International Standard for EMF in 1911.

The anode is an amalgam of cadmium with mercury, the cathode is of pure mercury, the electrolyte is a solution of cadmium sulphate and the depolariser is a paste of mercurous sulphate.

The cell is in an H-shaped glass vessel with the cadmium amalgam in one leg and the pure mercury in the other. Electrical connections to the cadmium amalgam and the mercury are made by platinum wires fused through the lower ends of the legs.

The original design was a saturated cadmium cell producing a convenient 1.0183 volt reference and had the advantage of having a lower temperature coefficient than the previously used Clark cell. (Reference cells must be applied in such a way that no current is drawn from them.)

The temperature coefficient can be reduced by shifting to an unsaturated design, the predominant type today. However, an unsaturated cell's output decreases by some 80 microvolts per year, which is compensated by periodical calibration against a saturated cell.

Rechargeable Battery

A rechargeable battery, also known as a storage battery, is a group of two or more secondary cells. These batteries can be restored to full charge by the application of electrical energy. In other words, they are electrochemical cells in which the electrochemical reaction that releases energy is readily reversible. Rechargeable electrochemical cells are therefore a type of accumulator. They come in many different designs using different chemicals. Commonly used secondary cell chemistries are lead and sulphuric acid, rechargeable alkaline battery (alkaline), nickel cadmium (NiCd), nickel hydrogen (NIH_2), nickel metal hydride (NiMH), lithium ion (Li-ion), and lithium ion polymer (Li-ion polymer).

Rechargeable batteries can offer economic and environmental benefits compared to disposable batteries. Some rechargeable battery types are available in the same sizes as disposable types. While the rechargeable cells have a higher first cost than disposable batteries, rechargeable batteries can be discharged and recharged many times. Proper selection of a rechargeable battery system can reduce toxic materials sent to landfill disposal compared to an equivalent series of disposable batteries. Some manufacturers of NiMH type rechargeable batteries claim a service life up to 3000 charge cycles for their batteries. Unlike nonrechargeable batteries (primary cells), secondary cells must be charged before use. Attempting to recharge nonrechargeable batteries has a small chance of causing a battery explosion.

Rechargeable batteries are susceptible to damage due to reverse charging if they are fully discharged. Fully integrated battery chargers that optimise the charging current are available.

Rechargeable batteries currently are used for applications such as automobile starters, portable consumer devices, tools, and uninterruptible power supplies. Emerging applications in hybrid electric vehicles and electric vehicles are driving the technology to improve cost, reduce weight, and increase lifetime. Rechargeable batteries have been known since the lead acid battery was invented in 1859.

Grid energy storage applications use rechargeable batteries for load levelling, where they store electric energy for use during peak load periods, and for renewable energy uses, such as storing power generated from photovoltaic arrays during the day to be used at night. By charging batteries during periods of low demand and returning energy to the grid during periods of high electrical demand, load-levelling helps eliminate the need for expensive peaking power plants and helps amortise the cost of generators over more hours of operation.

Charging and discharging

During charging, the positive active material is oxidised, producing electrons, and the negative material is reduced, consuming electrons. These electrons constitute the current flow in the external circuit. The electrolyte may serve as a simple buffer for ion flow between the electrodes, as in lithium-ion and nickel-cadmium cells, or it may be an active participant in the electrochemical reaction, as in lead-acid cells. The energy used to charge rechargeable batteries mostly comes from AC current (mains electricity) using an adapter unit. Most battery chargers can take several hours to charge a battery.

Reverse charging

Reverse charging, which damages batteries, is when a rechargeable battery is recharged with its polarity reversed. Reverse charging can occur under a number of circumstances, the two most important being:

1. When a battery is incorrectly inserted into a charger.
2. When multiple batteries are used in series in a device. When one battery completely discharges ahead of the rest, the other batteries in series may force the discharged battery to discharge to below zero voltage.

Chapter 14

Electrothermal Industries

INTRODUCTION

Many chemical products made at high temperatures demand the use of an electric furnace. Electric furnaces are capable of producing temperatures as high as 4100°C. This may be contrasted with the highest commercial combustion-furnace temperatures of about 1700°C. The effects of high temperature are two-fold: The speed of the reaction is increased, and new conditions of equilibrium are established. These new equilibrium conditions have resulted in the production of compounds unknown before the electric furnace. Silicon and calcium carbides are examples of new products thus formed. The electric furnace affords more exact control and more concentration of heat with less thermal loss than is possible with other types of furnaces. This favourable situation is caused by the lack of flue gases and by the high temperature gradient between the source of heat and the heated mass. The electric furnace is much cleaner and more convenient to operate than the combustion furnace. It is operated by alternating current of high amperage, usually with moderate voltage, whereas the electrolytic industries require direct current.

The three chief types of electric furnaces are arc, induction, and resistance. The heat in the arc furnace is produced by an electric arc between two or more electrodes, which are usually graphite or carbon, between the electrodes and the furnace charge or between two or more electrodes which are usually graphite or carbon and may or may not be consumed in the operation. The furnace itself is generally a cylindrical shell lined with a refractory material. Its use is not limited to those industries for which it is a necessity; some companies have rolling-mill operations for common-quality steels where electric-arc furnaces are the sole source of ingots.

The induction furnace may be applied only for conducting substances such as metals, where the electric energy is converted into heat by induced currents set up in the charge. The furnace can be considered as a transformer, with the secondary consisting of the metallic charge and the primary of heavy copper coils connected to the power source. Induction furnaces operate at frequencies

337

from 60 to 5,00,000 Hz, but those used in commercial-scale electrothermal processes do not usually use frequencies above 6000 Hz. The heating effect is obtained with lower field strengths as the frequency is increased. The charge should be placed around an iron core in the low-frequency furnace, but this core is unnecessary for high-frequency furnaces.

When the charged material furnishes the electric resistance required for the necessary heat, the furnace is direct-heated resistance; when high-resistance material is added to the charge for the purpose of creating heat, the furnace is indirect heated. In the electrochemical industry, the arc and the resistance furnaces are mostly used.

ABRASIVE

An abrasive is a material, often a mineral, that is used to shape or finish a workpiece through rubbing which leads to part of the workpiece being worn away. While finishing a material often means polishing it to gain a smooth, reflective surface it can also involve roughening as in satin, matte or beaded finishes.

Abrasives are used very extensively in a wide variety of industrial, domestic, and technological applications. This gives rise to a large variation in the physical and chemical composition of abrasives as well as the shape of the abrasive. Common uses for abrasives include grinding, polishing, buffing, honing, cutting, drilling, sharpening, and sanding.

Abrasives give rise to a form of wound called an abrasion or even an excoriation. Abrasions may arise following strong contract with surfaces made things such as concrete, stone, wood, carpet, and roads, though these surfaces are not intended for use as abrasives.

Abrasives generally rely upon a difference in hardness between the abrasive and the material being worked upon, the abrasive being the harder of the two substances. However, this is not necessary as any two solid materials that repeatedly rub against each other will tend to wear each other away.

Typically, materials used as abrasives are either hard minerals (rated at 7 or above on Mohs scale of mineral hardness) or are synthetic stones, some of which may be chemically and physically identical to naturally occurring minerals but which cannot be called minerals as they did not arise naturally. Abrasives may be classified as either natural or synthetic.

Many synthetic abrasives are effectively identical to a natural mineral, differing only in that the synthetic mineral has been manufactured rather than been mined. Impurities in the natural mineral may make it less effective.

Manufactured Abrasives

Abrasives are shaped for various purposes. Natural abrasives are often sold as dressed stones, usually in the from of a rectangular block. Both natural and synthetic abrasives are commonly available in a wide variety of shapes, often coming as bonded or coated abrasives, including blocks, belts, discs, wheels, sheets, rods and loose grains.

Bonded Abrasives

A bonded abrasive is composed of an abrasive material contained within a matrix, although very fine aluminium oxide abrasive may comprise sintered material. This matrix is called a binder and is often a clay, a resin, a glass or a rubber. This mixture of binder and abrasive is typically shaped into blocks, sticks, or wheels. The most usual abrasive used is aluminium oxide. Also common are silicon carbide, tungsten carbide and garnet. Artificial sharpening stones are often a bonded abrasive and are readily available as a two sided block, each side being a different grade of grit.

Bonded abrasives need to be trued and dressed after they are used. Dressing is cleaning the waste material (swarf and loose abrasive) from the surface and exposing fresh grit. Depending upon the abrasive and how it was used, dressing may involve the abrasive being simply placed under running water and brushed with a stiff brush for a soft stone or the abrasive being ground against another abrasive, such as aluminium oxide used to dress a grinding wheel.

Truing is restoring the abrasive to its original surface shape. Wheels and stones tend to wear unevenly, leaving the cutting surface no longer flat (said to be 'dished out' if it is meant to be a flat stone) or no longer the same diameter across the cutting face. This will lead to uneven abrasion and other difficulties.

Coated Abrasives

A coated abrasive comprises an abrasive fixed to a backing material such as paper, cloth, rubber, resin, polyester or even metal, many of which are flexible. Sandpaper is a very common coated abrasive. Coated abrasives are commonly the same minerals as are used for bonded abrasives.

Coated abrasives may be shaped for use in rotary and orbital sanders, for wrapping around sanding blocks, as handpads, as closed loops for use on belt grinders, as striking surfaces on matchboxes, on diamond plates and diamond steels. Diamond tools, though for cutting, are often abrasive in nature.

Other Abrasives and their Uses

Sand, glass beads, metal pellets and dry ice may all be used for a process called sandblasting (or similar, such as the use of glass beads which is 'bead blasting'). Dry ice will sublimate meaning that there is no residual abrasive left afterwards.

Cutting compound used on automotive paint is an example of an abrasive suspended in a liquid, paste or wax, as are some polishing liquids for silverware and optical media. The liquid, paste or wax acts as a binding agent that keeps the abrasive attached to the cloth which is used to as a backing to move the abrasive across the workpiece. On cars in particular, wax may serve as both a protective agent by preventing exposure of the paint of metal to air and also act as an optical filler to make scratches less noticeable. Toothpaste contains calcium carbonate or silica as a polishing agent to remove plaque and other matter from teeth as the hardness of calcium carbonate is less than that of tooth enamel but more than that of the contaminating agent.

Very fine rouge powder was commonly used for grinding glass, being somewhat replaced by modern ceramics, and is still used in jewellery making for a highly reflective finish.

Cleaning products may also contain abrasives suspended in a paste or cream. They are chosen to be reasonably safe on some linoleum, tile, metal or stone surfaces. However, many laminate surfaces and ceramic topped stoves are easily damaged by these abrasive compounds. Even ceramic/pottery tableware or cookware can damage these surfaces, particularly the bottom of the tableware which is often unglazed in part or in whole and acts as simply another bonded abrasive.

Metal pots and stoves are often scoured with abrasive cleaners, typically in the form of the aforementioned cream or paste or of steel wool.

Human skin is also subjected to abrasion in the form of exfoliation. Abrasives for this can be much softer and more exotic than for other purposes and may include things like almond and oatmeal. Dermabrasion and micro-dermabrasion are now rather commonplace cosmetic procedures which use mineral abrasives.

Choice of Abrasive

The shape, size and nature of the workpiece and the desired finish will influence the choice of the abrasive used. A bonded abrasive grind wheel may be used to commercially sharpen a knife (producing a hollow grind), but an individual may then sharpen the same knife with a natural sharpening stone or an even flexible coated abrasive (like a sandpaper) stuck to a soft, non-slip surface to make achieving a convex grind easier.

Similarly, a brass mirror may be cut with a bonded abrasive, have its surface flattened with a coated abrasive to achieve a basic shape, and then have finer grades of abrasive successively applied culminating in a wax paste impregnated with rouge to leave a sort of grainless finish called, in this case, a mirror finish.

Also, different shapes of adhesive may make it harder to abrade certain areas of the workpiece. Health hazards can arise from any dust produced (which may be ameliorated through the use of a lubricant) which could lead to silicosis (when the abrasive or workpiece is a silicate) and the choice of any lubricant. Besides water, oils are the most common lubricants. These may present inhalation hazards, contact hazards and, as friction necessarily produces heat, flammable material hazards.

An abrasive which is too hard or too coarse can remove too much material or leave undesired scratch marks. A finer or softer abrasive will tend to leave much finer scratch marks which may even be invisible to the naked eye (a grainless finish); a softer abrasive may not even significantly abrade a certain object. A softer or finer abrasive will take longer to cut as tends to cut less deeply than a coarser, harder material. Also, the softer abrasive may become less effective more quickly as the abrasive is itself abraded. This allows fine abrasives to be used in the polishing of metal and lenses where the series of increasingly fine scratches tends to take on a much more shiny or reflective appearance or greater transparency. Very fine abrasives may be used to coat the strop for a cut-throat razors, however, the purpose of stropping is not to abrade material but to straighten the burr on an edge.

The final stage of sharpening Japanese swords called polishing and may be a form of superfinishing.

Different chemical or structural modifications may be made to alter the cutting properties of the abrasive. Other very important considerations are price and availability. Diamond, for a long time considered the hardest substance in existence, is actually softer than fullerite and even harder aggregated diamond nanorods, both of which have been synthesised in laboratories but no commercial process has yet been developed. Diamond itself is expensive due to scarcity in nature and the cost of synthesising it. Bauxite is a very common ore which, along with corundum's reasonably high hardness, contributes to corundum's status as a common, inexpensive abrasive.

Thought must be given to the desired task about using an appropriately hard abrasive. At one end, using an excessively hard abrasive wastes money by wearing it down when a cheaper, less hard abrasive would suffice. At the other end, if too soft, abrasion does not take place in a timely fashion, effectively wasting the abrasive as well as any accruing costs associated with loss of time.

Other Instances of Abrasion

Aside from the aforementioned uses of shaping and finishing, abrasives may also be used to prepare surfaces for application of some sort of paint of adhesive. An excessively smooth surface may prevent paint and adhesives from adhering as strongly as an irregular surface could allow. Inflatable tyre repair kits (which, on bicycles particularly, are actually patches for the inner tube rather than the tyre) require use of an abrasive so that the self-vulcanising cement will stick strongly.

Inadvertently, people who use knives on glass or metal cutting boards are abrading their knife blades. The pressure at the knife edge can easily create microscopic (or even macroscopic) cuts in the board. This cut is a ready source of abrasive material as well as a channel full of this abrasive through which the edge slides. For this reason—without regard for the health benefits—wooden boards are much more desirable. A similar occurrence arises with glass-cutters. Glass-cutters have circular blades that are designed to roll not slide. They should never retrace an already effected cut.

Undesired abrasion may result from the presence of carbon in internal combustion engines. While smaller particles are readily transported by the lubrication system, larger carbon particles may abrade components with close tolerances. The carbon arises from the excessive heating of engine oil or from incomplete combustion. This soot may contain fullerenes which are noted for their extreme hardness—and small size and limited quantity which would tend to limit their effect.

SILICON CARBIDE

Silicon carbide (SiC) is a compound of silicon and carbon bonded together to form ceramics, but it also occurs in nature as the extremely rare mineral moissanite.

Manufacture

Due to the rarity of natural moissanite, silicon carbide is typically man-made. Most often it is used as an abrasive, and more recently as a semiconductor and diamond simulant of gem quality. The simplest manufacturing process is to combine silica sand and carbon in an Acheson graphite electric resistance furnace at a high temperature, between 1600° and 2500°C.

The material formed in the Acheson furnace varies in purity, according to its distance from the graphite resistor heat source. Colourless, pale yellow and green crystals have the highest purity and are found closest to the resistor. The colour changes to blue and black at greater distance from the resistor, and these darker crystals are less pure. Nitrogen and aluminium are common impurities, and they affect the electrical conductivity of SiC. The raw materials for the production of silicon carbide are sand and carbon. Carbon is obtained from anthracite, coke, pitch, or petroleum cokes, and sand contains 98 to 99.5 per cent silica. The equations usually given for the reactions involved are given below:

$$SiO_2(c) + 2C(amorph) \longrightarrow Si(c) + 2CO(g) \qquad \Delta H = +144.8 \text{ kcal}$$

$$Si(c) + C(amorph) \longrightarrow SiC(c) \qquad \Delta H = -30.5 \text{ kcal}$$

The total reaction obtained by combining these equations is

$$SiO_2(c) + 3C(amorph) \longrightarrow SiC(c) + 2CO(g) \qquad \Delta H = +114.3 \text{ kcal}$$

In the production of silicon carbide sand and carbon are mixed in an approximate molar ratio of 1:3 and charged into the furnace. Sawdust, if added, increases the porosity of the charge and permits the circulation of vapours and the escape of the carbon monoxide produced. The charge is built up in the furnace around a heating core of granular carbon. This core is in the center of the 30 to 50 ft-long furnace and connects the electrodes. The walls of the furnace are loose firebrick supported by iron castings and are taken away at the end of a charge to facilitate product removal. Excessive heat loss does not occur because the outside unreacted charge serves as an insulator. A typical initial current between the electrodes is 6000 A at 230 V, and the final current is 20,000 A at 75 V. This change in voltage is due to the decrease in the resistance of the charge as the reaction progresses.

The temperature at the core is 2200°C. The temperature should not become too high, or the silicon carbide will decompose with the volatilisation of silicon and the formation of graphite. Indeed, artificial graphite was so discovered. The energy efficiency is about 50 per cent, and the chemical conversion is from 70 to 80 per cent.

The time of the reaction is about 60 hours, 36 hours of heating and 24 hours of cooling. After cooling, the silicon carbide crystals are removed, with a yield of about 6 to 8 tons per furnace. The larger pieces of crystals are broken, washed, and cleaned by chemical treatment with sulphuric acid and caustic soda. The crystals are classified and screened, the finished product ranging from 6-mesh to fine powder. The outer unreacted part of the charge is combined with the next charge for the furnace. Part of the core used in the furnace can be made of coke suitable for graphite manufacture. After the run is completed, the graphite can be separated from the silicon carbide and converted to desired shapes.

Purer silicon carbide can be made by the more expensive process of chemical vapour deposition (CVD). Commercial large single crystal silicon carbide is grown using a physical vapour transport method commonly known as modified Lely method. Purer silicon carbide can also be prepared by the thermal decomposition of a polymer, poly (methylsilyne), under an inert atmosphere at low temperatures. Relative to the CVD process, the pyrolysis method is advantageous because the polymer can be formed into various shapes prior to thermalisation into the ceramic.

Properties

Silicon carbide exists in at least 70 crystalline forms. Alpha silicon carbide (α-SiC) is the most commonly encountered polymorph; it is formed at temperatures greater than 2000°C and has a hexagonal crystal structure (similar to Wurtzite). The beta modification (β-SiC), with a zinc blende crystal structure (similar to diamond), is formed at temperatures below 2000°C. Until recently, the beta form has had relatively few commercial uses, although there is now increasing interest in its use as a support for heterogeneous catalysts, owing to its higher surface area compared to the alpha form.

Silicon carbide has a density of 3.2 g/cm^3, and its high sublimation temperature (approximately 2700°C) makes it useful for bearings and furnace parts. Silicon carbide does not melt at any known pressure. It is also highly inert chemically. There is currently much interest in its use as a semiconductor material in electronics, where its high thermal conductivity, high electric field breakdown strength and high maximum current density make it more promising than silicon for high-powered devices. In addition, it has strong coupling to microwave radiation, which together with its high sublimation point, permits practical use in heating and casting metals. SiC also has a very low coefficient of thermal expansion (4.0×10^{-6}/K) and experiences no phase transitions that would cause discontinuities in thermal expansion.

Pure SiC is colourless. The brown to black colour of industrial product results from iron impurities. The rainbow-like lustre of the crystals is caused by a passivation layer of silicon dioxide that forms on the surface.

Uses

Circuit elements

Silicon carbide is used for blue LEDs, ultrafast, high-voltage Schottky diodes, MOSFETs and high temperature thyristors for high-power switching.

High-temperature applications

Due to its high thermal conductivity, SiC is also used as substrate for other semiconductor materials such as gallium nitride. Due to its wide band gap, SiC-based parts are capable of operating at high temperature (over 350°C), which together with good thermal conductivity of SiC makes SiC devices good candidates for elevated temperature applications.

Other uses

Lightning arrestors, ultraviolet detector, structural material, astronomy, disc brake, diesel particulate filter, ceramic membrane, cutting tools, heating elements, jewelery, and steel.

ALUMINIUM OXIDE

Aluminium oxide is an amphoteric oxide of aluminium with the chemical formula Al_2O_3. It is also commonly referred to as alumina or aloxite in the mining, ceramic and materials science communities. It is produced by the Bayer process from bauxite. Its most significant use is in the production of aluminium metal, although it is also used as an abrasive due to its hardness and as a refractory material due to its high melting point. Corundum is the most common naturally-occurring crystalline form of aluminium oxide. Much less-common rubies and sapphires are gem-quality forms of corundum with their characteristic colours due to trace impurities in the corundum structure. Aluminium oxide is an electrical insulator but has a relatively high thermal conductivity (40 W/mK). In its most commonly occurring crystalline form, called corundum or α-aluminium oxide, its hardness makes it suitable for use as an abrasive and as a component in cutting tools.

Aluminium oxide is responsible for metallic aluminium's resistance to weathering. Metallic aluminium is very reactive with atmospheric oxygen, and a thin passivation layer of alumina quickly forms on any exposed aluminium surface. This layer protects the metal from further oxidation. The thickness and properties of this oxide layer can be enhanced using a process called anodising. A number of alloys, such as aluminium bronzes, exploit this property by including a proportion of aluminium in the alloy to enhance corrosion resistance.

Manufacture

Aluminium hydroxide minerals are the main component of bauxite, the principal ore of aluminium. The bauxite ore is made up of a mixture of the minerals gibbsite [$Al(OH)_3$], boehmite [γ-$AlO(OH)$], and diaspore [α-$AlO(OH)$] along with iron oxides and hydroxides, quartz and clay minerals.

Bauxite is purified by the Bayer process:

$$Al_2O_3 + 3H_2O + 2NaOH \rightarrow 2NaAl(OH)_4$$

The Fe_2O_3 does not dissolve in the base. The SiO_2 dissolves as silicate $Si(OH)_6^{2-}$. Upon filtering, Fe_2O_3 is removed. When the Bayer liquor is cooled, $Al(OH)_3$ precipitates, leaving the silicates in solution. The mixture is then calcined (heated strongly) to give aluminium oxide:

$$2Al(OH)_3 + heat \rightarrow Al_2O_3 + 3H_2O$$

The formed Al_2O_3 is alumina. The alumina formed tends to be multi-phase, i.e. constituting several of the alumina phases rather than solely corundum. The production process can therefore be optimised to produce a tailored product. The type of phases present affects, for example, the solubility and pore structure of the alumina product which, in turn, affects the cost of aluminium production and pollution control.

Uses

Alumina is a medium for chemical chromatography, available in basic (pH 9.5), acidic (pH 4.5 when in water) and neutral formulations. Health and medical applications include it as a material in hip replacements. It is used in water filters (derived water treatment chemicals such as aluminium sulphate, aluminium chlorohydrate and sodium aluminate, are one of the few methods available to filter water-soluble fluorides out of water). It is also used in toothpaste formulations. It is widely used as a coarse or fine abrasive, including as a much less expensive substitute for industrial diamond. Many types of sandpaper use aluminium oxide crystals. In addition, its low heat retention and low specific heat make it widely used in grinding operations, particularly cutoff tools. Aluminium oxide is widely used in the fabrication of super-conducting devices, particularly single electron transistors and superconducting quantum interference devices (SQUID), where it is used to form highly resistive quantum tunnelling barriers.

Sapphire

Sapphire refers to gem varieties of the mineral corundum, an aluminium oxide (Al_2O_3), when it is a colour other than red, in which case the gem would instead be a ruby. Trace amounts of other elements such as iron, titanium, or chromium can give corundum blue, yellow, pink, purple, orange, or greenish colour, corundum are also sapphires.

Because it is a gemstone, sapphire is commonly worn as jewelry. Sapphire can be found naturally, or manufactured in large crystal boules. Because of its remarkable hardness, sapphire is used in many applications, including infrared optical components, watch crystals, high-durability windows, and wafers for the deposition of semiconductors, such as GaN nanorods and blue LEDs.

Natural sapphires

Sapphire is one of the two gem varieties of corundum, the other being the red ruby. Although blue is the most well known hue, sapphire is any colour of corundum except red, red corundum is known as ruby. Sapphire may also be colourless, and it also occurs in the non-spectral shades gray and black.

The cost of natural sapphire varies depending on their colour, clarity, size, cut, and overall quality as well as geographic origin. Significant sapphire deposits are found in Eastern Australia, Thailand, Sri Lanka, Madagascar, East Africa and in the United States at various locations (Gem Mountain) and in the Missouri River near Helena, Montana. Sapphire and rubies are often found together in the same area, but one gem is usually more abundant.

Blue sapphire

Colour in gemstones breaks down into three components: hue, saturation, and tone. Hue is most commonly understood as the colour of the gemstone. Saturation refers to the vividness or brightness or colourfulness of the hue, and tone is the lightness to darkness of the hue. Blue sapphire exists in various mixtures of its primary and secondary hues, various tonal levels (shades) and at various levels of saturation (brightness): the primary hue must, of course, be blue.

Blue sapphires are evaluated based upon the purity of their primary hue. Purple, violet and green are the normal secondary hues found in blue sapphires. Violet and purple can contribute to the overall beauty of the colour, while green is considered a distinct negative. Blue sapphires with no more than 15 per cent violet or purple are generally said to be of fine quality. Blue sapphires with any amount of green as a secondary hue are not considered to be fine quality. Gray is the normal saturation modifier or mask found in blue sapphires. Gray reduces the saturation or brightness of the hue and therefore has a distinctly negative effect.

Boron Carbide

Boron carbide (chemical formula B_4C) is an extremely hard ceramic material used in tank armour, bulletproof vests, and numerous industrial applications. With a hardness of 9.3 on the mohs scale, it is the fifth hardest material known behind boron nitride, diamond, ultrahard fullerite, and aggregated diamond nanorods.

Boron carbide is now produced industrially by the carbo-thermal reduction of B_2O_3 (boron oxide) in an electric arc furnace. The boric acid is caused to react with coke in a carbon resistnce furnace at 2600°C. The product is about 99 per cent B_4C. It finds specialised use as a powdered abrasive and in moulded shapes such as nozzles for sandblasting.

Its ability to absorb neutrons without forming long lived radionuclides makes the material attractive as an absorbent for neutron radiation arising in nuclear power plants. Nuclear applications of boron carbide include shielding, control rod and shut down pellets. Within control rods, boron carbide is often powdered, to increase its surface area.

Boron Nitride

Boron nitride (BN) is a binary chemical compound, consisting of equal numbers of boron and nitrogen atoms. Its empirical formula is therefore BN. Boron nitride is isoelectronic with carbon and, like carbon, boron nitrides exists as various polymorphic forms, one of which is analogous to diamond and one analogous to graphite. The diamond-like polymorph is one of the hardest materials known and the graphite-like polymorph is a useful lubricant. In addition, both of these polymorphs exhibit radar-absorptive properties.

Cubic Boron Nitride

Cubic boron nitride is extremely hard, although less so than diamond and some related materials. Also like diamond, cubic boron nitride is an electrical insulator and an excellent conductor of heat. This diamond-like polymorph, known as cubic boron nitride, c-BN, β-BN, or z-BN (after zinc blende crystalline structure), is widely used as an abrasive for industrial tools. Its usefulness arises from its insolubility in iron, nickel, and related alloys at high temperatures, whereas diamond is soluble in these metals to give carbides.

Boron Nitride Fibres

Hexagonal BN can be prepared in the form of fibres, structurally similar to carbon fibres, sometimes called white carbon fibre. They can be prepared by thermal decomposition of extruded borazine fibres with addition of boron oxide in nitrogen at 1800°C. The material also arises by the thermal decomposition of cellulose fibres impregnated with boric acid or ammonium tetraborate in an atmosphere of ammonia and nitrogen above 1000°C. Boron nitride fibres are used as reinforcement in composite materials, with the matrix materials ranging from organic resins to ceramics to metals.

CALCIUM CARBIDE

Calcium carbide (CaC_2) is manufactured by heating a lime and carbon mixture to 2000° to 2100°C (3632° to 3812°F) in an electric arc furnace. At those temperatures, the lime is reduced by carbon to calcium carbide and carbon monoxide (CO), according to the following reaction:

$$CaO + 3C \rightarrow CaC_2 + CO$$

Lime for the reaction is usually made by calcining limestone in a kiln at the plant site. The sources of carbon for the reaction are petroleum coke, metallurgical coke, and anthracite coal. Because impurities in the furnace charge remain in the calcium carbide product, the lime should contain no more han 0.5 per cent each of magnesium oxide, aluminium oxide, and iron oxide, and 0.004 per cent phosphorus. Also, the coke charge should be low in ash and sulphur. Analyses indicate that 0.2 to 1.0 per cent ash and 5 to 6 per cent sulphur are typical in petroleum coke. About 991 kilograms (2185 pounds) of lime, 683 kg (1506 lb) of coke, and 17 to 20 kg (37 to 44 lb) of electrode paste are required to produce 1 megagram (Mg) (2205 lb) of calcium carbide. Moisture is removed from coke in a coke dryer, while limestone is converted to lime in a lime kiln. Fines from coke drying and lime operations are removed and may be recycled. The two charge materials are then conveyed to an electric arc furnace, the primary piece of equipment used to produce calcium carbide.

There are three basic types of electric arc furnaces: the open furnace, in which the CO burns to carbon dioxide (CO_2) when it contacts the air above the charge; the closed furnace, in which the gas is collected from the furnace and is either used as fuel for other processes or flared; and the semi-covered furnace, in which mix is fed around the electrode openings in the primary furnace cover resulting in mix seals. Electrode paste composed of coal tar pitch binder and anthracite coal is fed into a steel casing where it is baked by heat from the electric arc furnace before being introduced into the furnace. The baked electrode exits the steel casing just inside the furnace cover and is consumed in the calcium carbide production process. Molten calcium carbide is tapped continuously from the furnace into chills and is allowed to cool and solidify. Then, the solidified calcium carbide goes through primary crushing by jaw crushers, followed by secondary crushing and screening for size. To prevent explosion hazards from acetylene generated by the reaction of calcium carbide with ambient moisture, crushing and screening operations may be performed in either an air-swept environment before the calcium carbide has completely cooled, or in an inert atmosphere. The calcium carbide product is used primarily in generating acetylene and in desulphurising iron.

Emissions and Controls

Emissions from calcium carbide manufacturing include particulate matter (PM), sulphur oxides (SO_x), CO, CO_2, and hydrocarbons. Particulate matter is emitted from a variety of equipment and operations in the production of calcium carbide including the coke dryer, lime kiln, electric furnace, tap fume vents, furnace room vents, primary and secondary crushers, and conveying equipment.

Impurities present in the carbide include phosphide, which produces phosphine when hydrolysed. This reaction was an important part of the industrial revolution in chemistry.

Pure calcium carbide is a colourless solid. The common crystalline form at room temperature is a distorted rock salt structure with the C_2^{2-} units lying parallel.

Applications

Manufacture of acetylene

It is produced by reacting calcium carbide with water:

$$CaC_2 + 2H_2O \rightarrow C_2H_2 + Ca(OH)_2$$

This reaction is the basis of the industrial manufacture of acetylene, and is the major industrial use of calcium carbide.

Production of calcium cyanamide

Calcium carbide reacts with nitrogen at high temperature to form calcium cyanamide:

$$CaC_2 + N_2 \rightarrow CaCN_2 + C$$

Calcium cyanamide is used as fertiliser. It is hydrolysed to cyanamide, H_2NCN.

Steelmaking

Calcium carbide is used:
1. In the desulphurisation of iron (pig iron, cast iron and steel).
2. As a fuel in steelmaking to extend the scrap ratio to liquid iron depending on economics.
3. As a powerful deoxidiser at ladle treatment facilities.

Carbide lamps

Calcium carbide is used in carbide lamps, in which water drips on carbide and the formed acetylene is ignited. These lamps were unusable in coal mines where the presence of the flammable gas methane made them a serious hazard. The presence of flammable gases in coal mines led to the miner safety lamp. However, carbide lamps were used extensively in slate, copper and tin mines, but most have now been replaced by electric lamps. Carbide lamps are still used by some cavers exploring caves and other underground areas, though they are increasingly being replaced in this use by LED lights. They were also used extensively as head lights in early automobiles, although in this application they are also obsolete, having been replaced entirely by electric lamps.

Other uses

Other uses are: (i) in the ripening of fruit, and (ii) it is used in toy cannons.

Fused Quartz

Fused quartz and fused silica are types of glass containing primarily silica in amorphous (non-crystalline) form. They are manufactured using several different processes. Note that glasses formed by the traditional melt-quench methods (heating the material to melting temperatures, then rapidly cooling to the solid glass phase), are often referred to as vitreous, as in vitreous silica. The term vitreous is synonymous with glass, when used in the melt-quench context.

Fused quartz is manufactured by melting naturally occurring quartz crystals of high purity at approximately 2000°C, using either an electrically heated furnace (electrically fused) or a gas/oxygen-fuelled furnace (flame fused). Fused quartz is normally transparent.

Fused quartz can also form naturally. The naturally occurring form is usually referred to as metaquartzite and is formed under metamorphic conditions. An increase in heat causes the crystals within the quartz to become fused together.

Fused silica is produced using high purity silica sand as the feedstock, and is normally melted using an electric furnace, resulting in a material that is translucent or opaque. (This opacity is caused by very small air bubbles trapped within the material.)

Synthetic fused silica is made from a silicon-rich chemical precursor usually using a continuous flame hydrolysis process which involves chemical gasification of silicon, oxidation of this gas to silicon dioxide, and thermal fusion of the resulting dust (although there are alternative processes). This results in a transparent glass with an ultra-high purity and improved optical transmission in the deep ultraviolet. One common method involves adding silicon tetrachloride to a hydrogen-oxygen flame, however, use of this precursor results in environmentally unfriendly by-products including chlorine and hydrochloric acid. To eliminate these by-products, new processes have been developed using an alternative feedstock, which has also resulted in a higher purity fused silica with further improved deep ultraviolet transmission.

Fumed silica is manufactured by a similar flame hydrolysis process to synthetic fused silica, however, it is in the form of a fine powder/dust and is typically used in applications such as fillers for rubbers and plastics, coatings, adhesives, cements, sealants, cosmetics, pharmaceuticals, inks and abrasives.

The optical and thermal properties are superior to those of other types of glass due to its purity (or rather, its lack of impurities). For these reasons, it finds use in situations such as semiconductor fabrication and laboratory equipment. It has better ultraviolet transmission than most other glasses, and so is used to make lenses and other optics for the ultraviolet spectrum. Its low coefficient of thermal expansion also makes it a useful material for precision mirror substrates.

Fused quartz is a noncrystalline form of silicon dioxide (SiO_2), which is also called silica. (The crystalline form of this material is quartz).

Applications

Specially prepared fused silica is also the key starting material used to make optical fibre for telecommunications.

Because of its strength and high melting point (compared to ordinary glass), fused silica is used as the envelope of halogen lamps, which must operate at a high envelope temperature to achieve their combination of high brightness and long life.

The combination of strength, thermal stability, and UV transparency makes it an excellent substrate for projection masks for photolithography. Due to the thermal stability and composition it is used in the semiconductor fabrication furnaces.

Fused quartz has nearly ideal properties for fabricating first surface mirrors such as those used in telescopes. The material behaves in a predictable way and allows the optical fabricator to put a very smooth polish onto the surface and produce the desired figure with fewer testing iterations. In some instances, fused quartz has been used to make the individual elements of special purpose lens.

Fused silica as an industrial raw material is used to make various refractory shapes such as crucibles, trays, shrouds, and rollers for many high temperature thermal processes including steel making, foundries, and glass manufacture. Refractory shapes made from fused silica have excellent thermal shock resistance and are chemically inert to most elements and compounds including virtually all acids, regardless of concentration. Translucent fused silica tubes are commonly used to sheathe electric elements in room heaters, industrial furnaces and other similar applications.

Physical Properties

The extremely low coefficient of thermal expansion, about 0.55 ppm/°C (20°–320°C), accounts for its remarkable ability to undergo large, rapid temperature changes without cracking. UV grade synthetic fused silica has a very low metallic impurity content making it transparent deeper into the ultraviolet. An optic with a thickness of 1 cm will have a transmittance of about 50 per cent at a wavelength of 170 nm, which drops to only a few per cent at 160 nm. However, its infrared transmission is limited by strong water absorptions at 2.2 μm and 2.7 μm.

IR grade fused quartz which is electrically fused, has a greater presence of metallic impurities, limiting its UV transmittance wavelength to around 250 nm, but a much lower water content, leading to excellent infrared transmission up to 3.6 μm wavelength. All grades of transparent fused quartz/ fused silica have near-identical physical properties.

The water content (and therefore infrared transmission of fused quartz and fused silica) is determined by the manufacturing process. Flame fused material always has a higher water content due to the combination of the hydrocarbons and oxygen fuelling the furnace forming hydroxyl [OH] within the material. An IR grade material typically has an [OH] content of <10 parts per million.

Chapter 15

Phosphorus Industries

INTRODUCTION

The use of artificial fertilisers, phosphoric acid, and phosphate salts and derivatives has increased greatly, chiefly because of aggressive and intelligent consumption promotion on the part of various manufacturers. However, before full consumption of these products could be achieved, more efficient and less expensive methods of production had to be developed. During recent decades, the various phosphate industries have made rapid strides in cutting the costs both of production and distribution and have thus enabled phosphorus, phosphoric acid, and its salts to be employed in wider fields and newer derivatives to be introduced. Supplementing the development of more efficient phosphorus industries have been the epoch-making pure chemical studies of phosphorus, in its old and in its new compounds.

These phosphates are not simple inorganic chemicals, as was assumed several decades ago, and their study has become a unique and complicated branch of chemistry that may some day be compared with the carbon (organic) or silicon branches of today. The properties of phosphorus chemicals are unique because of the important role of phosphorus in many biochemical processes, the ability of polyphosphates to complex or sequester many metal cations, and versatility in forming various types of organic and inorganic polymers.

ROCK PHOSPHATE

Rock phosphate is a general term for rock that contains a high concentration of phosphate minerals, which commonly belong to the apatite group. Phosphate rock minerals are the only significant global resources of phosphorus. Phosphorus is an essential element for plant and animal nutrition. Mined rock phosphate is primarily used in the production of phosphate fertilisers for agriculture. Phosphorus from rock phosphate is also used in animal feed supplements, food preservatives, anti-corrosion agents, cosmetics, fungicides, ceramics, water treatment and metallurgy.

Manufacturing Process

The flow sheet is given in Fig. 15.1.

Reactions

1. $2Ca_3(PO_4)_2 + 10C + 6SiO_2 \xrightarrow{250°-450°C} P_4 \text{ (yellow)} + CaSiO_3 + 10CO$

2. $P_4 \text{ (yellow)} \xrightarrow[\text{heating}]{20°C} P_4 \text{ (red)}$

Raw Material

Low grade rock, coke reducing agent, and silica as flux. Material requirements:
One ton of yellow phosphorus

Calcium phosphate		8 T
Sand		13.5 T
Coke		1.7 T
Carbon electrode	18 to 25 kg	
Consumption electricity	12,000–15,000 kwh	
Cooling water	200–250 tons	

Process Description

Rock is ground, mixed with coke and converted into nodules to prevent entrainment of fines with phosphorus and to have better temperature distribution and the mixture is then fed to an electric arc furnace. The reaction is highly endothermic.

The energy is supplied from electric arc of a 250–300 volt AC three phase. The temperature maintained in the order of 1400°–1500°C where the mass gets fused and reduction takes place. About 20 per cent of the fluorine present in the rock is converted to SiF_4 (silicon fluoride) and volatilised. This in presence of water vapour gives silica and hydrofluorosilic acid by the reaction:

$$3SiF_4 + 2H_2O \longrightarrow 2H_2SiF_4 + SiO_2$$

The gases and dust coming out from the furnace are sent to an electrostatic precipitator to remove dust and carbon particles. The gases are then cooled and sent to a condenser where yellow phosphorus gets collected under water. The carbon monoxide is used as a fuel for preparing the furnace raw material.

This yellow phosphorus can be converted to red phosphorus in a converted vessels at a temperature of 250°–450°C in absence of air.

The reaction is exothermic and the mass comes out as solid red phosphorus which is washed with sodium carbonate and water to remove yellow variety and then dried.

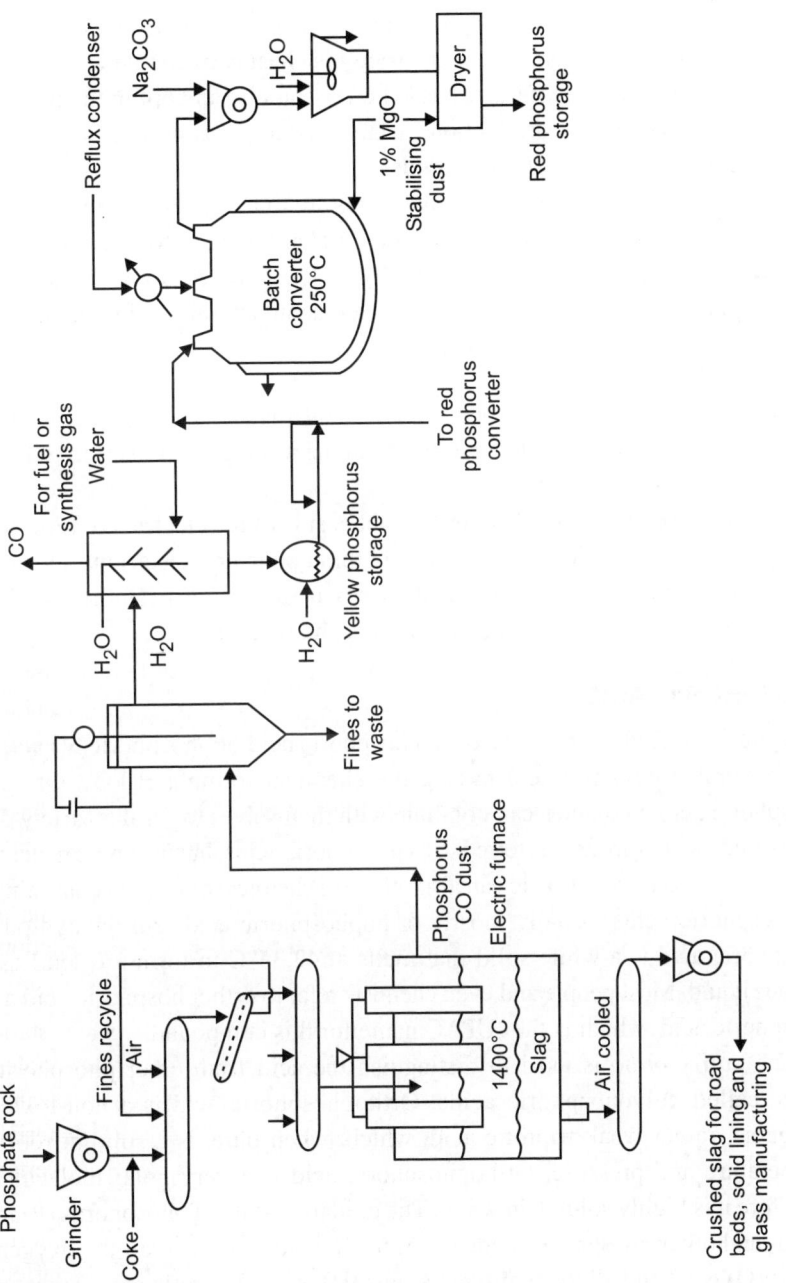

Fig. 15.1. Production of elemental phosphorus.

Engineering Problems

1. The process consumes significant amount of electricity and therefore, is only economical where electricity is cheaply available.
2. The design of the furnace has to be very precise and optimum in order to get economic production. A large reduction zone is essential for complete release of phosphorus.
3. Exclusion of air and safety measures are to be given priority.
4. Transportation of phosphorus is a problem as, it has to be transported under water. Therefore, normally at the plant it is converted into phosphoric acid which is then used for manufacturing various products.

Economics

1. Although low grade rock can be used (it is an advantage in electric process) but it should be enriched at the mining plant itself to reduce transportation cost.
2. About 60 to 70 per cent of the total cost is of the electric arc furnace.
3. In a country like India, where both sulphur and electricity are not available relative merits and demerits are to be closely studied for the wet process for manufacture of phosphoric acid.

PHOSPHORIC ACID

Phosphoric acid, also known as orthophosphoric acid or phosphoric(V) acid, is a mineral (inorganic) acid having the chemical formula H_3PO_4. Ortho-phosphoric acid molecules can combine with themselves to form a variety of compounds which are also referred to as phosphoric acids, but in a more general way. The term phosphoric acid can also refer to a chemical or reagent consisting of phosphoric acids, usually mostly orthophosphoric acid. Pure anhydrous phosphoric acid is a white solid that melts at 42.35°C to form a colourless, viscous liquid. Most people and even chemists refer to orthophosphoric acid as phosphoric acid, which is the IUPAC name for this compound.

The prefix *ortho* is used to distinguish the acid from other phosphoric acids, called polyphosphoric acids. Orthophosphoric acid is a non-toxic, inorganic, rather weak triprotic acid, which, when pure, is a solid at room temperature and pressure. Orthophosphoric acid is a very polar molecule; therefore it is highly soluble in water. The oxidation state of phosphorus (P) in ortho- and other phosphoric acids is +5; the oxidation state of all the oxygen atoms (O) is −2 and all the hydrogen atoms (H) is +1. Triprotic means that an orthophosphoric acid molecule can dissociate up to three times, giving up an

H^+ each time, which typically combines with a water molecule, H_2O, as shown in these reactions:

$$H_3PO_{4(s)} + H_2O_{(l)} \rightarrow H_3O^+_{(aq)} + H_2PO^-_{4(aq)} \; K_{a1} = 7.5 \times 10^{-3}$$

$$H_2PO_4^-{}_{(aq)} + H_2O_{(l)} \rightarrow H_3O^+{}_{(aq)} + HPO_4^{2-}{}_{(aq)} \; K_{a2} = 6.2 \times 10^{-8}$$

$$HPO_4^{2-}{}_{(aq)} + H_2O_{(l)} \rightarrow H_3O^+{}_{(aq)} + PO_4^{3-}{}_{(aq)} \; K_{a3} = 2.14 \times 10^{-13}$$

The anion after the first dissociation, $H_2PO_4^-$, is the dihydrogen phosphate anion. The anion after the second dissociation, HPO_4^{2-}, is the hydrogen phosphate anion. The anion after the third dissociation, PO_4^{3-}, is the phosphate or orthophosphate anion. For each of the dissociation reactions shown above, there is a separate acid dissociation constant, called K_{a1}, K_{a2}, and K_{a3} given at 25°C. Associated with these three dissociation constants are corresponding $pK_{a1}=2.12$, $pK_{a2}=7.21$, and $pK_{a3}=12.67$ values at 25°C. Even though all three hydrogen (H) atoms are equivalent on an orthophosphoric acid molecule, the successive K_a values differ since it is energetically less favourable to lose another H^+ if one (or more) has already been lost and the molecule/ion is more negatively-charged.

Because the triprotic dissociation of orthophosphoric acid, the fact that its conjugate bases (the phosphates mentioned above) cover a wide pH range, and, because phosphoric acid/phosphate solutions are, in general, nontoxic, mixtures of these types of phosphates are often used as buffering agents or to make buffer solutions, where the desired pH depends on the proportions of the phosphates in the mixtures. Similarly, the nontoxic, anion salts of triprotic organic citric acid are also often used to make buffers. Phosphates are found pervasively in biology, especially in the compounds derived from phosphorylated sugars, such as DNA, RNA, and adenosine triphosphate (ATP).

Upon heating orthophosphoric acid, condensation of the phosphoric units can be induced by driving off the water formed from condensation. When one molecule of water has been removed for each two molecules of phosphoric acid, the result is pyrophosphoric acid ($H_4P_2O_7$). When an average of one molecule of water per phosphoric unit has been driven off, the resulting substance is a glassy solid having an empirical formula of HPO_3 and is called metaphosphoric acid. Metaphosphoric acid is a singly anhydrous version of orthophosphoic acid and is sometimes used as a water- or moisture-absorbing reagent. Further dehydrating is very difficult, and can be accomplished only by means of an extremely strong desiccant (and not by heating alone).

It produces phosphoric anhydride, which has an empirical formula P_2O_5, although an actual molecule has a chemical formula of P_4O_{10}. Phosphoric anhydride is a solid, which is very strongly moisture-absorbing and is used as a desiccant.

Manufacture of Phosphoric Acid

Electric process

Elemental phosphorus as described before, is first product in an electric arc furnace and then this is oxidised with oxygen or air to give phosphorus pentaoxide in a stainless steel or acid resistant bricks lined reactor. The phosphorus pentaoxide formed is removed by water spray to get phosphoric acid.

The 85 per cent acid is given sulphuric acid wash to remove entrained calcium salts as $CaSO_4$, powdered silica added to remove hydrofluoric acid with a countercurrent scrubbing with hydrogen sulphide to remove arsenic as AsS_4.

The sludge is removed in a sand filter and acid sold as such or diluted to 50 per cent acid. The following reaction takes place.

$$2P + 5O_2 \longrightarrow P_2O_5; \qquad \Delta H = -720 \text{ kcal.}$$

$$P_2O_5 + 3H_2O \longrightarrow 2H_3PO_4; \qquad \Delta H = -45 \text{ kcal.}$$

Material requirement: 1 T of 100 per cent H_3PO_4 on 90 per cent yield basis.

Phosphorus	0.33 T
Air	1260 nm^3

Wet process (strong acid process)

1. Flow sheet (Fig. 15.2).
2. Chemical reactions:
   (a) $Ca_3(PO_4) + 3H_2SO_4 + 6H_2O \longrightarrow 2H_3PO_4 + 5(CaSO_4 \cdot 2H_2O)$

   $\qquad\qquad\qquad\qquad\qquad\qquad\qquad\qquad\qquad\qquad$ Gypsum

   (b) Side reaction:

   $$CaF_2 + H_2SO_4 + 2H_2 \longrightarrow 2HF + CaSO_4 + 2H_2O$$

   $$6HF + SiO_2 \longrightarrow H_2SiF_6 + 2H_2O$$
3. Raw materials: High grade enriched beneficiated rock and strong sulphuric acid.
4. Material requirements.
   1 T of 100 per cent H_3PO_4 on 90 per cent yield basis.
Phosphate rock	2.5 T
98 per cent sulphuric acid	2.0 T
5. Process description: Rock is ground to about 200 mesh and is mixed with recycle weak phosphoric acid. This is sent to a reaction vessel where strong sulphuric acid in controlled amounts is added.
   A single baffled reactor or series of 5/6 reactors can be designed with proper retention time distribution. The reaction is exothermic and

temperature is maintained by passing air across. The total retention time is about 4 to 6 hours. The gypsum acid slurry goes to a vacuum filter where weak phosphoric acid is removed from the gypsum as a precipitate. This is washed with water and the weak washings are used for mixing with ground rock as described above. The gypsum is dried to be either used as plaster of paris or as a fertiliser. Temperature control is necessary to prevent formation of $CaSO_4 \cdot \frac{1}{2} H_2O$ or anhydrous $CaSO_4$ which is difficult to filter. This gypsum can also be used for manufacture of ammonium sulphate. The dilute acid is concentrated to 50 per cent or more.

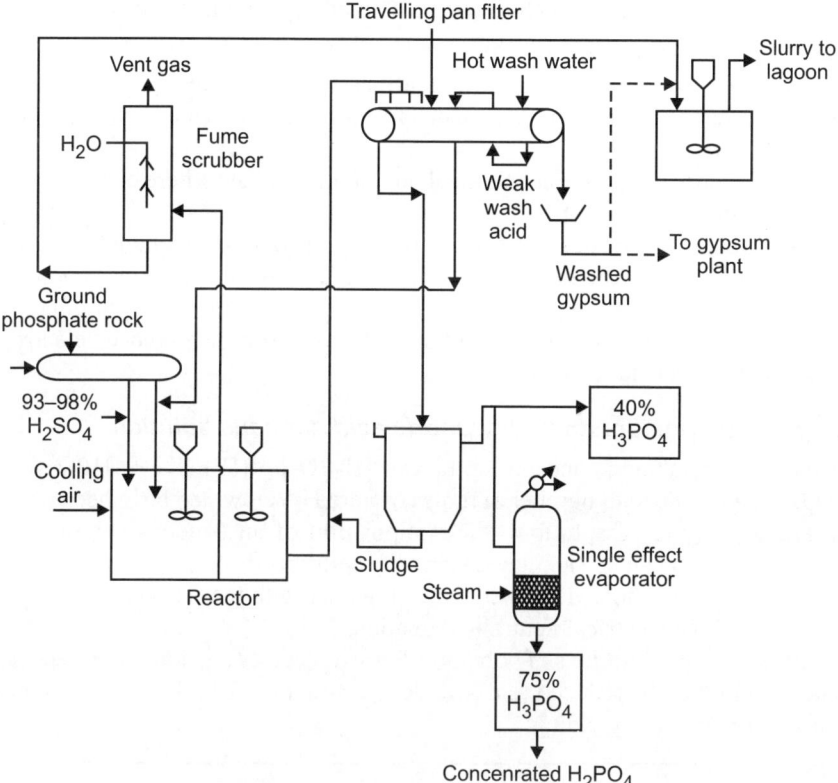

Fig. 15.2. Production of phosphoric acid.

6. Engineering problems:
    (a) Fineness of rock phosphate is of major importance as the whole economics depends upon this. The fine rock where on one hand gives greater production rate/yield, also consumes large amounts of power (used for grinding).

(b) If the temperature is higher than 100°C semihydrate $CaSO_4 \cdot \frac{1}{2}H_2O$ and anhydrous $CaSO_4$ are formed which are difficult to filter.

(c) Some sulphuric acid (1 to 1.5 per cent) is allowed to go with gypsum to make it easily filterable.

Comparison of the two processes

1. Strong acid process or wet process uses high grade rock whereas in electric process low grade can also be used.
2. The first cost of the plant in electric process is higher as compare to wet process.
3. Gypsum, a by-product of wet process is a useful material which can either be used as such or can be converted to ammonium sulphate.

$$(NH_4)_2CO_3 + CaSO_4 \longrightarrow CaCO_3 + (NH_4)_2SO_4$$

4. Wet process uses sulphuric acid based on imported sulphur which after some time would present a formidable problem in a country like India.
5. Electric process is economical only in those place where electricity is quite cheap.

Wet process inspite of some disadvantages is in vogue throughout the world and a 80 per cent of the total acid is produced by this process. For a country like India where sulphuric acid is entirely based on imported sulphur either a new process of manufacture of sulphuric acid is to be evolved or electric process should be adopted.

pH and composition of a phosphoric acid aqueous solution

For a given total acid concentration $[A] = [H_3PO_4] + [H_2PO_4^-] + [HPO_4^{2-}] + [PO_4^{3-}]$ ($[A]$ is the total number of moles of pure H_3PO_4 which have been used to prepare 1 litre of solution), the composition of an aqueous solution of phosphoric acid can be calculated using the equilibrium equations associated with the three reactions described above together with the $[H^+][OH^-] = 10^{-14}$ relation and the electrical neutrality equation.

Possible concentrations of polyphosphoric molecules and ions is neglected. The system may be reduced to a fifth degree equation for $[H^+]$ which can be solved numerically, yielding:

$[A]$ (mol/l)	pH	$[H_3PO_4]/$ $[A]$(%)	$[H_2PO_4^-]/$ $[A]$(%)	$[HPO_4^{2-}]/$ $[A]$(%)	$[PO_4^{3-}]/$ $[A]$(%)
1	1.08	91.7	8.29	6.20×10^{-6}	1.60×10^{-17}
10^{-1}	1.62	76.1	23.9	6.20×10^{-5}	5.55×10^{-16}
10^{-2}	2.25	43.1	56.9	6.20×10^{-4}	2.33×10^{-14}
10^{-3}	3.05	10.6	89.3	6.20×10^{-3}	1.48×10^{-12}

(Contd ...)

$[A]$ (mol/l)	pH	$[H_3PO_4]/$ $[A](\%)$	$[H_2PO_4^-]/$ $[A](\%)$	$[HPO_4^{2-}]/$ $[A](\%)$	$[PO_4^{3-}]/$ $[A](\%)$
10^{-4}	4.01	1.30	98.6	6.19×10^{-2}	1.34×10^{-10}
10^{-5}	5.00	0.133	99.3	0.612	1.30×10^{-8}
10^{-6}	5.97	1.34×10^{-2}	94.5	5.50	1.11×10^{-6}
10^{-7}	6.74	1.80×10^{-3}	74.5	25.5	3.02×10^{-5}
10^{-10}	7.00	8.24×10^{-4}	61.7	38.3	8.18×10^{-5}

For large acid concentrations, the solution is mainly composed of H_3PO_4. For $[A] = 10^{-2}$, the pH is closed to pK_{a1}, giving an equimolar mixture of H_3PO_4 and $H_2PO_4^-$. For $[A]$ below 10^{-3}, the solution is mainly composed of $H_2PO_4^-$ with $[HPO_4^{2-}]$ becoming non-negligible for very dilute solutions. $[PO_4^{3-}]$ is always negligible.

Chemical reagent

Pure 75–85 per cent aqueous solutions (the most common) are clear, colourless, odourless, non-volatile, rather viscous, syrupy liquids, but still pourable. Phosphoric acid is very commonly used as an aqueous solution of 85 per cent phosphoric acid or H_3PO_4.

Because it is a concentrated acid, an 85 per cent solution can be corrosive, although nontoxic when diluted. Because of the high percentage of phosphoric acid in this reagent, at least some of the orthophosphoric acid is condensed into polyphosphoric acids in a temperature-dependent equilibrium, but, for the sake of labelling and simplicity, the 85 per cent represents H_3PO_4 as if it were all orthophosphoric acid.

Other percentages are possible too, even above 100 per cent, where the phosphoric acids and water would be in an unspecified equilibrium, but the overall elemental mole content would be considered specified. When aqueous solutions of phosphoric acid and/or phosphate are dilute, they are in or will reach an equilibrium after a while where practically all the phosphoric/ phosphate units are in the ortho- form.

Preparation of Hydrogen Halides

Phosphoric acid reacts with halides to form the corresponding hydrogen halide gas (steamy fumes are observed on warming the reaction mixture). This is a common practice for the laboratory preparation of hydrogen halides.

$$NaCl(s) + H_3PO_4(l) \rightarrow NaH_2PO_4(s) + HCl(g)$$

$$NaBr(s) + H_3PO_4(l) \rightarrow NaH_2PO_4(s) + HBr(g)$$

$$NaI(s) + H_3PO_4(l) \rightarrow NaH_2PO_4(s) + HI(g)$$

Uses

Rust removal

Phosphoric acid may be used by direct application to rusted iron, steel tools, or surfaces to convert iron(III) oxide (rust) to a water-soluble phosphate compound. It is usually available as a greenish liquid, suitable for dipping (acid bath), but is more generally used as a component in a gel, commonly called naval jelly. As a thick gel, it may be applied to sloping, vertical, or even overhead surfaces. Care must be taken to avoid acid burns of the skin and especially the eyes, but the residue is easily diluted with water. When sufficiently diluted, it can even be nutritious to plant life, containing the essential nutrients phosphorus and iron.

It is sometimes sold under other names, such as rust remover or rust killer. It should not be directly introduced into surface water such as creeks or into drains, however. After treatment, the reddish-brown iron oxide will be converted to a black iron phosphate compound coating that may be scrubbed off. Multiple applications of phosphoric acid may be required to remove all rust. The resultant black compound can provide further corrosion resistance (such protection is somewhat provided by the superficially similar Parkerising and blued electrochemical conversion coating processes). After application and removal of rust using phosphoric acid compounds, the metal should be oiled (if to be used bare, as in a tool) or appropriately painted, by using a multiple coat process of primer, intermediate, and finish coats.

Processed food

Food-grade phosphoric acid (often labelled as E number E338.) is used to acidify foods and beverages such as various colas, but not without controversy regarding its health effects. It provides a tangy or sour taste and, being a mass-produced chemical, is available cheaply and in large quantities. The low cost and bulk availability is unlike more expensive natural seasonings that give comparable flavours, such as citric acid which is obtainable from lemons and limes. (However, most citric acid in the food industry is not extracted from citrus fruit, but fermented by *Aspergillus niger* mould from scrap molasses, waste starch hydrolysates and phosphoric acid.)

Biological effects on bone calcium and kidney health

Phosphoric acid, used in many soft drinks (primarily cola), has been linked to lower bone density in epidemiological studies. For example, a study using dual-energy X-ray absorptiometry rather than a questionnaire about breakage, provides reasonable evidence to support the theory that drinking cola results in lower bone density.

On the other hand, a study funded by Pepsi suggests that low intake of phosphorus leads to lower bone density. The study does not examine the effect of phosphoric acid, which binds with magnesium and calcium in the digestive tract to form salts that are not absorbed, but, rather, it studies general phosphorus intake.

Cola consumption has also been linked to chronic kidney disease and kidney stones through medical research. This study differentiated between the effects of cola (generally contains phosphoric acid), non-cola carbonated beverages (substitute citric acid) and coffee (control for caffeine), and found that drinking 2 or more colas per day more than doubled the incidence of kidney disease.

Medical

Phosphoric acid is used in dentistry and orthodontics as an etching solution, to clean and roughen the surfaces of teeth where dental appliances or fillings will be placed. Phosphoric acid is also an ingredient in over-the-counter anti-nausea medications that also contain high levels of sugar (glucose and fructose). It should not be used by diabetics without consultation with a doctor. This acid is also used in teeth whiteners to eliminate any plaque that may be on your teeth.

Kiln Phosphoric Acid

Kiln phosphoric acid (KPA) process technology is the most recent technology. Called the improved hard process, this technology will both make low grade phosphate rock reserves commercially viable and will increase the P_2O_5 recovery from existing phosphate reserves. This will significantly extend the commercial viability phosphate reserves.

Other Applications

Other applications of kiln phosphoric acid are:
1. Phosphoric acid is used as the electrolyte in phosphoric acid fuel cells. It is also used as an external standard for phosphorus-31, nuclear magnetic resonance (NMR).
2. Phosphoric acid is used as a cleaner by construction trades to remove mineral deposits, cementitious smears, and hard water stains. It is also used as a chelant in some household cleaners aimed at similar cleaning tasks.
3. Hot phosphoric acid is used in microfabrication to etch silicon nitride (Si_3N_4). It is highly selective in etching Si_3N_4 instead of SiO_2, silicon dioxide.
4. Phosphoric acid is used as a flux by hobbyists (such as model railroaders) as an aid to soldering.

5. Phosphoric acid is also used in hydroponics pH solutions to lower the pH of nutrient solutions. While other types of acids can be used, phosphorus is a nutrient used by plants, especially during flowering, making phosphoric acid particularly desirable. General hydroponics pH down liquid solution contains phosphoric acid in addition to citric acid and ammonium bisulphate with buffers to maintain a stable pH in the nutrient reservoir.

6. Phosphoric acid is used as a pH adjuster in cosmetics and skin-care products.

7. Phosphoric acid is used as a chemical oxidising agent for activated carbon production.

8. Phosphoric acid is also used for high pressure liquid chromotography (HPLC).

9. Phosphoric acid can be used as an additive to stabilise acidic aqueous solutions within a wanted and specified pH range.

PHOSPHORUS

Phosphorus is the chemical element that has the symbol P and atomic number 15. A multivalent nonmetal of the nitrogen group, phosphorus is commonly found in inorganic phosphate rocks. Due to its high reactivity, phosphorus is never found as a free element in nature on earth.

Phosphorus is a component of DNA, RNA, ATP, and also the phospholipids which form all cell membranes. It is thus an essential element for all living cells. The most important commercial use of phosphorus-based chemicals is the production of fertilisers.

Phosphorus compounds are also widely used in explosives, nerve agents, friction matches, fireworks, pesticides, toothpaste and detergents.

Phosphorus is an excellent example of an element that exhibits allotropy, as its various allotropes have strikingly different properties.

The two most common allotropes are white phosphorus and red phosphorus. A third form, scarlet phosphorus, is obtained by allowing a solution of white phosphorus in carbon disulphide to evaporate in sunlight. A fourth allotrope, black phosphorus, is obtained by heating white phosphorus under very high pressures (12,000 atmospheres). In appearance, properties and structure it is very like graphite, being black and flaky, a conductor of electricity and has puckered sheets of linked atoms. Another allotrope is diphosphorus — which is highly reactive.

Both phosphorus and arsenic have many allotropes, but only the white and red forms predominate. White phosphorus and yellow arsenic both have four atoms arranged in a tetrahedral structure in which each atom is bound to the other three atoms by a single bond. This form of the elements is the least stable, most reactive, more volatile, less dense, and more toxic than the other allotropes. The toxicity of white phosphorus led to its discontinued use in matches. The crystal melts at 44°C and has a density of 1.83 kg/l. The liquid boils at 280°C. White phosphorus (P_4) exists as individual molecules made up of four atoms in a tetrahedral arrangement, resulting in very high ring strain and instability. It contains six single bonds.

White phosphorus is a white, waxy transparent solid. This allotrope is thermodynamically unstable at normal condition and will gradually change to red phosphorus. This transformation, which is accelerated by light and heat, makes white phosphorus almost always contain some red phosphorus and therefore appear yellow. For this reason, it is also called yellow phosphorus. It glows greenish in the dark (when exposed to oxygen), is highly flammable and pyrophoric (self-igniting) upon contact with air as well as toxic (causing severe liver damage on ingestion). Because of pyrophoricity white phosphorus is used as an additive in napalm. The odour of combustion of this form has a characteristic garlic smell, and samples are commonly coated with white (di)phosphorus pentoxide, which consists of P_4O_{10} tetrahedra with oxygen inserted between the phosphorus atoms and at their vertices. White phosphorus is insoluble in water but soluble in carbon disulphide. The white allotrope can be produced using several different methods. In one process, calcium phosphate, which is derived from phosphate rock, is heated in an electric or fuel-fired furnace in the presence of carbon and silica. Elemental phosphorus is then liberated as a vapour and can be collected under phosphoric acid. This process is similar to the first synthesis of phosphorus from calcium phosphate in urine.

In red phosphorus one of the bonds in P_4 described above has been broken, and one additional bond is formed with a neighbouring tetrahedron. Red phosphorus may be formed by heating white phosphorus to 250°C (482°F) or by exposing white phosphorus to sunlight. Phosphorus after this treatment exists as an amorphous network of atoms which reduces strain and gives greater stability; further heating results in the red phosphorus becoming crystalline. Red phosphorus does not catch fire in air at temperatures below 240°C, whereas white phosphorus ignites at about 30°C.

SODIUM PHOSPHATES

Sodium phosphate is a generic term for the salts of sodium and phosphate. They are: (i) monosodium phosphate (NaH_2PO_4), (ii) disodium phosphate (Na_2HPO_4), and (iii) trisodium phosphate (Na_3PO_4).

Uses

Sodium phosphates are used as food additives. Sodium phosphates are added to many foods as an emulsifier to prevent oil separation. Some examples are processed cheeses, processed meats, ready-made meals and tinned (canned) soups. Sodium phosphates are also commonly added to powdered soups, boullions and gravy mixtures. Sodium phosphates can also be used as a leavening agent. Some examples of these foods include the batter coating on breaded fish or chicken, and commercially baked cakes. Adding sodium phosphates to food increases the shelf-life of the food; maintaining the texture and appearance of the food. Sodium phosphate (trisodium phosphate) is also an ingredient of cleaning products, e.g. sugar soap.

Oral sodium phosphates for bowel preparation for colonoscopy carry a risk of kidney injury under the form of phosphate nephropathy.

PYROPHOSPHATES

Tetrasodium pyrophosphate (TSPP) ($Na_4P_2O_7$), is used as a water softener and as soap and detergent builder. It is manufactured by reacting phosphoric acid and soda ash to yield a disodium phosphate (DSP) solution, which may be dried to give anhydrous Na_2HPO_4 or crystallised to give $Na_2HPO_4 \cdot 2H_2O$ or $Na_2HPO_4 \cdot 7H_2O$. These compounds are calcined at a high temperature in an oil-or gas-fired rotary kiln to yield TSPP in a plant such as that shown in Fig. 15.3. The reactions which take place are given below:

$$2Na_2HPO_4 \longrightarrow Na_4P_2O_7 + H_2O$$

$$2Na_2HPO_4 \cdot 2H_2O \longrightarrow Na_4P_2O_7 + 5H_2O$$

A nonhygroscopic sodium acid pyrophosphate is used extensively as a chemical leavening agent in making doughnuts, cakes, and packaged biscuit doughs. It is manufactured by partially dehydrating monosodium acid orthophosphate at a temperature between 25° and 250°C over the course of 6 to 12 hours.

$$2NaH_2PO_4 \longrightarrow Na_2H_2P_2O_7 + H_2O$$

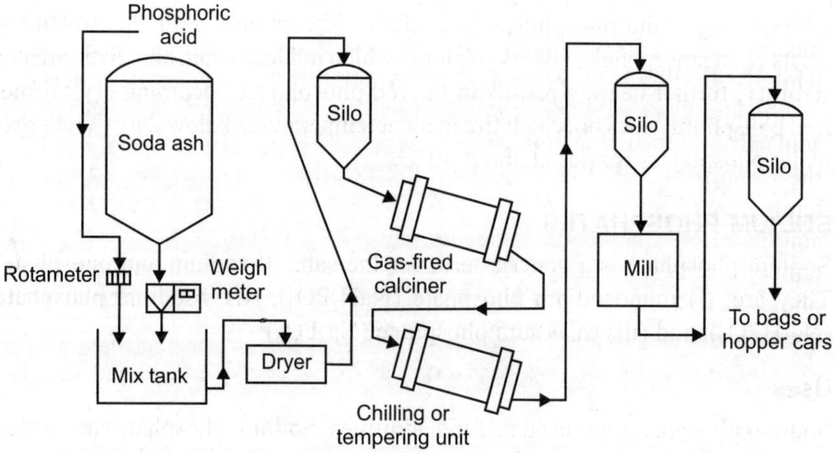

Fig. 15.3. Oil or gas-fired rotary kiln to yield TSPP in a plant.

PHOSPHORUS FIRE-RETARDANT CHEMICALS

In recent years, there has been an obvious increase in the use of fire-retardants both in fire-proofing various textiles and for combatting forest fires. Phosphorus compounds are most suitable for these purposes. The water resistant fire retardants commonly used for cotton textiles are a combination of trisazirdinly phosphorus oxide (commonly referred to in the literature as APO) and tetrakishydroxymethyl phosphonium chloride (THPC). APO is prepared by reacting phosphoryl trichloride and ethyleneimine:

$$POCl_3 + \begin{matrix} CH_2 \!-\!\!-\! CH_2 \\ \diagdown \quad \diagup \\ NH \end{matrix} \xrightarrow{\text{base}} \left[\begin{matrix} CH_2 \!-\!\!-\! CH_2 \\ \diagdown \quad \diagup \\ NH \end{matrix} \right]_3 PO$$

THPC is readily made from phosphine and formaldehyde:

$$4CH_2O + PH_3 + HCl \longrightarrow (HOCH_2)_4PCl$$

APO is a very reactive molecule and presumably reacts with both the hydroxyl groups of cellulose and THPC to form a polymeric material. For combatting forest and brush fires, mixtures based on $(NH_4)2HPO_4$ or $(NH_4)2SO_4$, thickening agents, colouring matter, and corrosion inhibitors are most commonly used. It is believed that the phosphorus compounds act as catalysts to produce non-combustible gas and char. Furthermore, phosphorus compounds which can yield phosphoric acid from thermal degradation are effective in suppressing glow reactions.

BAKING POWDER

Baking powder is a dry chemical leavening agent used in cooking, mainly baking. It is most often found in quick breads like pancakes, waffles, and muffins. When dissolved in water the baking powder's ingredients react and emit carbon dioxide gas which expands, producing bubbles to leaven the mixture. Baking powder is used instead of yeast because its action is instantaneous, while yeast takes two to three hours to produce its leavening action.

Most modern baking powders are double acting, that is, they contain two acid salts, one which reacts at room temperature, producing a rise as soon as the dough or batter is prepared, and another which reacts at a higher temperature, causing a further rise during baking. Common low-temperature acid salts include cream of tartar, calcium phosphate, and citrate. High-temperature acid salts are usually aluminium salts, such as calcium aluminium phosphate. Baking powders that contain only the low-temperature acid salts are called single acting.

Usage

Generally (in countries where the cup is used as a standard measure in cookery) one teaspoon (5 ml) of baking powder is used to raise a mixture of one cup

(200–250 ml) of flour, one cup of liquid, and one egg. However, if the mixture is acidic, baking powder's additional acids will remain unconsumed in the chemical reaction and often lend an unpleasant chemical taste to food. High acidity can be caused by ingredients like buttermilk, lemon, yoghurt, citrus or honey. When excessive acidity is present, some of the baking powder is replaced with baking soda. For example, one cup of flour, one egg, and one cup of buttermilk requires only $\frac{1}{2}$ teaspoon of baking powder—the remaining leavening is caused by buttermilk acids reacting with $\frac{1}{4}$ teaspoon of baking soda.

Moisture and heat can cause baking powder to lose its effectiveness over time, and commercial varieties have a somewhat arbitrary expiration date printed on the container. Regardless of the expiration date, the effectiveness can be tested by placing a teaspoon of the powder into a small container of water. If it fizzes energetically, it's still active and usable.

Substituting in recipes

Baking powder is generally just baking soda mixed with an acid, and a number of kitchen acids may be mixed with baking soda to simulate commercial blends of baking powder. Vinegar (dilute ethanoic acid), especially white vinegar, is also a common acidifier in baking; for example, many heirloom chocolate cake recipes call for a tablespoon or two of vinegar. Where a recipe already uses buttermilk or yoghurt, baking soda can be used without cream of tartar (or with less). Alternatively, lemon juice can be substituted for some of the liquid in the recipe, to provide the required acidity to activate the baking soda.

In older times, when chemically manufactured baking soda was not available, ash water was used instead, especially in confectionery. Wood ash is also weakly alkaline. To prepare ash water, one used a fistful of ash from the fireplace in a big pot of water. Ash from solid woods, such as the olive tree, is preferred, whereas resinous woods, like pine, cannot be used. The ash water is given a boil, then left overnight to settle. The water is then filtered through a cloth and is ready to use. Many traditional recipes call for ash water instead of baking soda, because of some unique qualities: for example, ash water dripped on hot vegetable oils congeals into a gel-like mixture.

Baking powders are available both with and without aluminium compounds. Some people prefer not to use baking powder with aluminium because they believe it gives food a vaguely metallic taste, and because of a possible (but controversial) link between aluminium consumption and Alzheimer's disease.

Chapter 16

Potassium Industries

INTRODUCTION

Potassium salts are present in the soil in order to have normal plant growth, hence they are essential components of the fertilisers used in food production throughout the world. The yardstick by which potassium compounds are measured is the content of potash as K_2O. Various potassium salts are also basically important as raw materials for many commodities such as soaps, detergents, glass, dyes, gunpowder, and pyrotechnics.

The largest domestic production of potassium salts has come from deep Permian sedimentary deposits of sylvinite [a natural mixture of sylvite (KCl) and halite (NaCl)] and langbeinite ($K_2SO_4 \cdot 2MgSO_4$) near Carlsbad, Mexico. The sylvinite is mined and treated to yield high-grade potassium chloride, and langbeinite is processed to make potassium sulphate. Another domestic source of potassium salts is Searles Lake at Trona, California, which is a deposit of solid sodium salts permeated by a saturated complex brine. This brine is processed to separate high-grade potassium chloride and borax, together with numerous other saline products.

Some companies evaporate the Solidus Marsh brines by solar heat of salt lake crystallising out the sodium and potassium chloride. The two salts are repulped and separated by froth and flotation to yield potassium chloride. Other sources contribute in a minor way to potash production include alcohol fermentation, using molasses as a raw materials and a by-product from natural salt brine.

Potassium is a chemical element. It has the symbol K. Potassium was first isolated from potash, hence the name. Elemental potassium is a soft silvery-white metallic alkali metal that oxidises rapidly in air and is very reactive with water, generating sufficient heat to ignite the evolved hydrogen.

Potassium in nature occurs only as ionic salt. As such, it is found dissolved in seawater, and as part of many minerals. Potassium ion is necessary for the function of all living cells, and is thus present in all plant and animal tissues. It is found in especially high concentrations in plant cells, and in a mixed diet, it is most highly concentrated in fruits.

In many respects, potassium and sodium are chemically similar, although they have very different functions in organisms in general, and in animal cells in particular.

POTASSIUM CHLORIDE

In most of the countries potassium chloride is produced as a fertiliser grade of about 97 per cent purity. The chemical grade or 99.9 per cent potassium chloride, is the basis of the manufacture of most potassium salts.

The chemical compound potassium chloride (KCl) is a metal halide salt composed of potassium and chlorine. In its pure state it is odourless. It has a white or colourless vitreous crystal, with a crystal structure that cleaves easily in three directions. Potassium chloride crystals are face-centered cubic. Potassium chloride is occasionally known as muriate of potash, particularly when used as a fertiliser. Potash varies in colour from pink or red to white depending on the mining and recovery process used. White potash, sometimes referred to as soluble potash, is usually higher in analysis and is used primarily for making liquid starter fertilisers. KCl is used in medicine, scientific applications, food processing and in judicial execution through lethal injection. It occurs naturally as the mineral sylvite and in combination with sodium chloride as sylvinite.

Properties

Potassium chloride can react as a source of chloride ion. As with any other soluble ionic chloride, it will precipitate insoluble chloride salts when added to a solution of an appropriate metal ion:

$$KCl(aq) + AgNO_3(aq) \rightarrow AgCl(s) + KNO_3(aq)$$

Although potassium is more electropositive than sodium, KCl can be reduced to the metal by reaction with metallic sodium at 850°C because the potassium is removed by distillation.

$$KCl(l) + Na(l) \rightarrow NaCl(l) + K(g)$$

This method is the main method for producing metallic potassium. Electrolysis (used for sodium) fails because of the high solubility of potassium in molten KCl. As with other compounds containing potassium, KCl in powdered form gives a lilac flame test result.

MANUFACTURE OF POTASSIUM CHLORIDE

From Sylvinite by Fractional Crystallisation

Sylvinite (42.7 per cent KCl, 56.6 per cent NaCl), is the raw material for the potassium salts. In the oldest plant which probably works with the highest-grade ores, the sodium and potassium chlorides are separated by fractional crystallisation; the others use flotation processes (Fig. 16.1).

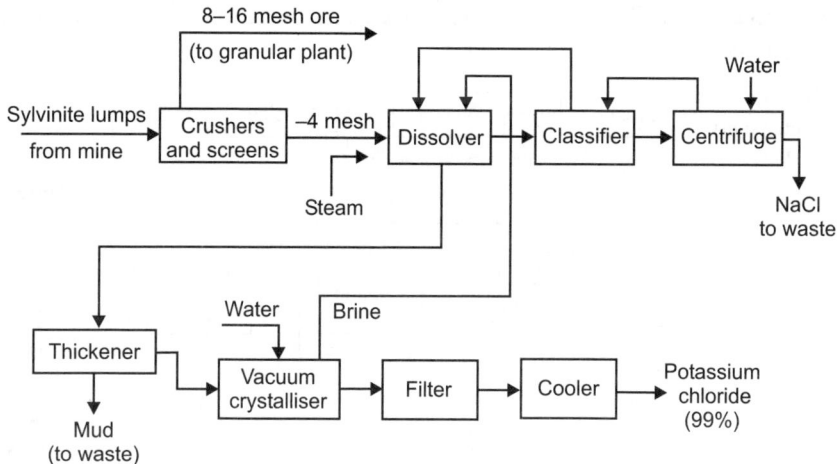

Fig. 16.1. Flow diagram for manufacture of potassium chloride from sylvinite by fractional crystallisation.

The feed for the fractional crystallisation process is –4 mesh ore. The 8 to 16 mesh material from the grinding operation is separated on tables to yield a granular 50 per cent K_2O product. For fertiliser use, the potassium content of all potassium salts is expressed as the equivalent percentage of K_2O. For instance, 50 per cent K_2O is equivalent to 80 per cent potassium chloride, and 60 per cent K_2O to 97 per cent potassium chloride. The fines (–4 mesh) are fed to a tank where they are partially dissolved in mother liquor returned from further along in the process. This mother liquor, when heated in the dissolver, remains saturated with respect to sodium chloride but has the capacity to dissolve more potassium chloride. Undissolved sodium chloride is removed from the dissolver, dewatered, washed, and sent to waste. The solution, saturated with both salts at 110°C, is clarified in a thickener. The underflow, containing insoluble mud, is concentrated and washed in counter-current decantation units.

The hot clarified solution is then sent to a vacuum crystalliser, where potassium chloride crystallises at 32°C. To prevent sodium chloride from crystallising, water is added to replace that which evaporates. The cooled slurry is sent to a filter from which the mother liquor is recirculated to the dissolving tank. The cake goes to a rotary cooler, and then to storage.

From Sylvinite by Flotation

Standard 60 per cent (K_2O) muriate of potash can be produced by separating potassium chloride from sodium chloride by froth flotation of finely ground sylvinite ore (Fig. 16.2). The ore is obtained by shaft mining and is broken from the mine face by drilling and blasting. About 375 grams of blasting powder is required per metric ton of ore. The ore is crushed underground to

13 cm maximum size, hoisted to the surface, and then ground and screened to smaller size.

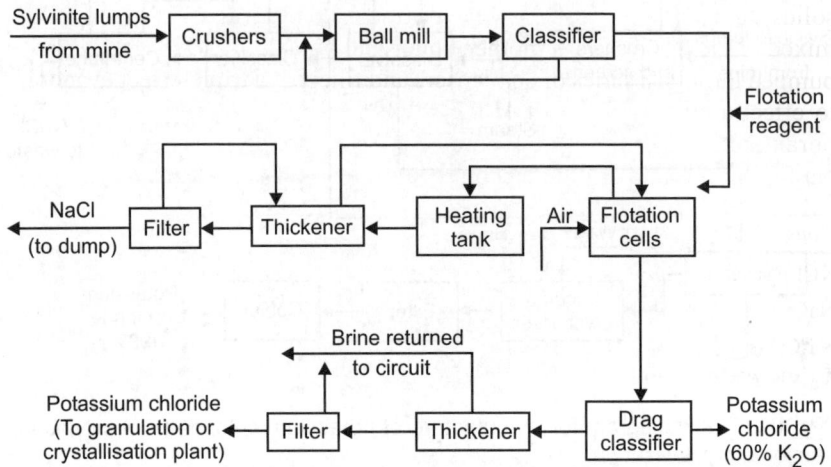

Fig. 16.2. Flow diagram for manufacture of potassium chloride from sylvinite by flotation.

Twelve-centimeter lumps of sylvinite, hoisted from the mine, are crushed to fine granules, pulped with brine, and further ground in a wet ball mill. The milled product is sent to a spiral classifier from which the underflow is returned to the grinding circuit. The flotation reagent mixture is added to the overflow on the way to the flotation cells. A typical flotation reagent mixture is 0.1 kg tallow amine and 0.11 to 0.12 kg of polyalkylglycol/metric ton of ore. In some cases, cornstarch is also added to the flotation-cell feed to adjust its gravity and viscosity. In the cell, air is intimately mixed into the feed. The finely divided sodium chloride particles are carried into the froth and overflow from the cells. The froth breaks and is pumped through a heating tank, a thickener, and a filter in series. Tailings of sodium chloride are sent to the dump; the final filtrate is recycled to the main stream.

The potassium chloride underflow from the flotation cell is pumped to a drag classifier from which a wet solid phase is removed and dewatered by filtration. The product is the standard 60 per cent muriate of potash. The overflow from the classifier is thickened and filtered. The filter cake may be melted in a furnace and flaked on a cold drum to a granular grade 60 per cent K_2O product (97 per cent KCl), or it may be sent to a crystallisation plant for the manufacture of a chemical-grade product (99.9 per cent KCl).

In some of the newer plants, sylvinite is floated and sodium salts removed in the underflow. This reversal from the process described is accomplished by proper selection of cell-feed gravity and the flotation agent.

From Searles Lake Brine

Potassium chloride has been recovered from brines. The brine (34.68 per cent solids, sp. gr. 1.3, pH 9.38) has the following composition. Raw lake brine is mixed with soda products mother liquor and borax mother liquors, and then pumped through a series of condensers and filters to a triple-effect evaporator to effect a crude separation of the soluble salts into a potassium chloride-borax solution, solid crude sodium chloride, burkeite crystals ($Na_2CO_3 \cdot Na_2SO_4$), and solid dilithium sodium phosphate crystals (Li_2NaPO_4).

Constituent	Per cent	Constituent	Per cent
KCl	4.70	Na_3PO_4	0.16
NaCl	16.35	NaF	0.01
Na_2CO_3	4.70	Misc. solids	0.30
Na_2SO_4	6.96	Water	65.32
$Na_2B_4O_7$	1.50	–	–

The hot, concentrated liquor, saturated with potassium chloride and borax, is clarified and then cooled quickly to 38°C in a three-stage vacuum crystalliser (Fig. 16.3.) The rate of crystallisation of potassium chloride is so much faster than that of borax that an excellent separation is effected. To prevent crystallisation of sodium chloride, enough additional water is added to replace that which evaporates during the vacuum cooling.

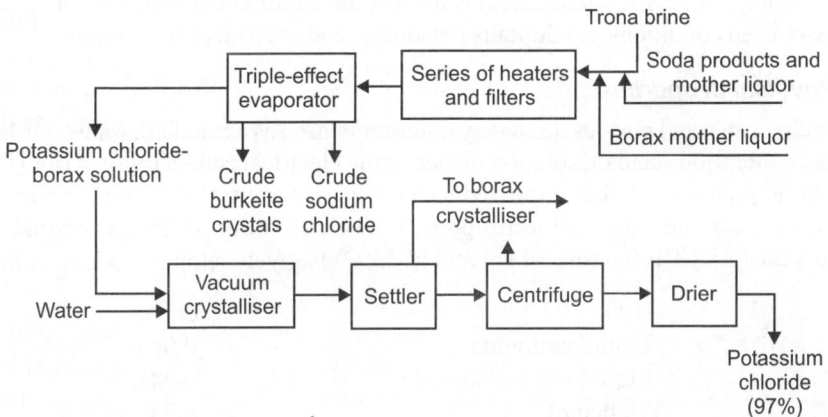

Fig. 16.3. Flow diagram for manufacture of potassium chloride from Searles lake brine.

The suspension of solid potassium chloride crystals in mother liquor goes to a cone settler. The clear overflow liquor is sent to borax crystallisers; the underflow potassium chloride sludge is further concentrated in continuous centrifuges, and then dried in a rotary dryer to yield potassium chloride crystals of 97 per cent purity.

Uses

The majority of the potassium chloride produced is used for making fertiliser, since the growth of many plants is limited by their potassium intake. As a chemical feedstock it is used for the manufacture of potassium hydroxide and potassium metal. It is also used in medicine, scientific applications, food processing, and as a sodium-free substitute for table salt (sodium chloride).

Potassium chloride is also used as the third of a three drug combination in lethal injection. Additionally, KCl is used to terminate the life of the unborn fetus in induced abortion procedures by lethal injection to the heart, which induces cardiac arrest. It is sometimes used in water as a completion fluid in petroleum and natural gas operations, as well as being an alternative to sodium chloride in household water softener units.

Potassium chloride has also been used to create heat packs which employ exothermic chemical reactions, but these are no longer being created due to cheaper and more efficient methods such as the oxidation of metals.

Biological and medical properties

Potassium is vital in the human body and oral potassium chloride is the common means to replenish it, although it can also be diluted and given intravenously (of course, in concentrations much lower than those used in executions). It can be used as a salt substitute for food, but due to its weak, bitter, unsalty flavour, it is usually mixed with regular salt (sodium chloride), for this purpose to improve the taste. Medically it is used in the treatment of hypokalemia and associated conditions, for digitalis poisoning, and as an electrolyte replenisher.

Physical properties

In chemistry and physics it is a very commonly used as a standard, for example as a calibration standard solution in measuring electrical conductivity of (ionic) solutions, since carefully prepared KCl solutions have well-reproducible and well-repeatable measurable properties. The solubility of KCl in various solvents(g KCl/100 grams of solvent at 25°C) is given below:

H_2O	36
Liquid ammonia	0.04
Liquid sulphur dioxide	0.041
Methanol	0.53
Formic acid	19.2
Sulpholane	0.004
Acetonitrile	0.0024
Acetone	0.000091
Formamide	6.2
Acetamide	2.45
Dimethylformamide	0.017–0.05

POTASSIUM SALTS

Potassium Sulphate

Potassium sulphate is also called sulphate of potash or archaically known as potash of sulphur is a flammable white crystalline salt which is soluble in water. The chemical is commonly used in fertilisers, providing both potassium and sulphur.

The mineral form of potassium sulphate, namely arcanite, is relatively rare. Natural resources of potassium sulphate are minerals abundant in the Stassfurt salt. These are co-crystalisations of potassium sulphate and sulphates of magnesium, calcium and sodium. The minerals are:

1. Kainite ($MgSO_4 \cdot KCl \cdot H_2O$)
2. Schönite ($K_2SO_4 \cdot MgSO_4 \cdot 6H_2O$)
3. Leonite ($K_2SO_4 \cdot MgSO_4 \cdot 4H_2O$)
4. Langbeinite ($K_2SO_4 \cdot 2MgSO_4$).
5. Glaserite ($K_3Na(SO_4)_2$).
6. Polyhalite ($K_2SO_4 \cdot MgSO_4 \cdot 2CaSO_4 \cdot 2H_2O$).

From some of the minerals like kainite, the potassium sulphate can be separated, because the corresponding salt is less soluble in water. With potassium chloride kieserite $MgSO_4 \cdot 2H_2O$ can be transformed and then the potassium sulphate can be dissolved in water.

Manufacture

1. Potassium sulphate can be synthesised by the decomposition of potassium chloride with sodium sulphate.
2. The Hargreaves method is basically the same process with different starting materials. Sulphur dioxide, oxygen and water (the starting materials for sulphuric acid) are reacted with potassium chloride. Hydrochloric acid evaporates off.
3. Potassium sulphate is produced by mixing.
4. Potassium chloride and sulphuric acid (with molar ratio):

$$2KCl + H_2SO_4 \rightarrow 2HCl + K_2SO_4$$

Properties

The anhydrous crystals form a double six-sided pyramid, but are in fact classified as rhombic. They are transparent, very hard and have a bitter, salty taste. The salt is soluble in water, but insoluble in solutions of potassium hydroxide (sp. gr. 1.35), or in absolute ethanol. It melts at 1078°C.

Pottassium Hydroxide

Potassium hydroxide is the inorganic compound with the formula KOH. Along with sodium hydroxide, this colourless solid is a prototypical strong base. It has many industrial and niche applications. Most applications exploit its reactivity toward acids and its corrosive nature. KOH is noteworthy as the precursor to most soft and liquid soaps as well as numerous potassium-containing chemicals.

Potassium hydroxide is usually sold as translucent pellets, which will become tacky in air because KOH is hygroscopic. Consequently, KOH characteristically contains varying amounts of water. Its dissolution in water is strongly exothermic, meaning the process gives off significant heat. Concentrated aqueous solutions are sometimes called potassium lyes.

Manufacture

KOH is made by boiling a solution of potassium carbonate (potash) with calcium hydroxide (slaked lime), leading to a metathesis reaction which caused calcium carbonate to precipitate, leaving potassium hydroxide in solution:

$$Ca(OH)_2 + K_2CO_3 \rightarrow CaCO_3 + 2KOH$$

Filtering off the precipitated calcium carbonate and boiling down the solution gives potassium hydroxide ('calcinated or caustic potash').

This method was used potash extracted from wood ashes using slaked lime. It was the most important method of producing potassium hydroxide until the late 19th century, when it was largely replaced by the modern method of electrolysis of potassium chloride solutions, analogous to the method of manufacturing sodium hydroxide:

$$2KCl + 2H_2O \rightarrow 2KOH + Cl_2 + H_2$$

Hydrogen gas forms as a by-product on the cathode; concurrently, an anodic oxidation of the chloride ion takes place, forming chlorine gas as a by-product. Separation of the anodic and cathodic spaces in the electrolysis cell is essential for this process.

Properties

At higher temperatures, solid KOH crystallises in the NaCl motif. The OH group is either rapidly or randomly disordered so that the OH^- group is effectively a spherical anion of radius 1.53 Å (between Cl^- and F^- in size). At room temperature the OH^- groups are ordered and the environment about the K^+ centres is distorted with K^+—OH^- distances ranging from 2.69 to 3.15 Å, depending on the orientation of the OH group. KOH forms a series of crystalline hydrates, namely the monohydrate $KOH \cdot H_2O$, the dihydrate $KOH \cdot 2H_2O$, and the tetrahydrate $KOH \cdot 4H_2O$.

Solubility

Approximately 121 grams of KOH will dissolve in 100 ml of water at room temperature (compared with 100 grams of NaOH in the same volume). Lower alcohols such as methanol, ethanol, and propanols are also excellent solvents. The solubility in ethanol is about 40 grams KOH/100 ml.

Because of its high affinity for water, KOH serves as a desiccant in the laboratory. It is often used to dry basic solvents, especially amines and pyridines; distillation of these basic liquids from a slurry of KOH yields the anhydrous reagent.

Thermal stability

Like NaOH, KOH exhibits high thermal stability. KOH sublimes unchanged at 400°C; the gaseous species is dimeric. Even at high temperatures, dehydration does not occur. Because of its high stability and relatively low melting point, it is often melt-cast as pellets or rods forms that have low surface area and convenient handling properties.

Reactions

As a base

KOH is highly basic, forming strongly alkali solutions in water and other polar solvents. These solutions are capable of deprotonating many acids, even weak ones. In analytical chemistry, titrations using solutions of KOH are used to assay acids.

As a nucleophile in organic chemistry

KOH, like NaOH, serves as a source of OH^-, a highly nucleophilic anion that attacks polar bonds in both inorganic and organic materials. In perhaps its most well-known reaction, aqueous KOH saponifies esters:

$$KOH + RCO_2R' \rightarrow RCO_2K + R'OH$$

When R is a long chain, the product is called a potassium soap. This reaction is manifested by the 'greasy' feel that KOH gives when touched — fats on the skin are rapidly converted to soap and glycerol. Molten KOH is used to displace halides and other leaving groups. The reaction is especially useful for aromatic reagents to give the corresponding phenols.

Reactions with inorganic compounds

Complementary to its reactivity toward acids, KOH attacks anhydrides, defined in the broadest sense. Thus, SiO_2 and CO_2 are attacked by KOH to give the silicates and bicarbonate, respectively:

$$KOH + CO_2 \rightarrow KHCO_3$$

Precursor to other potassium compounds

Many radium salts are prepared by neutralisation reactions involving KOH. The potassium salts of carbonate, cyanide, permanganate, phosphate, and various silicates are prepared by treating either the oxides or the acids with KOH. The high solubility of potassium phosphate is desirable in fertilisers.

Manufacture of biodiesel

Although more expensive than using sodium hydroxide, KOH works well in the manufacture of biodiesel by saponification of the fats in vegetable oil. Glycerine from potassium hydroxide-processed biodiesel is useful as an inexpensive food supplement for livestock, once the toxic methanol is removed.

Manufacture of soft soaps

The saponification of fats with KOH is used to prepare the corresponding potassium soaps, which are softer than the more common sodium hydroxide-derived soaps. Because of their softness and greater solubility, potassium soaps require less water to liquify, and can thus contain more cleaning agent than liquified sodium soaps.

As an electrolyte

Aqueous potassium hydroxide is employed as the electrolyte in alkaline batteries based on nickel-cadmium and manganese dioxide-zinc. Potassium hydroxide is preferred over sodium hydroxide because its solutions are more conductive.

Niche applications

KOH attracts numerous specialised applications, which virtually all rely on its basic or degradative properties. KOH is widely used in the laboratory for the same purposes. In chemical synthesis, the selection of KOH vs. NaOH is guided by the solubility for the resulting salt. Its corrosivity is sometimes used in cleaning and disinfection of resistant surfaces and materials. It is often the main active ingrediant in chemical 'cuticle removers'.

Potassium Carbonate

Potassium carbonate is a white salt, soluble in water (insoluble in alcohol), which forms a strongly alkaline solution. It can be made as the product of potassium hydroxide's absorbent reaction with carbon dioxide. It is deliquescent, often appearing a damp or wet solid. It has sp. gr. 2.428 (19°C), m.p. 891°C. It can exist as tri-, di-, and monohydrates, all of which loose their water of crystallisation on heating above 130°C. In its reactions it closely resembles sodium carbonate.

Manufacture

Potassium carbonate is prepared commercially by the electrolysis of potassium chloride. The resulting potassium hydroxide is then carbonated using carbon dioxide to form potassium carbonate, which is often used to produce other potassium compounds.

$$2KOH + CO_2 \rightarrow K_2CO_3 + H_2O$$

Potassium carbonate is also prepared by reacting alkylamines or ion exhange resins with potassium chloride and carbon dioxide.

Uses

Potassium carbonate is used in the production of soap and glass. Mixed with water it causes an exothermic reaction that results in a temperature change, producing heat.

Potassium carbonate is being used as the electrolyte in many cold fusion experiments. Aqueous potassium carbonate is also used as a fire suppressant in extinguishing deep fat fryers. Potassium carbonate is used in reactions to maintain anhydrous conditions without reacting with the reactants and product formed. It may also be used to predry some ketones, alcohols, and amines prior to distillation. In the laboratory, it may be used as a mild drying agent where other drying agents such as calcium chloride may be incompatible. However, it is not suitable for acidic compounds.

Potassium Nitrate

Potassium nitrate is a chemical compound with the chemical formula KNO_3. A naturally occurring mineral source of nitrogen, KNO_3 constitutes a critical oxidising component of black powder/gunpowder. In the past it was also used for several kinds of burning fuses, including slow matches. It is slightly hygroscopic, pungent, saline taste, sp. gr. 2.1062, m.p. 337°C, b.p. decomposes at 400°C.

Potassium nitrate is the oxidising component of black powder. Before the large-scale industrial fixation of nitrogen through the Haber process, major sources of potassium nitrate were the deposits crystallising from cave walls and the draining of decomposing organic material. Dung-heaps were a particularly common source. Ammonia from the decomposition of urea and other nitrogenous materials would undergo bacterial oxidation to produce nitrate. It was and is also used as a component in some fertilisers. When used by itself as a fertiliser, it has an NPK rating of 13-0-38 (indicating 13.9 per cent, 0 per cent, and 38.7 per cent of nitrogen, phosphorus, and potassium, by mass, respectively). However, potassium nitrate and other nitrates do successfully combat high blood pressure and are used medically to relieve angina.

Manufacture

It is manufactured by adding potassium chloride to sodium nitrate.

$$NaNO_3 + KCl \rightleftharpoons KNO_3 + NaCl$$

Because sodium chloride is the least soluble of these substances in hot water, much of it is precipitated from solution and filtered of. On cooling the mixture, most of the potassium nitrate formed crystallises out because it is much less soluble in cold water than in hot water.

Uses

Potassium nitrate is also used as a fertiliser, in amateur rocket propellant, and in several fireworks such as smoke bombs. It is commonly used in pre-rolled cigarettes to maintain an even burn of the tobacco.

As a fertiliser, it is used as a source of nitrogen and potassium, two of the plant macro nutrients for plants. The other macro nutrients are carbon, oxygen, and hydrogen.

Potassium nitrate is also the main component (usually about 98 per cent) of tree stump remover; it accelerates the natural decomposition of the stump.

Potassium nitrate is also commonly used in the heat treatment of metals as a solvent in the post-wash. The oxidising, water solubility and low cost make it an ideal short-term rust inhibitor.

It has also been used in the manufacture of ice cream and can be found in some toothpastes for sensitive teeth. Recently, the use of potassium nitrate in toothpastes for treating sensitive teeth has increased dramatically, despite the fact that it has not been conclusively shown to help dental hypersensitivity.

Potassium nitrate is also one of the three components of black powder, along with powdered charcoal (substantially carbon) and sulphur, where it acts as an oxidiser. When subjected to the flame test it produces a lilac flame due to the presence of potassium.

Potassium Bitartrate

Potassium bitartrate crystallises in wine casks during the fermentation of grape juice. In wines bottled before they are fully ripe, it can precipitate on the side of the bottle in a sort of crust, thus forming what is called crusted wine. It is soluble in boiling water, pleasant slightly acid taste, insoluble in alcohol, sp. gr. 1.984 (18°C). This crude form (known as beeswing) is collected and purified to produce the white, odourless, acidic powder used for many culinary and other household purposes.

Manufacture

It is prepared from wine lees by extraction with water and crystallisation.

Applications

In food

In food, potassium bitartrate is used for: (i) stabilising egg whites, increasing their heat tolerance and volume, (ii) preventing sugar syrups from crystallising,

(iii) reducing discolouration of boiled vegetables, (iv) frequent combination with baking soda (which needs an acid ingredient to activate it) in formulations of baking powder, and (v) commonly used in combination with potassium chloride in sodium-free salt substitutes.

Household use

It can be used with white vinegar to make a paste-like cleaning agent. It is a common ingredient in Playdoh and gingerbread house icing.

Chapter 17

Nitrogen Industries

INTRODUCTION

Humans have increased their supply of food by feeding nitrogen compounds back into the soil. Chemists and chemical engineers have found ways of making nitrogen derivatives out of air in an economical way. The first successful process, the arc process, required much cheap electric energy. However, the present solution to the nitrogen-fixation problem was obtained by reacting nitrogen with low-cost hydrogen to make ammonia under conditions requiring less power and low conversion expense. This made the arc process obsolete. Ammonia has now become one of our heavy chemical, produced in enormous tonnage around the globe, and at such low prices as to dominate the world supply of nitrogen fertilisers and most nitrogen compounds. This is probably one of the most important chemical engineering achievements in history.

The chemical nitrogen industries include not only nitrogen fixed by humans, but also by-product ammonia from coke ovens and such natural nitrogen deposits both of which are subjected to manufacturing processes. Ammonia is used in heat treating and paper pulping, nitric acid for nitro compounds, high explosives, and propellants. Hydrogen cyanide, made largely from ammonia, is used mainly for acrylates and methacrylates, although a little is used for fumigation. Acrylonitrile uses substantial amounts of nitrogen compounds. Hydroxylamine and hydrazine also are smaller-volume nitrogen consumers. Amines of fatty acids enjoy wide use as surface-active agents and in ore flotation.

CYANAMIDE

Cyanamide (CN_2H_2) is a white, crystalline compound. The term can also refer to a salt of this compound, having one or both of the hydrogen atoms replaced by another element or radical, such as in the most common case of calcium cyanamide ($CaCN_2$), a compound used as a fertiliser and as a source of other compounds of nitrogen.

Cyanamide can be prepared by hydrolysis of calcium cyanamide in presence of carbon dioxide by Frank-Caro process.

Its employment as a chemical raw material has become increasingly important, however, the largest chemical use being for the preparation of the dimer dicyandiamide. This dimer is polymerised to produce the trimer melamine, which has applications in plastic resins. Other commercial derivatives are produced by way of a crude calcium cyanide (48 to 50 per cent expressed as NaCN) from a high temperature melt of cyanamide with excess carbon and NaCl. This is used directly for cyanidation of ores and to manufacture ferrocyanides. The use of cyanamide fertilisers is enjoying a minor resurgence, because these chemicals act as pesticides to snails and larvae and also enrich the soil with nitrogen. Cyanamide compounds are used to fire-proof shingles and cotton fabrics.

Reactions and Energy Changes

The essential reactions for the production of calcium cyanamide are:

$$CaCO_3(s) \longrightarrow CaO(s) + CO_2(g) \qquad \Delta H = +43.5 \text{ kcal} \quad \dots (17.1)$$

$$CaO(s) + 3C(\text{amorph}) \longrightarrow CaC_2(s) + CO(g) \qquad \Delta H = +103.0 \text{ kcal} \quad \dots (17.2)$$

$$CaC_2(s) + N_2(g) \longrightarrow CaCN_2(s) + C(\text{amorph}) \qquad \Delta H = -68.0 \text{ kcal} \quad \dots (17.3)$$

Various catalysts or fluxes are used to increase the rate of reaction or cause it to proceed at lower temperatures. Some companies use calcium fluoride, and reaction-rate studies have shown that it reduces the temperature of optimum reactivity and increases the velocity of the reaction by 4.5 times at 1000°C. Equation 17.3 takes place at 900° to 1000°C. Equation 17.2 is carried out in two 20,000- and two 10,000 kW furnaces at about 2000° to 2200°C. Under normal conditions Eq. 17.2 yields up to 90 per cent calcium carbide. Considerable overall energy is needed, principally to secure the high temperature for Eq. 17.3 to start when it is self-sustaining and for making the calcium carbide in Eq. 17.2. The source of carbon is usually coke. Coal is required to burn the limestone and to dry the raw materials (Fig. 17.1).

The following physical operations and chemical conversions are needed to commercialise the reactions on which Fig. 17.1 is based on:

1. Limestone, coal and coke are pulverised, separately.
2. Limestone is calcined to quicklime.
3. Coke is pulverised, dried and mixed with quicklime.
4. Carbide is formed in an electric furnace at nearly 2000° to 2200°C and run out molten.
5. Carbide is cooled, crushed, and finely ground.
6. Air is liquefied by compressing, cooling and expansion.
7. Nitrogen is separated from oxygen by liquid rectification.
8. Calcium carbide is nitrified over the course of 40 hrs with 99.9 per cent nitrogen at about 1000°C.
9. Calcium cyanamide is pulverised and treated with a small amount of water to hydrate residual CaO and CaC_2. It may be oiled to reduce dust.

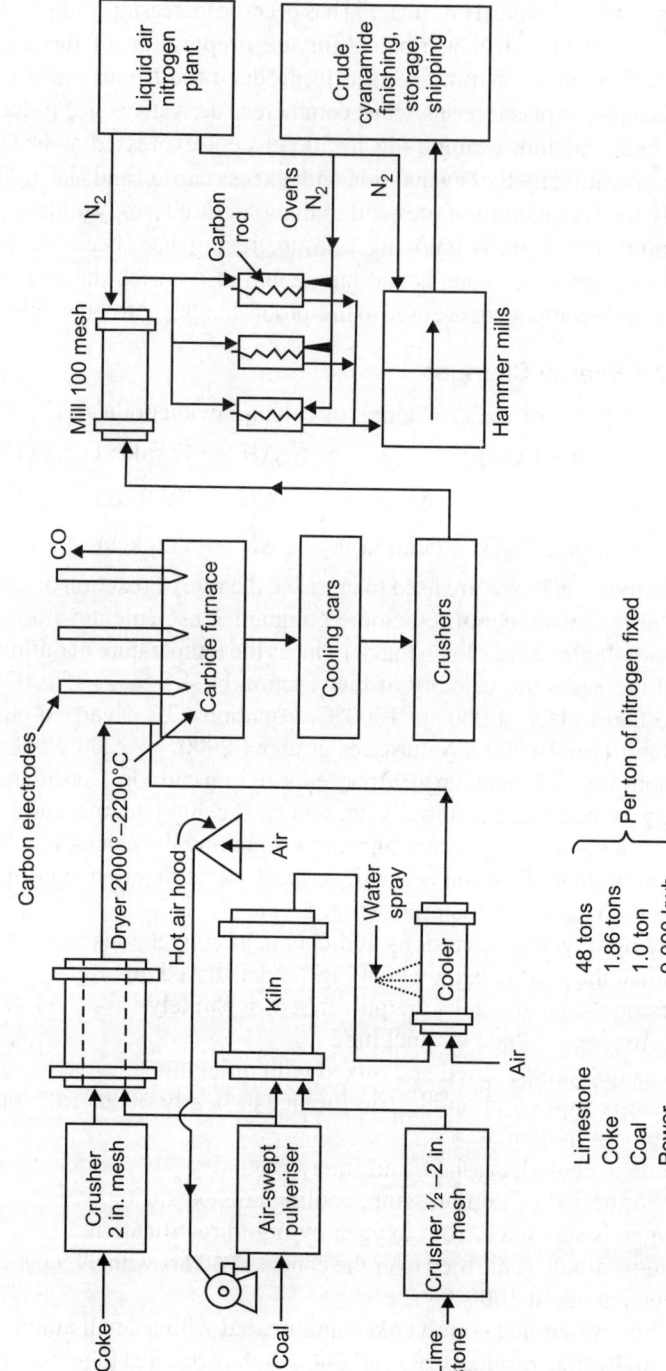

Fig. 17.1. Flow diagram for the manufacture of calcium carbide and calcium cyanamide.

Limestone	48 tons	
Coke	1.86 tons	Per ton of nitrogen fixed
Coal	1.0 ton	
Power	9,000 kwh	

Uses

Since mid-1960s, there have been developed procedures to produce stabilised cyanamide for industry use. Cyanamide is used as a plant growth modulator and has many uses in chemical industry. Aqueous solutions of cyanamide with high concentration may undergo explosive polymerisation when heated. Stability of its solution can be increased by addition of a dicarboxylic acid such as adipic acid.

AMMONIA

Ammonia is a compound with the formula NH_3. It is normally encountered as a gas with a characteristic pungent odour. Ammonia contributes significantly to the nutritional needs of terrestrial organisms by serving as a precursor to foodstuffs and fertilisers. Ammonia, either directly or indirectly, is also a building block for the synthesis of many pharmaceuticals. Although in wide use, ammonia is both caustic and hazardous.

Ammonia, as used commercially, is often called anhydrous ammonia. This term emphasises the absence of water in the material. Because NH_3 boils at $-33.34°C$, the liquid must be stored under high pressure or at low temperature. Its heat of vapourisation is, however, sufficiently great that NH_3 can be readily handled in ordinary beakers in a fume hood.

Household ammonia or ammonium hydroxide is a solution of NH_3 in water. The strength of such solutions is measured in units of baume (density), with 26 degrees baume (about 30 weight per cent ammonia at $15.5°C$) being the typical high concentration commercial product. Household ammonia ranges in concentration from 5 to 10 weight percent ammonia.

Manufacture

Because of its many uses, ammonia is one of the most highly produced inorganic chemicals. Dozens of chemical plants worldwide produce ammonia.

Today, the typical modern ammonia-producing plant first converts natural gas (i.e. methane) or liquified petroleum gas (such gases are propane and butane) or petroleum naphtha into gaseous hydrogen. The processes used in producing the hydrogen begins with removal of sulphur compounds from the natural gas (because sulphur deactivates the catalysts used in subsequent steps). Catalytic hydrogenation converts organosulphur compounds into gaseous hydrogen sulphide:

$$H_2 + RSH \rightarrow RH + H_2S(g)$$

The hydrogen sulphide is then removed by passing the gas through beds of zinc oxide where it is absorbed and converted to solid zinc sulphide:

$$H_2S + ZnO \rightarrow ZnS + H_2O$$

Catalytic steam reforming of the sulphur-free feedstock is then used to form hydrogen plus carbon monoxide:

$$CH_4 + H_2O \rightarrow CO + 3H_2$$

In the next step, the water gas shift reaction is used to convert the carbon monoxide into carbon dioxide and more hydrogen:

$$CO + H_2O \rightarrow CO_2 + H_2$$

The carbon dioxide is then removed either by absorption in aqueous ethanolamine solutions or by adsorption in pressure swing adsorbers (PSA) using proprietary solid adsorption media.

The final step in producing the hydrogen is to use catalytic methanation to remove any small residual amounts of carbon monoxide or carbon dioxide from the hydrogen:

$$CO + 3H_2 \rightarrow CH_4 + H_2O$$
$$CO_2 + 4H_2 \rightarrow CH_4 + 2H_2O$$

To produce the desired end-product ammonia, the hydrogen is then catalytically reacted with nitrogen (derived from process air) to form anhydrous liquid ammonia. This step is known as the ammonia synthesis loop (also referred to as the Haber-Bosch process):

$$3H_2 + N_2 \rightarrow 2NH_3$$

As is evident from the reaction, there is a decrease in total volume, therefore, reaction is carried out at high pressures to obtain high equilibrium conversion. Although the reaction is exothermic demanding a low temperature for reaction but the reaction is quite slow at lower temperature, therefore, a compromise of high temperature has to be made.

In this case the temperature is about 500° to 600°C. Further, the rate is enhanced by using iron as catalyst with catalyst promoters to promote the activity of iron. The reaction is highly exothermic and design of the converter has to be very precise so that temperature at various points are optimum for economic consideration.

Out of the two raw materials H_2 and N_2, nitrogen essentially comes from the atmospheric air either in the pure form after air fractionation or directly with air. The other, hydrogen may be obtained from hydrocarbon (natural gas, naphtha, etc.), coke oven gas, water gas, producer gas and also from electrolysis of water. The steps in the ammonia manufacture are as follows: (i) production of synthesis gas (1 mole of N_2 and 3 moles of hydrogen), (ii) NH_3 formation, and (iii) Refrigeration and storage of NH_3.

Production of Synthesis Gas

1. The source of H_2 is normally some hydrocarbon which is either partially oxidised with air or steam reformed (than reformed with air) to give N_2, H_2, CO, $CO_2 + H_2O$, etc.

Partial oxidation:

$C_nH_{2n} + 2O_2$ (Air) Ni catalyst $\longrightarrow CO + CO_2 + H_2O$ (Exothermic reaction)

Steam reforming:

$C_nH_{2n} + 2H_2O$ (Steam) $\longrightarrow CO + H_2$ (Endothermic reaction)

2. Shift conversion of CO to CO_2 with steam.

$$CO(g) + H_2O(g) \xrightarrow[400°C]{FeO + Cr_2O_2} CO_2(g) + H_2O(g); \ \Delta H = -9.2 \ \text{kcal.}$$

3. Purification by removal of CO_2 using mono-ethanolamine (MEA) or hot potassium carbonate or sulphinol, recently developed by Shell Oil Co., USA.

4. Removal of residual amounts of CO_2, CO to less than 10 ppm by methanation over Ni catalyst followed by N_2 wash.

$$CO_2 + 4H_2 \longrightarrow CH_4 + 2H_2O$$

Figure 17.2 shows these steps producing synthesis gas (1 mole of N_2 and 3 moles of H_2 which is then sent to convertor to get NH_3).

Ammonia Synthesis

The synthesis gas (3 moles H_2 per 1 mole. N_2) is compressed to the operating pressure 100–1000 atmosphere depending upon the process. It is then sent through filter to remove compression oil and then to the convertor.

Ammonia synthesis reactors/converters are of various types depending upon the design. These convertors/reactors are long cylinderical tubes which are either forged or made of seamless tubing. A design of the convertor has to be optimum in respect of temperature which is controlled by heat exchangers between cold incoming synthesis gas and out going product rich gas.

It is a tricky problem as the amount of heat released during the exothermic reactions are different at different points.

A higher temperature, than desired would decrease the activity of catalyst which in turn would effect the conversion. Choice of suitable material of construction to withstand a very high pressure, is also a matter of great importance.

Catalysts

The catalyst has no effect on the position of chemical equilibrium; rather, it provides an alternative pathway with lower activation energy and hence increases the reaction rate, while remaining chemically unchanged at the end of the reaction. The first Haber-Bosch reaction chambers used osmium and uranium as catalysts. In industrial practice, the iron catalyst is prepared by exposing a mass of magnetite, an iron oxide, to the hot hydrogen feedstock.

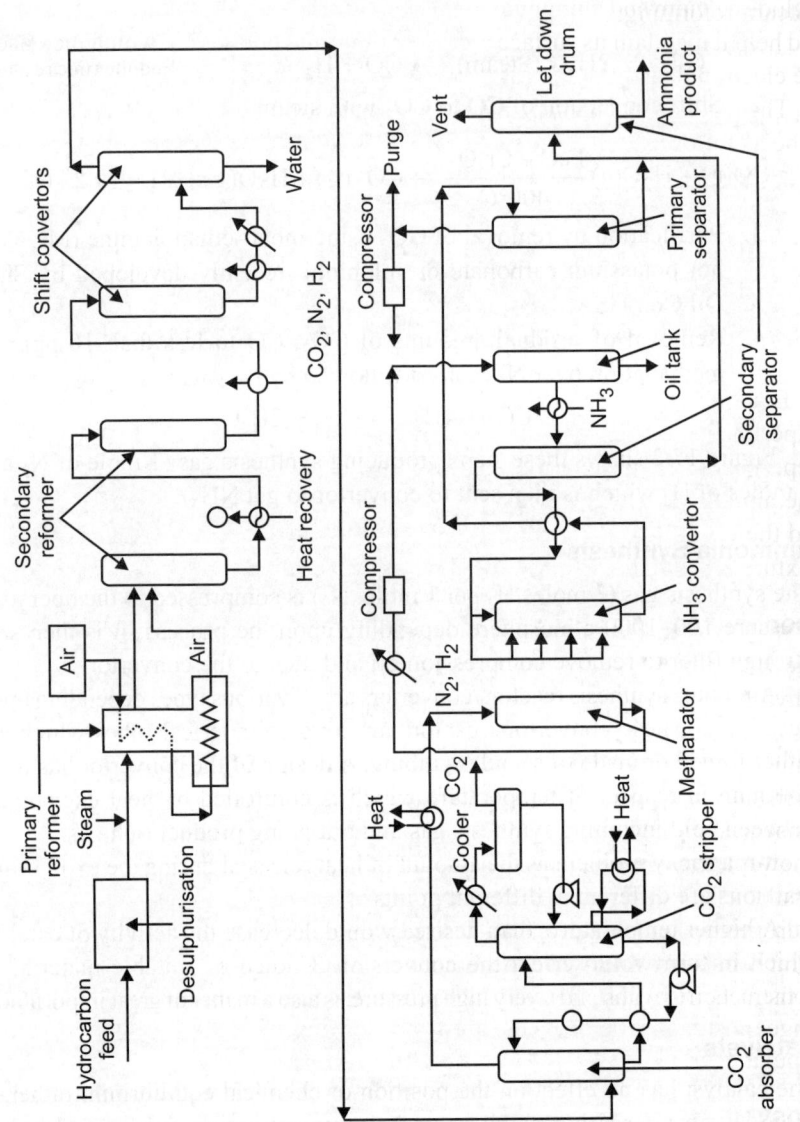

Fig. 17.2. Ammonia synthesis process.

This reduces some of the magnetite to metallic iron, removing oxygen in the process. However, the catalyst maintains most of its bulk volume during the reduction, and so the result is a highly porous material whose large surface area aids its effectiveness as a catalyst. Other minor components of the catalyst include calcium and aluminium oxides, which support the porous iron catalyst and help it maintain its surface area over time, and potassium, which increases the electron density of the catalyst and so improves its activity.

The reaction mechanism, involving the heterogeneous catalyst, is believed to be as follows:

$$N_2(g) \rightarrow N_2(adsorbed) \quad \text{... (17.4)}$$
$$N_2(adsorbed) \rightarrow 2N(adsorbed) \quad \text{... (17.5)}$$
$$H_2(g) \rightarrow H_2(adsorbed) \quad \text{... (17.6)}$$
$$H_2(adsorbed) \rightarrow 2H(adsorbed) \quad \text{... (17.7)}$$
$$N(adsorbed) + 3H(adsorbed) \rightarrow NH_3(adsorbed) \quad \text{... (17.8)}$$
$$NH_3(adsorbed) \rightarrow NH_3(g) \quad \text{... (17.9)}$$

Equation 17.8 occurs in three steps, forming NH, NH_2, and then NH_3. Experimental evidence points to Eq. 17.5 as being the slow, rate-determining step. A major contributor to the elucidation of this mechanism is Gerhard Ertl. The ammonia formed is removed by condensation, first with water cooling and then by ammonia refrigeration. The unconverted nitrogen and hydrogen mixture is sent back to allow an yield of about 90 per cent.

Economic and Environmental Aspects

The Haber process now produces nitrogen fertiliser, mostly in the form of anhydrous ammonia, ammonium nitrate, and urea. 3–5 per cent of world natural gas production is consumed in the Haber process (~1–2 per cent of the world's annual energy supply). That fertiliser is responsible for sustaining one-third of the earth's population, as well as various deleterious environmental consequences. Generation of hydrogen using electrolysis of water, using renewable energy, is not currently competitive cost-wise with hydrogen from fossil fuels, such as natural gas, and is responsible for 4 per cent of current hydrogen production. Notably, the rise of this industrial process led to the nitrate crisis in Chile, when the industrials who owned the nitrate mines (most of them British) left the country—since the natural nitrate mines were no longer profitable—closing the mines and leaving a large unemployed Chilean population behind.

Biosynthesis

In certain organisms, ammonia is produced from atmospheric N_2 by enzymes called nitrogenases. The overall process is called nitrogen fixation. Although it is unlikely that biomimetic methods will be developed that are competitive with the Haber process, intense effort has been directed toward understanding the mechanism of biological nitrogen fixation. The scientific interest in this problem is motivated by the unusual structure of the active site of the enzyme, which consists of an Fe_7MoS_9 ensemble. Ammonia is also a metabolic product of amino acid deamination. Ammonia excretion is common in aquatic animals. In humans, it is quickly converted to urea, which is much less toxic. This urea is a major component of the dry weight of urine. Most reptiles, including birds, as well as insects and snails solely excrete uric acid as nitrogenous waste.

Properties

Ammonia is a colourless gas with a characteristic pungent smell. It is lighter than air, its density being 0.589 times that of air. It is easily liquefied due to the strong hydrogen bonding between molecules; the liquid boils at $-33.3°C$, and solidifies at $-77.7°C$ to white crystals. Liquid ammonia has a very high standard enthalpy change of vapourisation (23.35 kJ/mol, cf. water 40.65 kJ/mol, methane 8.19 kJ/mol, phosphine 14.6 kJ/mol) and can therefore be used in laboratories in non-insulated vessels without additional refrigeration.

It is miscible with water. Ammonia in an aqueous solution can be expelled by boiling. The aqueous solution of ammonia is basic. The maximum concentration of ammonia in water (a saturated solution) has a density of 0.880 g/cm^3 and is often known as 880 Ammonia. Ammonia does not burn readily or sustain combustion, except under narrow fuel-to-air mixtures of 15–25 per cent air. When mixed with oxygen, it burns with a pale yellowish-green flame. At high temperature and in the presence of a suitable catalyst, ammonia is decomposed into its constituent elements. Ignition occurs when chlorine is passed into ammonia, forming nitrogen and hydrogen chloride; if ammonia is present in excess, then the highly explosive nitrogen trichloride (NCl_3) is also formed.

The ammonia molecule readily undergoes nitrogen inversion at room temperature; a useful analogy is an umbrella turning itself inside out in a strong wind. The energy barrier to this inversion is 24.7 kJ/mol, and the resonance frequency is 23.79 GHz, corresponding to microwave radiation of a wavelength of 1.260 cm. The absorption at this frequency was the first microwave spectrum to be observed.

Basicity

One of the most characteristic properties of ammonia is its basicity. It combines with acids to form salts; thus with hydrochloric acid it forms ammonium chloride; with nitric acid, ammonium nitrate, etc. However, perfectly dry ammonia will not combine with perfectly dry hydrogen chloride: moisture is necessary to bring about the reaction.

$$NH_3 + HCl \rightarrow NH_4Cl$$

The salts produced by the action of ammonia on acids are known as the ammonium salts and all contain the ammonium ion (NH_4^+). Anhydrous ammonia is often used for the production of methamphetamine.

Acidity

Although ammonia is well-known as a base, it can also act as an extremely weak acid. It is a protic substance, and is capable of formation of amides (which contain the NH_2^- ion), for example when solid lithium nitride is added to liquid ammonia, forming a lithium amide solution:

$$Li_3N_{(s)} + 2\ NH_{3\ (l)} \rightarrow 3\ Li^+_{(am)} + 3\ NH^-_{2\,(am)}$$

In this Brønsted-Lowry acid-base reaction, ammonia serves as an acid.

Combustion

The combustion of ammonia to nitrogen and water is exothermic:

$$4NH_3 + 3O_2 \rightarrow 2N_2 + 6H_2O\ (g)\ \Delta H_r^\circ = -1267.20\ kJ/mol$$

The standard enthalpy change of combustion, ΔH_c°, expressed per mole of ammonia and with condensation of the water formed, is -382.81 kJ/mol. Dinitrogen is the thermodynamic product of combustion: all nitrogen oxides are unstable with respect to nitrogen and oxygen, which is the principle behind the catalytic converter. However, nitrogen oxides can be formed as kinetic products in the presence of appropriate catalysts, a reaction of great industrial importance in the production of nitric acid.

$$4NH_3 + 5O_2 \rightarrow 4NO + 6H_2O$$

The combustion of ammonia in air is very difficult in the absence of a catalyst (such as platinum gauze), as the temperature of the flame is usually lower than the ignition temperature of the ammonia-air mixture. The flammable range of ammonia in air is 16–25 per cent.

Formation of other compounds

In organic chemistry, ammonia can act as a nucleophile in substitution reactions. Amines can be formed by the reaction of ammonia with alkyl halides, although the resulting $-NH_2$ group is also nucleophilic and secondary and tertiary amines are often formed as by-products.

An excess of ammonia helps minimise multiple substitution, and neutralises the hydrogen halide formed. Methylamine is prepared commercially by the reaction of ammonia with chloromethane, and the reaction of ammonia with 2-bromopropanoic acid has been used to prepare racemic alanine in 70 per cent yield.

Ethanolamine is prepared by a ring-opening reaction with ethylene oxide: the reaction is sometimes allowed to go further to produce diethanolamine and triethanolamine. Amides can be prepared by the reaction of ammonia with a number of carboxylic acid derivatives. Acyl chlorides are the most reactive, but the ammonia must be present in at least a two-fold excess to neutralise the hydrogen chloride formed. Esters and anhydrides also react with ammonia to form amides. Ammonium salts of carboxylic acids can be dehydrated to amides so long as there are no thermally sensitive groups present: temperatures of 150°–200°C are required. The hydrogen in ammonia is capable of replacement by metals, thus magnesium burns in the gas with the formation of magnesium nitride Mg_3N_2, and when the gas is passed over heated sodium or potassium, sodamide, $NaNH_2$, and potassamide, KNH_2, are formed. Where necessary in substitutive nomenclature, IUPAC recommendations prefer the name azane to ammonia: hence chloramine would be named chloroazane in substitutive nomenclature, not chloroammonia.

Uses

Fertiliser

Major quanity of ammonia is used as fertilisers either as its salts or as solutions. Consuming more than 1 per cent of all man-made power, the production of ammonia is a significant component of the world energy budget.

Precursor to nitrogenous compounds

Ammonia is directly or indirectly the precursor to most nitrogen-containing compounds. Practically all synthetic and all inorganic nitrogen compounds are prepared from ammonia. An important derivative is nitric acid. This key material is generated via the Ostwald process by oxidisation of ammonia with air over a platinum catalyst at 700°–850°C, ~9 atmosphere. Nitric oxide is an intermediate:

$$NH_3 + 2O_2 \rightarrow HNO_3 + H_2O$$

Nitric acid is used for the production of fertilisers, explosives, and natural organonitrogen other chemical compounds.

Minor and Emerging Uses

Refrigeration-R717

Ammonia's thermodynamic properties made it one of the refrigerants commonly used prior to the discovery of dichlorodifluoromethane. Ammonia's toxicity complicates this application. Anhydrous ammonia is widely used in industrial refrigeration applications because of its high energy efficiency and low cost. Ammonia is used less frequently in commercial applications, such as in grocery store freezer cases and refrigerated displays due to its earlier mentioned toxicity.

For remediation of gaseous emissions

Ammonia used to scrub SO_2 from the burning of fossil fuels, the resulting product is converted to ammonium sulphate for use as fertiliser. Ammonia neutralises the nitrogen oxides (NO_x) pollutants emitted by diesel engines. This technology, called SCR (selective catalytic reduction), relies on a vanadia-based catalyst.

As a fuel

Liquid ammonia was used as the fuel of the rocket airplane, the X-15. Although not as powerful as other fuels, it left no soot in the reusable rocket engine and its density approximately matches that for the oxidiser, liquid oxygen, which simplified the aircraft's design.

As a vehicle fuel

Ammonia is proposed as a practical and clean alternative to fossil fuel for internal combustion engines. The biggest obstacle is the enormous increase in production required since present production, although the second most produced chemical, is a very small fraction of world petroleum usage. Ammonia has no more serious issues, as an alternative vehicle fuel compared to petrol or diesel, including toxicity, flammability, use in engines, pollution, energy density.It does require twice the storage volume of petrol/diesel. It can run in existing engines. It is already widely produced and distributed, and can be manufactured from renewable energy sources, coal or nuclear power. The main down side is that overall it is significantly less efficient than batteries. There are prototype solid state processes to use electricity to convert nitrogen and water directly to ammonia, which are claimed to be cheaper, more efficient and capable of much smaller scale application, i.e. to otherwise stranded assets such as remote wind turbines.

The calorific value of ammonia is 22.5 MJ/kg (9690 Btu/lb) which is about half that of diesel. In a normal engine, in which the water vapour is not condensed, the calorific value of ammonia will be about 21 per cent less than this figure.

Textile

Liquid ammonia is used for treatment of cotton materials, give a properties like mercerisation using alkalies. And also used for pre-washing of wool.

Poison treatment

Solutions of ammonia in water can be applied on the skin to lessen the effects of acidic animal poisons, especially insect poison and jellyfish poison.

Ammonia's Role in Biological Systems and Human Disease

Ammonia is an important source of nitrogen for living systems. Although atmospheric nitrogen abounds, few living creatures are capable of utilising this nitrogen. Nitrogen is required for the synthesis of amino acids, which are the building blocks of protein. Some plants rely on ammonia and other nitrogenous wastes incorporated into the soil by decaying matter. Others, such as nitrogen-fixing legumes, benefit from symbiotic relationships with rhizobia which create ammonia from atmospheric nitrogen.

Ammonia is important for normal animal acid/base balance. After formation of ammonium from glutamine, α-ketoglutarate may be degraded to produce two molecules of bicarbonate which are then available as buffers for dietary acids. Ammonium is excreted in the urine resulting in net acid loss. Ammonia may itself diffuse across the renal tubules, combine with a hydrogen ion, and thus allow for further acid excretion.

Excretion

Ammonium ions are a toxic waste product of the metabolism in animals. In fishes and aquatic invertebrates, it is excreted directly into the water. In mammals, sharks, and amphibians, it is converted in the urea cycle to urea, because it is less toxic and can be stored more efficiently. In birds, reptiles, and terrestrial snails, metabolic ammonium is converted into uric acid, which is solid, and can therefore be excreted with minimal water loss.

Theoretical role in alternative biochemistry

Ammonia has been proposed as a possible replacement for water as a bodily solvent in the theoretical alternative biochemistries of life-forms that do not use carbon for cellular structure and water as a solvent to dissolve bodily solutes and allow essential parts of metabolic processes to occur. It has been suggested that ammonia would be most favourable for life-forms that live in temperatures below the freezing point of water.

Liquid ammonia as a solvent

Liquid ammonia is the best-known and most widely studied non-aqueous ionising solvent. Its most conspicuous property is its ability to dissolve alkali metals to form highly coloured, electrically conducting solutions containing solvated electrons.

Apart from these remarkable solutions, much of the chemistry in liquid ammonia can be classified by analogy with related reactions in aqueous solutions.

Safety Precautions

Toxicity and storage information

Hydrochloric acid sample releasing HCl fumes which are reacting with ammonia fumes to produce a white smoke of ammonium chloride.

The toxicity of ammonia solutions does not usually cause problems for humans and other mammals, as a specific mechanism exists to prevent its build-up in the bloodstream. Ammonium compounds should never be allowed to come in contact with bases (unless in an intended and contained reaction), as dangerous quantities of ammonia gas could be released.

Solutions of ammonia (5–10 per cent by weight) are used as household cleaners, particularly for glass. These solutions are irritating to the eyes and mucous membranes (respiratory and digestive tracts), and to a lesser extent the skin. Caution should be used that the chemical is never mixed into any liquid containing bleach, or a poisonous gas may result. Mixing with chlorine-containing products or strong oxidants, for example household bleach can lead to hazardous compounds such as chloramines. Ammonia is also used for the production of plastics, fibres, explosives, and intermediates for dyes and pharma-ceuticals.

AMMONIUM COMPOUNDS

Ammonium Sulphate

Ammonium sulphate, $(NH_4)_2SO_4$, is an inorganic chemical compound commonly used as a fertiliser. It contains 21 per cent nitrogen as ammonium ions and 24 per cent sulphur as sulphate ions. Ammonium sulphate occurs naturally as the rare mineral mascagnite in volcanic fumaroles and due to coal fires on some dumps. The principal raw materials for production of ammonium sulphate are gysum and coke and the following reaction take place:

$$(NH_4)_2CO_3(aq) + CaSO_4 \cdot 2H_2O(s) \rightarrow CaCO_3(s) + 2H_2O + (NH_4)_2SO_4(aq)$$

Properties

Ammonium sulphate is not soluble in alcohol or liquid ammonia. The compound is slightly hygroscopic and absorbs water from the air at relative humidity > 81 per cent (at ca. 20°C). Ammonium sulphate is prepared commercially by reacting ammonia with sulphuric acid (H_2SO_4). Ammonium sulphate is prepared commercially from the ammoniacal liquor of gas-works and is purified by recrystallisation. It forms large rhombic prisms, has a somewhat saline taste and is easily soluble in water.

Uses

It is used largely as an artificial fertiliser for alkaline soils. In the soil the sulphate ion is released and forms bisulphate, lowering the pH balance of the soil (as do other sulphate compounds such as aluminium sulphate), while contributing essential nitrogen for plant growth.

It is also used as an agricultural spray adjuvant for water soluble insecticides, herbicides, and fungicides. There it functions to bind iron and calcium cations that are present in both well water and plant cells. It is particularly effective as an adjuvant for 2,4-D (amine), glyphosate, and gluphosinate herbicides. It is also used in the preparation of other ammonium salts.

In biochemistry, ammonium sulphate precipitation is a common method for purifying proteins by precipitation. As such, ammonium sulphate is also listed as an ingredient for many United States vaccines per the Center for Disease Control. Ammonium sulphate is also a food additive.

Ammonium Nitrate

The chemical compound ammonium nitrate, the nitrate of ammonia with the chemical formula NH_4NO_3, is a white powder at room temperature and standard pressure. It is commonly used in agriculture as a high-nitrogen fertiliser, and it has also been used as an oxidising agent in explosives, including improvised explosive devices. Ammonium nitrate, when mixed with water, will create a cold substance that can be used as a cold pack.

Manufacture

The processes involved in the production of ammonium nitrate in industry, although simple in chemistry, challenge technology: The acid-base reaction of ammonia with nitric acid gives a solution of ammonium nitrate:

$$HNO_3(aq) + NH_3(g) \rightarrow NH_4NO_3(aq)$$

For industrial production, this is done using anhydrous ammonia gas and concentrated nitric acid. This reaction is violent and very exothermic. After the solution is formed, typically at about 83 per ent concentration, the excess water is evaporated to an ammonium nitrate (AN) content of 95 per cent to 99.9 per cent concentration (AN melt), depending on grade. The AN melt is then made into prills or small beads in a spray tower, or into granules by spraying and tumbling in a rotating drum. The prills or granules may be further dried, cooled, and then coated to prevent caking. These prills or granules are the typical AN products in commerce (Fig. 17.3).

The Haber process combines nitrogen and hydrogen to produce ammonia, part of which can be oxidised to nitric acid and combined with the remaining ammonia to produce the nitrate. Another production method is used in the so-called Odda process.

Disasters

Ammonium nitrate decomposes into gases including oxygen when heated (non-explosive reaction); however, ammonium nitrate can be induced to decompose explosively by detonation. Large stockpiles of the material can be a major fire risk due to their supporting oxidation, and may also detonate.

Ammonium nitrate decomposes at temperatures above 210°C. Pure AN is stable and will stop decomposing once the heat source is removed, but when catalysts are present (combustible materials, acids, metal ions, chlorides) the reaction can become self-sustaining (known as self-sustaining decomposition, SSD). This is a well-known hazard with some types of NPK fertilisers, and is responsible for the loss of several cargo ships.

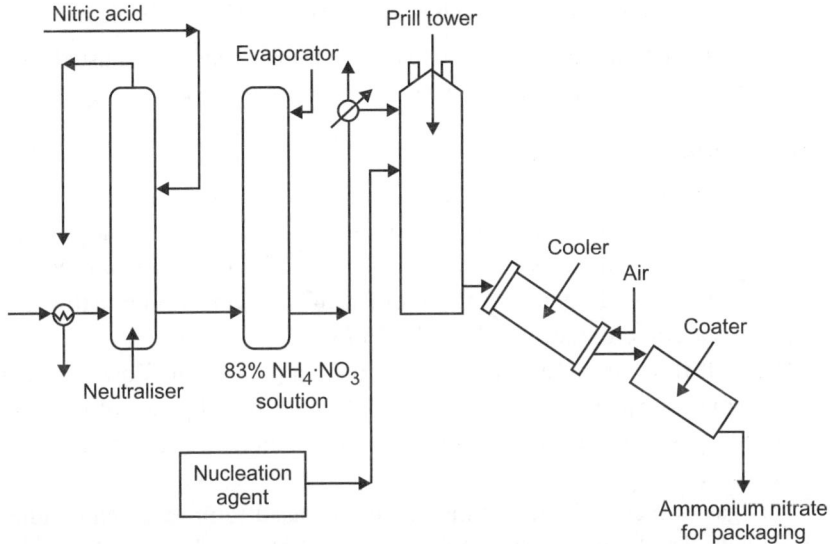

Fig. 17.3. Flow sheet for manufacture of ammonium nitrate.

UREA

Urea is also known by the International Nonproprietary Name (rINN) carbamide, as established by the World Health Organisation. For example, the medicinal compound hydroxyurea (old British approved name) is now hydroxycarbamide. Other names include carbamide resin, isourea, carbonyl diamide, and carbonyldiamine

It was the first organic compound to be artificially synthesised from inorganic starting materials, in 1828 by Friedrich Wöhler, who prepared it from silver isocyanate through a reaction with ammonium chloride:

$$AgNCO + NH_4Cl \rightarrow (NH_2)_2CO + AgCl$$

Although Wöhler was attempting to prepare ammonium cyanate, by forming urea, he inadvertently discredited vitalism, the theory that the chemicals of living organisms are fundamentally different from inanimate matter, thus starting the discipline of organic chemistry.

Carbamate Process

The process involves the following reaction.

$$\text{1.} \quad CO_2 + 2NH_3 \longrightarrow NH_4.COO.NH_2; \qquad \Delta H° = -37.4 \text{ kcal.}$$
$$\text{2.} \quad NH_4 COO.NH_2 \longrightarrow NH_2.CO.NH_2 + H_2O; \qquad \Delta H° = -10.0 \text{ kcal}$$

Undesirable side reaction

$$2NH_2.CO.NH_2 \longrightarrow NH_2.CO.NH_2.CO.NH_2.H_2O$$

The first reaction is easily carried to completion but the second usually has a conversion of 40–60 per cent depending upon process conditions.

Raw Materials

The main raw materials are: (i) ammonia, and (ii) carbon dioxide.

Ammonia and carbon dioxide are compressed and fed continuously into an autoclave maintained between 135° and 200°C and at pressure usually between 1.05 to 3.50 kg/cm² to form a melt of urea, ammonium carbonate, ammonia, carbon dioxide and water (Fig. 17.4).

The melt flows from the autoclaves through a pressure let down valve to the carbamate stripper where it is separated into a liquid phase of urea and water and a gaseous phase of ammonia and carbon dioxide with small amount of water.

This gas is recycled back to the process or used to produce ammonium compounds. The liquid product of the stripper can be used in the preparation of urea, ammonia solution or it can be processed further to produce solid urea.

The main variables affecting the reaction are temperature, pressure, feed composition and reaction time. Urea is formed only in liquid phase which is maintained by application of heat and pressure.

The pressure cannot be very high because of the exothermic nature of reaction. There are basically two types of processes for the manufacture of urea: (i) partial recycle, and (ii) total recycle.

There are various processes used through out the world but Monte-Catini process is the most conly (see Figs 17.4 and 17.5).

This process uses feed of 2 moles of ammonia and 1 mole of carbon dioxide. They form ammonium carbamate which is then converted into urea and water in an autoclave (stainless steel) at 360°C and 180 kg/cm². The temperature is controlled by cooling water in the jacket.

The effluent from the autoclave urea-ammonium carbamate, ammonia, carbon dioxide and water is compressed down to 28 kg/cm² and then passes through a shell and tube heater and then enters the carbamate condenser.

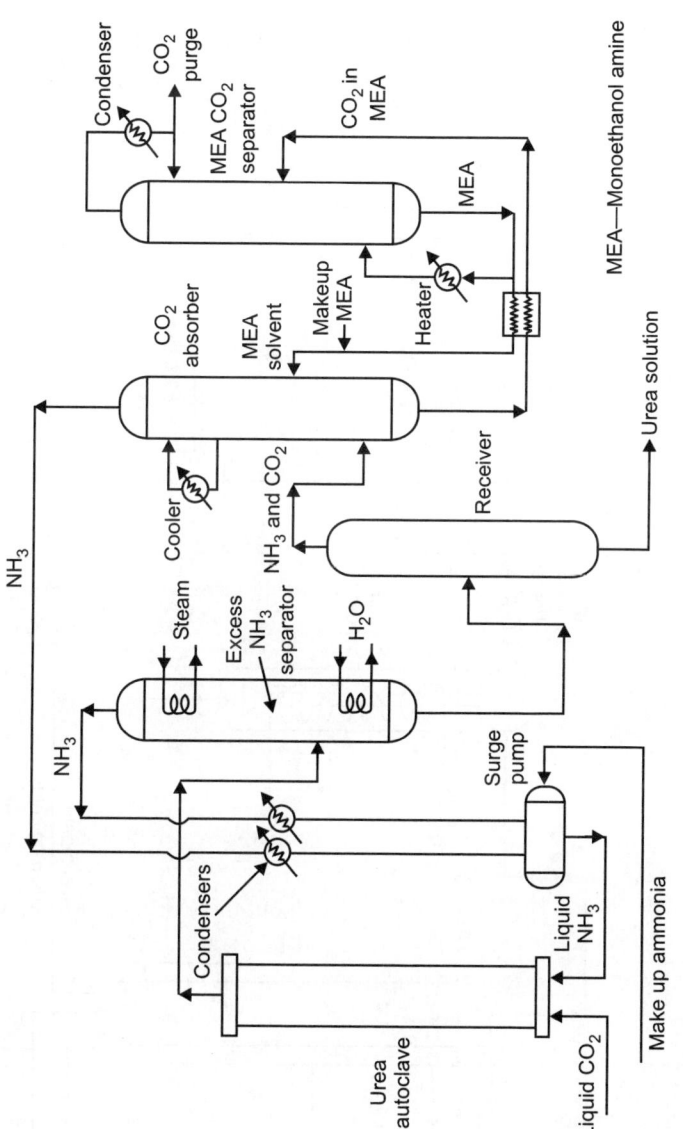

Fig. 17.4. Production of urea solution.

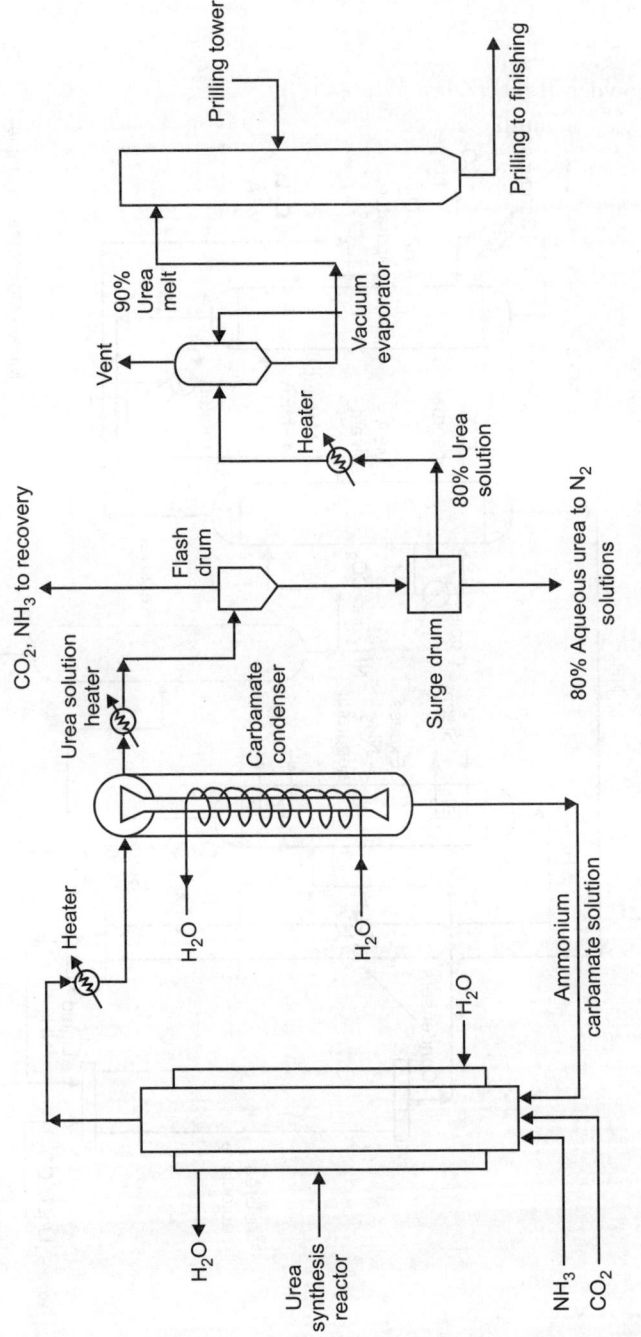

Fig. 17.5. Urea manufacture (Monte-Catini process).

The condenser is a two compartment pressure vessel lined with stainless steel. The upper compartment is a gas liquid separater and the lower one is the condenser provided with stainless steel water cooled coil. The unreacted ammonia and carbon dioxide in the separator pass through the downcomer pipe of the condenser and after being cooled by the water cooled coil, condense and react to form carbamate. This liquid stream is then pumped from the bottom of the condenser back to the autoclave. In a total recycle process, a second carbamate condenser is used which is operated at atmospheric pressure. The liquid urea solution from the top of the carbamate condenser is flashed to atmospheric pressure in the flash drum and flows through a heater to a stainless steel evaporator. The liquid stream contains about 80 per cent urea and can be compounded into nitrogen solution or prills. The unreacted ammonia and carbon dioxide leaving the separater enter the second carbamate condenser to be recycled to the autoclave. The 80 per cent aqueous urea solution is sent to a vacuum evaporator to obtain molten urea containing less than one per cent water. The molten mass is then sprayed into a prilling tower. To avoid formation of biurate the temperature must be kept just above the melting point for processing times of 1–2 seconds in this phase of the operation.

Engineering Problems

Autoclave variables

To get optimum yield it is very necessary to have a close control over temperature, pressure, NH_3/CO_2 ratio and feed rate.

The variables affect the reaction as:

1. There is an increase in production with increase in pressure.
2. Production increases with temperature to maximum of 175°–200°C then falls off sharply. The operating pressure should be above the dissociation pressure of carbonate. With excess ammonia higher pressure and temperature can be maintained to get better yields, but because of various reasons (e.g. increased cost of equipment, increased biurate formation and high corrosion rates) this is normally not done.

Carbamate decomposition and recycle

The main drawback in the biurate formation is that it is harmful when urea is used as cattle feed. This is checked by maintaining low temperature and low retention time. In recycle process, the unrelated off gasses must be recirculated or used elsewhere.

Prilling

Here also the problem of biurate formation is there. Prilling should be accomplished at a temperature just above the melting point of urea with minimum of retention time.

Corrosion

Because of this, much has been done for finding a suitable material of construction. At lower temperature and high NH_3/CO_2 ratio the corrosion is less but otherwise very severe. Special materials make the autoclave very costly. Stainless steel, haste-alloy, tantalum, etc. are used for making autoclave.

It is found in mammalian and amphibian urine as well as in some fish. Birds and reptiles excrete uric acid, comprising a different form of nitrogen metabolism that requires less water.

Urea is highly soluble in water and is, therefore, an efficient way for the human body to expel excess nitrogen. Its high solubility is due to extensive hydrogen bonding with water: up to eight hydrogen bonds may form—two from the oxygen atom, one from each hydrogen atom and one from each nitrogen atom. The urea molecule is planar and retains its full molecular point symmetry, due to conjugation of one of each nitrogen's P orbital to the carbonyl double bond. Each carbonyl oxygen atom accepts four N-H-O hydrogen bonds, a very unusual feature for such a bond type. This dense (and energetically favourable) hydrogen bond network is probably established at the cost of efficient molecular packing.

Function

In humans

Urea is, in essence, a waste product. It is found in and extracted from urine. However, it also plays a very important role in that it helps set up the countercurrent system in the nephrons. The countercurrent system in the nephrons allows for reabsorption of water and critical ions. Urea is reabsorbed in the inner medullary collecting ducts of the nephrons, thus raising the osmolarity in the medullary interstitium surrounding the thin ascending limb of the Loop of Henle. It is dissolved in blood (reference range of 2.5–7.5 mmol/litre) and excreted by the kidney as a component of urine. In addition, a small amount of urea is excreted (along with sodium chloride and water) in sweat.

Non-humans

Most organisms have to deal with the excretion of nitrogen waste originating from protein and amino acid catabolism. In aquatic organisms the most common form of nitrogen waste is ammonia, while land-dwelling organisms convert the toxic ammonia to either urea or uric acid. In general, birds and saurian reptiles excrete uric acid, whereas the remaining species, including mammals, excrete urea.

Urea can be irritating to skin and eyes. Too high concentrations in the blood can cause damage to organs of the body. Low concentrations of urea, such as are found in typical human urine, are not dangerous with additional water ingestion within a reasonable time-frame. Many animals (e.g. dogs) have a much more concentrated urine and it contains a higher urea amount than normal human urine; this can prove dangerous as a source of liquids for consumption in a life-threatening situation (such as in a desert).

It has been found that urea can cause algal blooms to produce toxins, and urea in runoff from fertilisers may play a role in the increase of toxic blooms.

Repeated or prolonged contact with urea in fertiliser form on the skin may cause dermatitis. The substance also irritates the eyes, the skin, and the respiratory tract. The substance decomposes on heating above melting point, producing toxic gases, and reacts violently with strong oxidants, nitrites, inorganic chlorides, chlorites and perchlorates, causing fire and explosion hazard.

Uses

Agricultural use

Urea is used as a nitrogen-release fertiliser, as it hydrolyses back to ammonia and carbon dioxide, but its most common impurity, biuret, must be present at less than 2 per cent, as it impairs plant growth. It is also used in many multi-component solid fertiliser formulations.

Storage of urea fertiliser

Like most nitrogen products, urea absorbs moisture from the atmosphere. Therefore it should be stored either in closed/sealed bags on pallets, or, if stored in bulk, under cover with a tarpaulin. As with most solid fertilisers, it should also be stored in a cool, dry, well-ventilated area.

Industrial use

Urea has the ability to form loose compounds, called clathrates, with many organic compounds. The organic compounds are held in channels formed by interpenetrating helices comprising of hydrogen-bonded urea molecules. This behaviour can be used to separate mixtures, and has been used in the production of aviation fuel and lubricating oils. As the helices are interconnected, all helices in a crystal must have the same handedness. This is determined when the crystal is nucleated and can thus be forced by seeding. This property has been used to separate racemic mixtures.

Medical use

Drug use

Urea is used in topical dermatological products to promote rehydration of the skin. If covered by an occlusive dressing, 40 per cent urea preparations may also be used for nonsurgical debridement of nails. This drug is also used as an earwax removal aid. Like saline, urea injection is used to perform abortions. It is also the main component of an alternative medicinal treatment referred to as urine therapy.

Textile use

Urea in textile laboratories are frequently used both in dyeing and printing as an important auxiliary, which provides solubility to the bath and retains some moisture required for the dyeing or printing process.

Ionic liquid

Choline chloride, in mixture with urea, is used as a deep eutectic solvent, a type of ionic liquid.

Automobile systems

A number of modern diesel engines, including that found on the current Mercedes-Benz ML320, use an injector containing a water-based urea solution to capture particulate emissions. The solution is injected into the exhaust system and releases ammonia. This reacts with the nitrogen oxide emissions and is converted into nitrogen and water within the catalytic converter.

Ureas

The term urea or carbamide is also used for the class of chemical compounds sharing the same functional group RR N-CO-NRR based on a carbonyl group flanked by two organic amine residues. They can be accessed in the laboratory by reaction of phosgene with primary or secondary amines, proceeding through an isocyanate intermediate.

Reactions

Urea reacts with alcohols to form urethanes. Urea reacts with malonic esters to make barbituric acids.

NITRIC ACID

Nitric acid (HNO_3), also known as aqua fortis and spirit of nitre, is a highly corrosive and toxic strong acid that can cause severe burns. Colourless when pure, older samples tend to acquire a yellow cast due to the accumulation of oxides of nitrogen. If the solution contains more than 86 per cent nitric acid, it is referred to as fuming nitric acid. Fuming nitric acid is characterised as white fuming nitric acid and red fuming nitric acid, depending on the amount of nitrogen dioxide present.

Properties

Pure anhydrous nitric acid (100 per cent) is a colourless liquid with a density of 1522 kg/m³ which solidifies at –42°C to form white crystals and boils at

83°C. When boiling in light, even at room temperature, there is a partial decomposition with the formation of nitrogen dioxide following the reaction:

$$4HNO_3 \rightarrow 2H_2O + 4NO_2 + O_2 \ (72°C)$$

which means that anhydrous nitric acid should be stored below 0°C to avoid decomposition. The nitrogen dioxide (NO_2) remains dissolved in the nitric acid colouring it yellow, or red at higher temperatures. While the pure acid tends to give off white fumes when exposed to air, acid with dissolved nitrogen dioxide gives off reddish-brown vapours, leading to the common name 'red fuming acid' or 'fuming nitric acid'.

Fuming nitric acid is also referred to as 16-molar nitric acid—as the most concentrated form of nitric acid at standard temperature and pressure (STP). Nitric acid is miscible with water and distillation gives an azeotrope with a concentration of 68 per cent HNO_3 and a boiling temperature of 120.5°C at 1 atm, which is the ordinary concentrated nitric acid of commerce.

Two solid hydrates are known; the monohydrate ($HNO_3 \cdot H_2O$) and the trihydrate ($HNO_3 \cdot 3H_2O$).

Nitrogen oxides (NO_x) are soluble in nitric acid and this property influences more or less all the physical characteristics depending on the concentration of the oxides. These mainly include the vapour pressure above the liquid and the boiling temperature, as well as the colour mentioned above.

Nitric acid is subject to thermal or light decomposition with increasing concentration and this may give rise to some non-negligible variations in the vapour pressure above the liquid because the nitrogen oxides produced dissolve partly or completely in the acid.

Acidic properties

Being a typical acid, nitric acid reacts with alkalies, basic oxides, and carbonates to form salts, such as ammonium nitrate. Due to its oxidising nature, nitric acid generally does not donate its proton (that is, it does not liberate hydrogen) on reaction with metals and the resulting salts are usually in the higher oxidised states. For this reason, heavy corrosion can be expected and should be guarded against by the appropriate use of corrosion resistant metals or alloys.

Nitric acid has an acid dissociation constant (pK_a) of –1.4: in aqueous solution, it almost completely (93 per cent at 0.1 mol/l) ionises into the nitrate ion NO_3^- and a hydrated proton, known as a hydronium ion, H_3O^+.

$$HNO_3 + H_2O \rightarrow H_3O^+ + NO_3^-$$

Oxidising properties

Reactions with metals

Being a powerful oxidising agent, nitric acid reacts violently with many organic materials and the reactions may be explosive. Depending on the acid concentration, temperature and the reducing agent involved, the end products can be variable. Reaction then takes place with all metals except the precious metal series and certain alloys. This characteristic has made it a common agent to be used in acid tests. As a general rule, oxidising reactions occur primarily with the concentrated acid, favouring the formation of nitrogen dioxide (NO_2).

$$Cu + 4H^+ + 2NO_3^- \rightarrow Cu^{2+} + 2NO_2 + 2H_2O$$

The acidic properties tend to dominate with dilute acid, coupled with the preferential formation of nitrogen oxide (NO).

$$3Cu + 8HNO_3 \rightarrow 3Cu(NO_3)_2 + 2NO + 4H_2O$$

Since nitric acid is an oxidising agent, hydrogen (H_2) is rarely formed. Only magnesium (Mg), manganese (Mn) and calcium (Ca) react with cold, dilute nitric acid to give hydrogen:

$$Mg_{(s)} + 2HNO_{3\ (aq)} \rightarrow Mg(NO_3)_{2\ (aq)} + H_{2\ (g)}$$

Nitric acid has the highest distinction (amongst all acids) of attacking and dissolving all metals on the periodic table except gold and platinum.

Passivation

Although chromium (Cr), iron (Fe) and aluminium (Al) readily dissolve in dilute nitric acid, the concentrated acid forms a metal oxide layer that protects the metal from further oxidation, which is called passivation. Typical passivation concentrations range from 18 to 22 per cent by weight.

Reactions with non-metals

Reaction with non-metallic elements, with the exception of silicon and halogens, usually oxidises them to their highest oxidation states as acids with the formation of nitrogen dioxide for concentrated acid and nitric oxide for dilute acid.

$$C + 4HNO_3 \rightarrow CO_2 + 4NO_2 + 2H_2O$$

or

$$3C + 4HNO_3 \rightarrow 3CO_2 + 4NO + 2H_2O$$

Industrial Production

Nitric acid is made by reacting nitrogen dioxide (NO_2) with water.

$$3NO_2 + H_2O \rightarrow 2HNO_3 + NO$$

Normally, the nitric oxide produced by the reaction is reoxidised by the oxygen in air to produce additional nitrogen dioxide (Fig. 17.6). Dilute nitric acid may be concentrated by distillation up to 68 per cent acid, which is a maximum boiling azeotrope containing 32 per cent water. In the laboratory, further concentration involves distillation with sulphuric acid which acts as a dehydrating agent. Such distillations must be done with all-glass apparatus at reduced pressure, to prevent decomposition of the acid. Industrially, strong nitric acid is produced by dissolving additional nitrogen dioxide in 68 per cent nitric acid in an absorbtion tower. Dissolved nitrogen oxides are either stripped in the case of white fuming nitric acid, or remain in solution to form red fuming nitric acid.

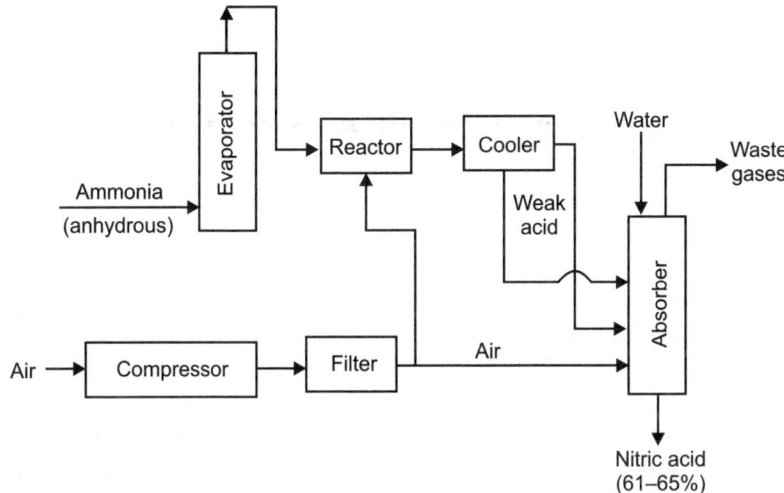

Fig. 17.6. Flow diagram for manufacture of nitrogen and its compounds from ammonia.

Commercial grade nitric acid solutions are usually between 52 and 68 per cent nitric acid. Production of nitric acid is via the Ostwald process, named after German chemist Wilhelm Ostwald. In this process, anhydrous ammonia is oxidised to nitric oxide, which is then reacted with oxygen in air to form nitrogen dioxide. This is subsequently absorbed in water to form nitric acid and nitric oxide. The nitric oxide is cycled back for reoxidation. By using ammonia derived from the Haber process, the final product can be produced from nitrogen, hydrogen, and oxygen which are derived from air and natural gas as the sole feedstocks.

Manufacturing Plants

There are three types of plants available for the production of nitric acid. They all employ the same chemical reaction but differ in their operating conditions. Plants may be named as:

1. Atmospheric pressure oxidation and atmospheric pressure absorption.
2. Pressure oxidation and pressure absorption.
3. Combined process, i.e. atmospheric pressure oxidation and pressure absorption.

Atmospheric pressure type

Firstly ammonia is vapourised either by using steam or some other energy source. The ammonia vapour is then mixed with hot air, heated by out-going nitrous oxide coming from burner. Ammonia content is 9.5 to 11 per cent in the ammonia air mixture.

The mixture is filtered and then sent through the bed of catalyst, i.e. 10 per cent rhodium, 90 per cent platinum gauge at a temperature of 750° to 800°C nitrous oxide from the preheater to heat the air is then passed through waste heat boiler where temperature is reduced to about 179° to 199°C. Then the gases are further passed through a series of coolers to condenses the water formed as much as possible. Here nitric acid begins to form. Water condensed here should contain smallest amount of nitric acid i.e. not more than 2 to per cent of nitric acid. The subsequent condensers produce approximately 20–22 per cent nitric acid. Condensates are then sent to the absorption system of the plant.

Nitrous gases coming from last condenser are sent through a series of absorption towers for scrubbing. The temperature of the nitrous gases approaches to that of cooling water. The make up water and weak condensate from first cooler are added at the top of absorption tower. To increase absorption efficiency the stronger condensate is added to the system at the point where the strength of the acid is approximately same. The acid in the first tower is cooled directly by evaporating ammonia. The acid is cooled by heat exchanger before reaching the top. The final product, viz. 45 to 52 per cent nitric acid is taken out from the second tower. It is desirable that the oxidation of nitrous oxide should be completed in first tower.

Pressure oxidation and pressure absorption

In this process following steps are involved:
1. Compression of ammonia.
2. Air compression.
3. Oxidation in a burner equipped with a platinum-rhodium catalyst.

Series of heat exchangers are used to reduce the temperature of nitrous gases after leaving the reactor. The nitrous gases are cooled to near the temperature of cooling water. Then the gases enter the absorption tower. In this case air compression may be done by either: (i) centrifugal compressor, and (ii) non-centrifugal compressor.

When centrifugal compressor is used, there is no need of preheating the air since air delivered from the compressor has the temperature of 230°–315°C which is suitably high for the satisfactory working of the ammonia oxidiser.

When compressors other than centrifugal are used, air is preheated in heat exchanger and heat exchangers also: (i) preheat the tail gas if power recovery is required, (ii) produce steam, and (iii) condense most of the water formed in the oxidation step, in the form of weak acid.

Atmospheric oxidation and pressure absorption

This process is called combination process. The first part of combination process involves the same method as in the atmospheric pressure type process. Then the cooled nitrous oxides from cooler condenser are compressed in the stainless steel turbo compressor from 2.1 to 7.4 kg/cm^2 pressure depending upon the plants. Compression helps in completion of the conversion of nitric acid to nitrogen peroxide due to the presence of air. The pressure absorption is done in plate tower in single stage. Although pressure vessels are costlier but since only one tower is required for the purpose the high pressure process has been found to be quite economical and is, therefore, in vogue. The compressed gases from the compressor are sent to cooler condensers where nitric acid of 55 to 62 per cent concentration is taken out. The nitric acid is then sent to the bleacher.

The remaining gases are sent to the series of packed towers where the gases are absorbed by acid which is circulated by centrifugal pump. The acid flows in the towers alternatively, counter-currently and cocurrently.

The nitric acid of 55 to 62 per cent concentration which is condensed before compression is added to the point where the strength of acid is the same as that of above. Power recovery is possible by preheating the tail gas from the absorption system by heat exchanger with hot nitrous gases from the burner. Then the tail gases are released to the atmosphere by an expander mounted on the shaft of gas compressor.

Uses

A solution of nitric acid and alcohol, Nital, is used for etching of metals to reveal the microstructure.

Commercially available aqueous blends of 5–30 per cent nitric acid and 15–40 per cent phosphoric acid are commonly used for cleaning food and dairy equipment primarily to remove precipitated calcium and magnesium compounds (either deposited from the process stream or resulting from the use of hard water during production and cleaning). Nitric acid is also used in explosives, and is key to the manufacture of nitroglycerine and RDX.

Clog remover

In a high medium concentration nitric acid is used as a cheap clog remover.

Woodworking

In a low concentration (approximately 10 per cent), nitric acid is often used to artificially age pine and maple. The colour produced is a grey-gold very much like very old wax or oil finished wood (wood finishing).

Other uses

Alone, it is useful in metallurgy and refining as it reacts with most metals, and in organic synthesis. When mixed with hydrochloric acid, nitric acid forms aqua regia, one of the few reagents capable of dissolving gold and platinum.

Safety

Nitric acid is a powerful oxidising agent, and the reactions of nitric acid with compounds such as cyanides, carbides, and metallic powders can be explosive. Reactions of nitric acid with many organic compounds, such as turpentine, are violent and hypergolic (i.e. self-igniting).

Concentrated nitric acid dyes human skin yellow due to a reaction with the keratin. These yellow stains turn orange when neutralised.

SODIUM NITRATE

Sodium nitrate is the chemical compound with the formula $NaNO_3$. This salt, also known as Chile saltpeter or Peru saltpeter (to distinguish it from ordinary saltpeter, potassium nitrate), is a white solid which is very soluble in water. The mineral form is also known as nitratine or soda nitre.

Sodium nitrate is used as an ingredient in fertilisers, pyrotechnics, as a food preservative, and as a solid rocket propellants, as well as in glass and pottery enamels; the compound has been mined extensively for those purposes. The mining of Chile saltpeter was such a profitable business that three nations, Chile, Peru, and Bolivia fought over the richest deposits in the War of the Pacific. The world's largest natural deposits of caliche ore were in the Atacama desert of Chile, and many deposits were mined for over a century, until the 1940s, when its value declined dramatically in the first decades of the twentieth century.

Chile still has the largest reserves of caliche, with active mines in such locations as Pedro de Valdivia, Maria Elena and Pampa Blanca. Sodium nitrate, potassium nitrate, sodium sulphate and iodine are all obtained by the processing of caliche. The former Chilean saltpeter mining communities of Humberstone and Santa Laura were declared UNESCO World Heritage sites in 2005. Sodium nitrate is also synthesised industrially by neutralising nitric acid with soda ash.

Applications

Sodium nitrate was used extensively as a fertiliser and a raw material for the manufacture of gunpowder in the late nineteenth century.

Sodium nitrate has antimicrobial properties when used as a food preservative. It is found naturally in leafy green vegetables.

It can be used in the production of nitric acid by combining it with sulphuric acid and subsequent separation through fractional distillation of the nitric acid, leaving behind a residue of sodium bisulphate. Hobbyist gold refiners use sodium nitrate to make a hybrid aqua regia that dissolves gold and other metals.

Less common applications include its use as a substitute oxidiser used in fireworks as a replacement for potassium nitrate commonly found in black powder and as a component in instant cold packs.

Because sodium nitrate can be used as a phase change material it may be used for heat transfer in solar power plants.

POTASSIUM NITRATE

Potassium nitrate is a chemical compound with the chemical formula KNO_3. A naturally occurring mineral source of nitrogen, KNO_3 constitutes a critical oxidising component of black powder/gunpowder. In the past it was also used for several kinds of burning fuses, including slow matches. Because potassium nitrate readily precipitates, urine was a significant source, through various means, from the late middle ages and early modern era through the 19th century.

Potassium nitrate is the oxidising component of black powder. Before the large-scale industrial fixation of nitrogen through the Haber process, major sources of potassium nitrate were the deposits crystallising from cave walls and the draining of decomposing organic material. Dung-heaps were a particularly common source: ammonia from the decomposition of urea and other nitrogenous materials would undergo bacterial oxidation to produce nitrate. It was and is also used as a component in some fertilisers. When used by itself as a fertiliser, it has an NPK rating of 13-0-38 (indicating 13.9 per cent, 0 per cent, and 38.7 per cent of nitrogen, phosphorus, and potassium, by mass, respectively). Potassium nitrate was once thought to induce impotence, and is still falsely rumored to be in institutional food (such as military fare) as an anaphrodisiac; these uses would be ineffective, since potassium nitrate has no such properties. However, potassium nitrate and other nitrates do successfully combat high blood pressure and are used medically to relieve angina.

Urine has also been used in the manufacture of saltpetre for gunpowder. In this process, stale urine placed in a container of straw hay is allowed to sour for many months, after which water is used to wash the resulting chemical salts from the straw. The process is completed by filtering the liquid through wood ashes and air-drying in the sun. Saltpetre crystals can then be collected and added to brimstone and charcoal to create black powder.

Uses

Potassium nitrate is also used as a fertiliser, in amateur rocket propellant, and in several fireworks such as smoke bombs.

In the process of food preservation, potassium nitrate has been a common ingredient of salted meat since the Middle ages, but its use has been mostly discontinued due to inconsistent results compared to more modern nitrate and nitrite compounds. Even so, saltpetre is still used in some food applications, such as charcuterie and the brine used to make corned beef. Sodium nitrate (and nitrite) have mostly supplanted saltpetre's culinary usage, as they are more reliable in preventing bacterial infection than saltpeter. All three give cured salami and corned beef their characteristic pink hue.

It is commonly used in pre-rolled cigarettes to maintain an even burn of the tobacco.

As a fertiliser, it is used as a source of nitrogen and potassium, two of the plant macronutrients for plants. The other macronutrients are carbon, oxygen, and hydrogen.

Potassium nitrate is also the main component (usually about 98 per cent) of tree stump remover; it accelerates the natural decomposition of the stump.

Potassium nitrate is also commonly used in the heat treatment of metals as a solvent in the post-wash. The oxidising, water solubility and low cost make it an ideal short-term rust inhibitor.

It has also been used in the manufacture of ice cream and can be found in some toothpastes for sensitive teeth. Recently, the use of potassium nitrate in toothpastes for treating sensitive teeth has increased dramatically, despite the fact that it has not been conclusively shown to help dental hypersensitivity.

Potassium nitrate is also one of the three components of black powder, along with powdered charcoal (substantially carbon) and sulphur, where it acts as an oxidiser. When subjected to the flame test it produces a lilac flame due to the presence of potassium.

Chapter 18

Sulphur and Sulphuric Acid

INTRODUCTION

One of the most important basic raw materials in the chemical process industries is sulphur. It exists in nature both in the free state and combined in ores such as pyrite (FeS_2). It is also an important constituent of petroleum and natural gas (as H_2S). The most important application of sulphur is in the manufacture of sulphuric acid.

SULPHUR

Sulphur is the chemical element that has the atomic number 16. It is denoted with the symbol S. It is an abundant multivalent non-metal. Sulphur, in its native form, is a yellow crystalline solid. In nature, it can be found as the pure element and as sulphide and sulphate minerals. It is an essential element for life and is found in two amino acids, cysteine and methionine.

Its commercial uses are primarily in fertilisers, but it is also widely used in black gunpowder, matches, insecticides and fungicides. Elemental sulphur crystals are commonly sought after by mineral collectors for their brightly coloured polyhedron shapes.

Characteristics

At room temperature, sulphur is a soft, bright-yellow solid. Elemental sulphur has only a faint odour, similar to that of matches. The odour associated with rotten eggs is due to hydrogen sulphide (H_2S) and organic sulphur compounds rather than elemental sulphur. Sulphur burns with a blue flame that emits sulphur dioxide, notable for its peculiar suffocating odour due to dissolving in the mucosa to form dilute sulphurus acid. Sulphur itself is insoluble in water, but soluble in carbon disulphide—and to a lesser extent in other non-polar organic solvents such as benzene and toluene. Sulphur forms stable compounds with all elements except the noble gases. Sulphur in the solid state ordinarily exists as cyclic crown-shaped S_8 molecules.

414 Chemical Process Industries

The crystallography of sulphur is complex. Depending on the specific conditions, the sulphur allotropes form several distinct crystal structures, with rhombic and monoclinic S_8 best known.

A noteworthy property of sulphur is that its viscosity in its molten state, unlike most other liquids, increases above temperatures of 200°C due to the formation of polymers. The molten sulphur assumes a dark red colour above this temperature. At higher temperatures, however, the viscosity is decreased as depolymerisation occurs.

Amorphous or plastic sulphur can be produced through the rapid cooling of molten sulphur. X-ray crystallography studies show that the amorphous form may have a helical structure with eight atoms per turn. This form is metastable at room temperature and gradually reverts back to crystalline form. This process happens within a matter of hours to days but can be rapidly catalysed.

Sulphur forms more than 30 solid allotropes, more than any other element. Besides S_8, several other rings are known. Removing one atom from the crown gives S_7, which is more deeply yellow than S_8. HPLC analysis of elemental sulphur reveals an equilibrium mixture of mainly S_8, but also S_7 and small amounts of S_6. Larger rings have been prepared, including S_{12} and S_{18}. By contrast, sulphur's lighter neighbour oxygen only exists in two states of allotropic significance: O_2 and O_3. Selenium, the heavier analogue of sulphur, can form rings but is more often found as a polymer chain.

Common naturally occurring sulphur compounds include the sulphide minerals, such as pyrite (iron sulphide), cinnabar (mercury sulphide), galena (lead sulphide), sphalerite (zinc sulphide) and stibnite (antimony sulphide), and the sulphates, such as gypsum (calcium sulphate), alunite (potassium aluminium sulphate), and barite (barium sulphate). It occurs naturally in volcanic emissions, such as from hydrothermal vents, and from bacterial action on decaying sulphur-containing organic matter.

Sulphur is present in many types of meteorites. Ordinary chondrites contain on average 2.1 per cent sulphur, and carbonaceous chondrites may contain as much as 6.6 per cent. Sulphur in meteorites is normally present entirely as troilite (FeS), but other sulphides are found in some meteorites, and carbonaceous chondrites contain free sulphur, sulphates, and possibly other sulphur compounds.

Extraction and Production

Extraction from natural resources

Sulphur is extracted by mainly two processes: the Sicilian process and the Frasch process. The Sicilian process, which was first used in Sicily, was used in ancient times to get sulphur from rocks present in volcanic regions. In this process, the sulphur deposits are piled and stacked in brick kilns built on sloping hillsides, and with airspaces between them. Then powdered sulphur is put on top of the sulphur deposit and ignited. As the sulphur burns, the heat melts the sulphur deposits, causing the molten sulphur to flow down the sloping hillside. The molten sulphur can then be collected in wooden buckets.

The second process used to obtain sulphur is the Frasch process. In this method, three concentric pipes are used: the outermost pipe contains superheated water, which melts the sulphur, and the innermost pipe is filled with hot compressed air, which serves to create foam and pressure. The resulting sulphur foam is then expelled through the middle pipe.

The Frasch process produces sulphur with a 99.5 per cent purity content, and which needs no further purification. The sulphur produced by the Sicilian process must be purified by distillation.

Manufacture from Hydrogen Sulphide

The concentrated hydrogen sulphide may then be converted to sulphur by oxidation. Details of the process differ from plant to plant, depending on the source of the hydrogen sulphide and other factors. The following description refers to a process commonly used by petroleum refineries, in which sulphur recovery plants are almost always necessary owing to pollution control laws. The reactions which take place are given below:

$$2H_2S + 3O_2 \longrightarrow 2SO_2 + 2H_2O$$
$$2H_2S + SO_2 \longrightarrow 3S + 2H_2O$$
$$2H_2S + O_2 \longrightarrow 2S + 2H_2O$$

In the Claus process one-third of the hydrogen sulphide feed to the plant is burned at 1000°C in a pressurised boiler, where 80 per cent of the total heat of reaction is removed by generating steam. The hydrogen sulphide is converted to sulphur dioxide by the first reaction shown above; the second reaction (Claus reaction) produces sulphur.

After the removal of the molten sulphur in a condenser, the remaining gases, now 2 parts hydrogen sulphide and 1 part sulphur dioxide, are passed through a series of two or three bauxite or alumina catalyst beds at 200°–260°C, with sulphur removal between each converter stage. Overhead conversion is 92–95 per cent with two catalytic stages, 95–96 per cent with three and 96–97 per cent with four.

Hydrogen or carbon dioxide in the feed gas to the plant reduces the sulphur yield by 0.25–2.5 per cent, by allowing the formation of carbonyl sulphide (COS) and carbon disulphide (CS_2). The composition of the tail gas from the final converter is about one-third sulphur dioxide and hydrogen sulphide, one-third carbonyl sulphide and carbon disulphide, and one-third sulphur. Although this tail gas may be burned in an incinerator to yield sulphur dioxide, which is then vented to the atmosphere, pollution regulations normally do not permit this, and the tail gases must be treated. In one treatment process (Beavon process) all sulphur compounds in the Claus tail gases are reconverted to hydrogen sulphide by hydrogenation in a converter containing a cobalt molybdate catalyst. No hydrogen needs to be added. The gas containing hydrogen sulphide from the condenser that follows the converter may then be treated by any appropriate process to remove the hydrogen sulphide.

Another treatment (Stratford process) uses sodium carbonate solution to react with the hydrogen sulphide, forming sodium hydrosulphide (NaHS) which in turn is oxidised to sulphur by sodium vanadate in solution. The finely divided sulphur froth is skimmed off, washed and dried. The effluent gas is now low enough in hydrogen sulphide to be vented.

Although what is described above refers to off-gases from petroleum refineries, the hydrogen sulphide content of other gaseous sources, such as sour natural gas, can be treated by similar methods. As a rule, differences between processes are in heat-recovery methods, or in the concentration of hydrogen sulphide in the raw material. In the biological route, hydrogen sulphide (H_2S) from natural gas or refinery gas is absorbed with a slight alkaline solution in a wet scrubber. Or the sulphide is produced by biological sulphate reduction. In the subsequent process step, the dissolved sulphide is biologically converted to elemental sulphur. This solid sulphur is removed from the reactor. This process has been built on commercial scale. The main advantages of this process are:

1. No use of expensive chemicals.
2. The process is safe as the H_2S is directly absorbed in an alkaline solution.
3. No production of a polluted waste stream.
4. Re-usable sulphur is produced.
5. The process occurs under ambient conditions.

Sulphur is one of the oldest pesticides used in agriculture. In organic production sulphur is the most important fungicide used. Biosulphur (biologically produced elemental sulphur with hydrophillic characteristics) can be used well for these applications.

The biosulphur product is different from other processes in which sulphur is produced because the sulphur is hydrophillic. Next to straightforward reuses as source for sulphuric acid production, it can also be applied as sulphur fertiliser. Hydrogen sulphide has the characteristic smell of rotten eggs. It has specific gravity 2.05, flash point (close cup) 205°C, m.p. 115°C. Dissolved in water, hydrogen sulphide is acidic and will react with metals to form a series of metal sulphides. Natural metal sulphides are common, especially those of iron.

Applications

One of the direct uses of sulphur is in vulcanisation of rubber, where polysulphides cross-link organic polymers. Sulphur is a component of gunpowder. It reacts directly with methane to give carbon disulphide, which is used to manufacture cellophane and rayon. Elemental sulphur is mainly used as a precursor to other chemicals. Approximately 85 per cent is converted to sulphuric acid (H_2SO_4), which is of such prime importance to the world's economies that the production and consumption of sulphuric acid is an indicator of a nation's industrial development. The principal use for the acid is the extraction of phosphate ores for the production of fertiliser manufacturing. Other applications of sulphuric acid include oil refining, waste-water processing, and mineral extraction. Sulphur compounds are also used in detergents, fungicides, dyestuffs, and agrichemicals. Sulphur is an ingredient in some acne treatments. An increasing application is as fertiliser. Standard sulphur is hydrophobic and therefore has to be covered with a surfactant by bacteria in the ground before it can be oxidised to sulphate. Sulphur also improves the use efficiency of other essential plant nutrients, particularly nitrogen and phosphorus. Biologically produced sulphur particles are naturally hydrophilic due to a biopolymer coating. This sulphur is therefore easier to disperse over the land (via spraying as a diluted slurry), and results in a faster release.

Sulphites, derived from burning sulphur, are heavily used to bleach paper. They are also used as preservatives in dried fruit.

Sulphur is used as a light-generating medium in the rare lighting fixtures known as sulphur lamps.

Sulphur is the only fungicide used in organically farmed apple production against the main disease apple scab under colder conditions. Sulphur is also a major fungicide in conventional culture of grapes, strawberry, many vegetables and several other crops. It has a good efficacy against a wide range of powdery mildew diseases as well as black spot.

Environmental Impact

The burning of coal and/or petroleum by industry and power plants generates sulphur dioxide (SO_2), which reacts with atmospheric water and oxygen to produce sulphuric acid (H_2SO_4). This sulphuric acid is a component of acid rain, which lowers the pH of soil and freshwater bodies, sometimes resulting in substantial damage to the environment and chemical weathering of statues and structures. Fuel standards increasingly require sulphur to be extracted from fossil fuels to prevent the formation of acid rain. This extracted sulphur is then refined and represents a large portion of sulphur production. In coal fired power plants, the flue gases are sometimes purified. In more modern power plants that use syngas the sulphur is extracted before the gas is burned.

SULPHURIC ACID

Sulphuric acid, H_2SO_4, is a strong mineral acid and is covalent. It is soluble in water at all concentrations. Sulphuric acid has many applications, and is one of the top products of the chemical industry. Principal uses include lead-acid batteries for cars and other vehicles, ore processing, fertiliser manufacturing, oil refining, waste-water processing, and chemical synthesis.

Most proteins contain sulphur (in the amino acids cysteine and methionine), which is metabolised by the body to sulphuric acid. Pure (undiluted) sulphuric acid is not encountered on earth, due to sulphuric acid's great affinity for water. Apart from that, sulphuric acid is a constituent of acid rain, which is formed by atmospheric oxidation of sulphur dioxide in the presence of water, i.e. oxidation of sulphurus acid. Sulphur dioxide is the main by-product produced when sulphur-containing fuels such as coal or oil are burned.

Sulphuric acid is formed naturally by the oxidation of sulphide minerals, such as iron sulphide. The resulting water can be highly acidic and is called acid mine drainage (AMD) or acid rock drainage (ARD). This acidic water is capable of dissolving metals present in sulphide ores, which results in brightly-coloured, toxic streams. The oxidation of iron sulphide pyrite by molecular oxygen produces iron(II), or Fe^{2+}:

$$2FeS_2 + 7O_2 + 2H_2O \rightarrow 2Fe^{2+} + 4SO_4^{2-} + 4H^+.$$

The Fe^{2+} can be further oxidised to Fe^{3+}, according to:

$$4Fe^{2+} + O_2 + 4H^+ \rightarrow 4Fe^{3+} + 2H_2O,$$

and the Fe^{3+} produced can be precipitated as the hydroxide or hydrous oxide. The equation for the formation of the hydroxide is

$$Fe^{3+} + 3H_2O \rightarrow Fe(OH)_3 + 3H^+.$$

The iron(III) ion (ferric iron, in casual nomenclature) can also oxidise pyrite. When iron(III) oxidation of pyrite occurs, the process can become rapid. pH values below zero have been measured in ARD produced by this process.

ARD can also produce sulphuric acid at a slower rate, so that the acid neutralisation capacity (ANC) of the aquifer can neutralise the produced acid. In such cases, the total dissolved solids (TDS) concentration of the water can be increased from the dissolution of minerals from the acid-neutralisation reaction with the minerals.

Extraterrestrial Sulphuric Acid

Atmosphere of Venus

Sulphuric acid is produced in the upper atmosphere of Venus by the sun's photochemical action on carbon dioxide, sulphur dioxide, and water colour. Ultraviolet photons of wavelengths less than 169 nm can photodissociate carbon dioxide into carbon monoxide and atomic oxygen. Atomic oxygen is highly reactive. When it reacts with sulphur dioxide, a trace component of the Venusian atmosphere, the result is sulphur trioxide, which can combine with water vapour, another trace component of Venus's atmosphere, to yield sulphuric acid.

$$CO_2 \rightarrow CO + O$$
$$SO_2 + O \rightarrow SO_3$$
$$SO_3 + H_2O \rightarrow H_2SO_4$$

In the upper, cooler portions of Venus's atmosphere, sulphuric acid exists as a liquid, and thick sulphuric acid clouds completely obscure the planet's surface when viewed from above. The main cloud layer extends from 45–70 km above the planet's surface, with thinner hazes extending as low as 30 and as high as 90 km above the surface.

The permanent Venusian clouds produce a concentrated acid rain, as the clouds on the atmosphere of earth produce water rains. Thus, it's exist a double combined cycle of sulphur dioxide and water, because when sulphuric drops fall down, they are heated up and release water vapour, becoming more and more concentrated.

And when they reach above 300°C, sulphuric acid begins to decompose in sulphur trioxide and water (both gaseous). Sulphur trioxide is highly reactive (like sulphuric acid) and become sulphuric dioxide and oxygen, which oxides traces of CO or surface rocks. Sulphuric dioxide and water (colour) continuously equilibrate their pressure from deep Venusian atmosphere to upper altitudes, where they will be transformed again in sulphuric acid, and the cycle is closed.

Physical Properties

Forms of sulphuric acid

Although nearly 100 per cent sulphuric acid can be made, this loses SO_3 at the boiling point to produce 98.3 per cent acid. The 98 per cent grade is more stable in storage, and is the usual form of what is described as concentrated sulphuric acid. Other concentrations are used for different purposes. Some common concentrations are:

1. 10 per cent, dilute sulphuric acid for laboratory use.
2. 33.53 per cent, battery acid (used in lead-acid batteries).
3. 62.18 per cent, chamber or fertiliser acid.
4. 73.61 per cent, tower or Glover acid.
5. 97 per cent, concentrated acid.

Different purities are also available. Technical grade H_2SO_4 is impure and often coloured, but is suitable for making fertiliser. Pure grades such as United States Pharmacopoeia (USP) grade are used for making pharmaceuticals and dyestuffs. When high concentrations of $SO_{3(g)}$ are added to sulphuric acid, $H_2S_2O_7$, called pyrosulphuric acid, fuming sulphuric acid or oleum or, less commonly, Nordhausen acid, is formed. Concentrations of oleum are either expressed in terms of %SO_3 (called %oleum) or as %H_2SO_4 (the amount made if H_2O were added); common concentrations are 40 per cent oleum (109 per cent H_2SO_4) and 65 per cent oleum (114.6 per cent H_2SO_4). Pure $H_2S_2O_7$ is a solid with melting point 36°C.

Polarity and conductivity

Anhydrous H_2SO_4 is a very polar liquid, having a dielectric constant of around 100. It has a high electrical conductivity, caused by dissociation through protonating itself, a process known as auto-protolysis.

Chemical Properties

Reaction with water

The hydration reaction of sulphuric acid is highly exothermic. If water is added to the concentrated sulphuric acid, it can boil and spit dangerously. One should always add the acid to the water rather than the water to the acid. The necessity for this safety precaution is due to the relative densities of these two liquids. Water is less dense than sulphuric acid, meaning water will tend to float on top of this acid. The reaction is best thought of as forming hydronium ions, by:

$$H_2SO_4 + H_2O \rightarrow H_3O^+ + HSO_4^-$$

and then

$$HSO_4^- + H_2O \rightarrow H_3O^+ + SO_4^{2-}$$

Because the hydration of sulphuric acid is thermodynamically favourable, sulphuric acid is an excellent dehydrating agent, and is used to prepare many dried fruits. The affinity of sulphuric acid for water is sufficiently strong that it will remove hydrogen and oxygen atoms from other compounds; for example, mixing starch $(C_6H_{12}O_6)_n$ and concentrated sulphuric acid will give elemental carbon and water which is absorbed by the sulphuric acid (which becomes slightly diluted):

$$(C_6H_{12}O_6)_n \rightarrow 6C + 6H_2O$$

The effect of this can be seen when concentrated sulphuric acid is spilled on paper; the cellulose reacts to give a burned appearance, the carbon appears much as soot would in a fire.

A more dramatic reaction occurs when sulphuric acid is added to a tablespoon of white sugar; a rigid column of black, porous carbon will quickly emerge. The carbon will smell strongly of caramel.

Other reactions

As an acid, sulphuric acid reacts with most bases to give the corresponding sulphate. For example, copper(II) sulphate. This blue salt of copper, commonly used for electroplating and as a fungicide, is prepared by the reaction of copper(II) oxide with sulphuric acid:

$$CuO + H_2SO_4 \rightarrow CuSO_4 + H_2O$$

Sulphuric acid can also be used to displace weaker acids from their salts. Reaction with sodium acetate, for example, displaces acetic acid:

$$H_2SO_4 + CH_3COONa \rightarrow NaHSO_4 + CH_3COOH$$

Similarly, reacting sulphuric acid with potassium nitrate can be used to produce nitric acid and a precipitate of potassium bisulphate. When combined with nitric acid, sulphuric acid acts both as an acid and a dehydrating agent, forming the nitronium ion NO_2^+, which is important in nitration reactions involving electrophilic aromatic substitution. This type of reaction, where protonation occurs on an oxygen atom, is important in many organic chemistry reactions, such as Fischer esterification and dehydration of alcohols.

Sulphuric acid reacts with most metals via a single displacement reaction to produce hydrogen gas and the metal sulphate. Dilute H_2SO_4 attacks iron, aluminium, zinc, manganese, magnesium and nickel, but reactions with tin and copper require the acid to be hot and concentrated. Lead and tungsten, however, are resistant to sulphuric acid. The reaction with iron (shown) is typical for most of these metals, but the reaction with tin is unusual in that it produces sulphur dioxide rather than hydrogen.

$$Fe(s) + H_2SO_4(aq) \rightarrow H_2(g) + FeSO_4(aq)$$
$$Sn(s) + 2H_2SO_4(aq) \rightarrow SnSO_4(aq) + 2H_2O(l) + SO_2(g)$$

Sulphuric acid undergoes electrophilic aromatic substitution with aromatic compounds to give the corresponding sulphonic acids.

Uses

Sulphuric acid is a very important commodity chemical. The major use (60 per cent of total production worldwide) for sulphuric acid is in the wet method for the production of phosphoric acid, used for manufacture of phosphate fertilisers as well as trisodium phosphate for detergents. In this method, phosphate rock is used, and more than 100 million tons are processed annually. This raw material is shown below as fluorapatite, though the exact composition may vary. This is treated with 93 per cent sulphuric acid to produce calcium sulphate, hydrogen fluoride (HF) and phosphoric acid. The HF is removed as hydrofluoric acid. The overall process can be represented as:

$$Ca_5F(PO_4)_3 + 5H_2SO_4 + 10H_2O \rightarrow 5CaSO_4 \cdot 2H_2O + HF + 3H_3PO_4.$$

Sulphuric acid is used in large quantities by the iron and steelmaking industry to remove oxidation, rust and scale from rolled sheet and billets prior to sale to the automobile and white-goods industry. Used acid is often recycled using a spent acid regeneration (SAR) plant. These plants combust spent acid with natural gas, refinery gas, fuel oil or other fuel sources. This combustion process produces gaseous sulphur dioxide (SO_2) and sulphur trioxide (SO_3) which are then used to manufacture new sulphuric acid. SAR plants are common additions to metal smelting plants, oil refineries, and other industries where sulphuric acid is consumed in bulk, as operating a SAR plant is much cheaper than the recurring costs of spent acid disposal and new acid purchases.

Ammonium sulphate, an important nitrogen fertiliser, is most commonly produced as a by-product from coking plants supplying the iron and steel making plants. Reacting the ammonia produced in the thermal decomposition of coal with waste sulphuric acid allows the ammonia to be crystallised out as a salt (often brown because of iron contamination) and sold into the agrochemicals industry.

Another important use for sulphuric acid is for the manufacture of aluminium sulphate, also known as paper maker's alum. This can react with small amounts of soap on paper pulp fibres to give gelatinous aluminium carboxylates, which help to coagulate the pulp fibres into a hard paper surface. It is also used for making aluminium hydroxide, which is used at water treatment plants to filter out impurities, as well as to improve the taste of the water. Aluminium sulphate is made by reacting bauxite with sulphuric acid:

$$Al_2O_3 + 3H_2SO_4 \rightarrow Al_2(SO_4)_3 + 3H_2O.$$

Sulphuric acid is used for a variety of other purposes in the chemical industry. For example, it is the usual acid catalyst for the conversion of cyclohexanoneoxime to caprolactam, used for making nylon. It is used for making hydrochloric acid from salt via the Mannheim process. Much H_2SO_4 is used in petroleum refining, for example as a catalyst for the reaction of isobutane with isobutylene to give isooctane, a compound that raises the octane rating of gasoline (petrol). Sulphuric acid is also important in the manufacture of dyestuffs solutions and is the acid in lead-acid (car) batteries. Sulphuric acid is also used as a general dehydrating agent in its concentrated form.

Sulphur-iodine cycle

The sulphur-iodine cycle is a series of thermochemical processes used to obtain hydrogen. It consists of three chemical reactions whose net reactant is water and whose net products are hydrogen and oxygen.

$$2H_2SO_4 \rightarrow 2SO_2 + 2H_2O + O_2 \qquad (830°C)$$
$$I_2 + SO_2 + 2H_2O \rightarrow 2\,HI + H_2SO_4 \qquad (120°C)$$
$$2HI \rightarrow I_2 + H_2 \qquad (320°C)$$

The sulphur and iodine compounds are recovered and reused, hence the consideration of the process as a cycle. This process is endothermic and must occur at high temperatures, so energy in the form of heat has to be supplied. The sulphur-iodine cycle has been proposed as a way to supply hydrogen for a hydrogen-based economy. It does not require hydrocarbons like current methods of steam reforming.

The sulphur-iodine cycle is currently being researched as a feasible method of obtaining hydrogen, but the concentrated, corrosive acid at high temperatures poses currently insurmountable safety hazards if the process were built on large-scale.

Safety

Laboratory hazards

The corrosive properties of sulphuric acid are accentuated by its highly exothermic reaction with water. Hence burns from sulphuric acid are potentially more serious than those of comparable strong acids (e.g. hydrochloric acid), as there is additional tissue damage due to dehydration and particularly due to the heat liberated by the reaction with water, i.e. secondary thermal damage. The danger is obviously greater with more concentrated preparations of sulphuric acid, but it should be remembered that even the normal laboratory dilute grade will char paper by dehydration if left in contact for a sufficient time. Fuming sulphuric acid (oleum) is not recommended for use in schools due to it being quite hazardous.

Preparation of the diluted acid can also be dangerous due to the heat released in the dilution process. It is essential that the concentrated acid is added to water and not the other way round, to take advantage of the relatively high heat capacity of water. Addition of water to concentrated sulphuric acid leads at best to the dispersal of a sulphuric acid aerosol, at worst to an explosion.

Industrial hazards

Although sulphuric acid is non-flammable, contact with metals in the event of a spillage can lead to the liberation of hydrogen gas. The dispersal of acid aerosols and gaseous sulphur dioxide is an additional hazard of fires involving sulphuric acid.

Sulphuric acid is not considered toxic besides its obvious corrosive hazard, and the main occupational risks are skin contact leading to burns and the inhalation of aerosols. Exposure to aerosols at high concentrations leads to immediate and severe irritation of the eyes, respiratory tract and mucous membranes: this ceases rapidly after exposure, although there is a risk of subsequent pulmonary edema if tissue damage has been more severe.

At lower concentrations, the most commonly reported symptom of chronic exposure to sulphuric acid aerosols is erosion of the teeth, found in virtually all studies.

Contact Process

The contact process is the current method of producing sulphuric acid in the high concentrations needed for industrial processes. Vanadium(V) oxide (vanadium pentoxide) is the catalyst employed.

This process was patented in 1831 by the British vinegar merchant Peregrine Phillips. In addition to being a far more economical process for producing concentrated sulphuric acid than the previous lead chamber process, the contact process also produces sulphur trioxide and oleum (Fig. 18.1).

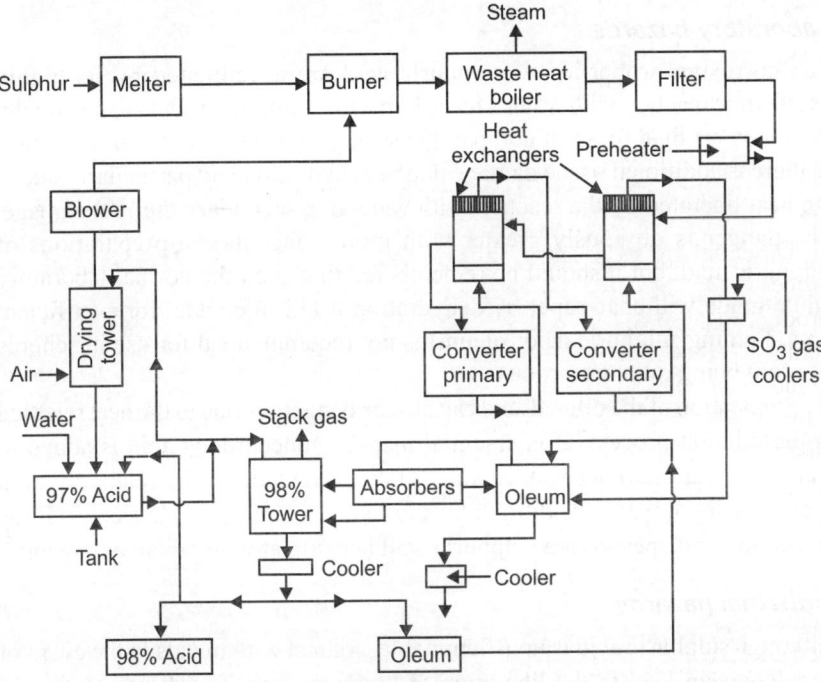

Fig. 18.1. Contact process which produced sulphur trioxide and oleum.

The process can be divided into three stages:
1. Preparation and purification of sulphur dioxide.
2. Catalytic oxidation (using vanadium pentoxide catalyst) of sulphur dioxide to sulphur trioxide.
3. Conversion of sulphur trioxide to sulphuric acid.
4. Purification of air and SO_2 is necessary to avoid catalyst poisoning (i.e. removing catalytic activities). The gas is then washed with water and dried by H_2SO_4.

To conserve energy, the mixture is heated by exhaust gases from the catalytic converter by heat exchangers. Sulphur dioxide and oxygen then react in the manner as follows:

$$2SO_2(g) + O_2(g) \rightarrow 2SO_3(g) \qquad \Delta H = -197 \text{ kJ mol}^{-1}$$

To increase the reaction rate, high temperatures (450°C), high pressures (10 atm), and vanadium(V) oxide (V_2O_5) are used to ensure a 99.5 per cent conversion. Platinum would be a more suitable catalyst, but it is very costly and easily poisoned. The catalyst only serves to increase the rate of reaction — it has no effect on how much SO_3 is produced.

Hot sulphur trioxide passes through the heat exchanger and is dissolved in concentrated H_2SO_4 in the absorption tower to form oleum:

$$H_2SO_4(l) + SO_3(g) \rightarrow H_2S_2O_7(l)$$

Note that directly dissolving SO_3 in water is impractical due to the highly exothermic nature of the reaction. Acidic vapour or mists are formed instead of a liquid. Oleum is reacted with water to form concentrated H_2SO_4. The average percentage yield of this reaction is around 30 per cent.

$$H_2S_2O_7(l) + H_2O(l) \rightarrow 2H_2SO_4(l)$$

Double contact double absorption (DCDA): The next step to the contact process is DCDA or double contact double absorption. In this process the product gases (SO_2) and (SO_3) are passed through absorption towers twice to achieve further absorption and conversion of SO_2 to SO_3 and production of higher grade sulphuric acid. SO_2 rich gases enter the catalytic converter, usually a tower with multiple catalyst beds, and get converted to SO_3, achieving the first stage of conversion. The exit gases from this stage contain both SO_2 and SO_3 which are passed through intermediate absorption towers where sulphuric acid is trickled down packed columns and SO_3 reacts with water increasing the sulphuric acid concentration. Though SO_2 too passes through the tower it is unreactive and comes out of the absorption tower. This stream of gas containing SO_2, after necessary cooling is passed through the catalytic converter bed column again achieving up to 99.8 per cent conversion of SO_2 to SO_3 and the gases are again passed through the final absorption column thus resulting not only achieving high conversion efficiency for SO_2 but also enabling production of higher concentration of sulphuric acid, H_2SO_4.

The industrial production of sulphuric acid involves proper control of temperatures and flow rates of the gases as both the conversion efficiency and absorption are dependent on these.

Catalysis

Catalysis is the process in which the rate of a chemical reaction is increased by means of a chemical substance known as a catalyst. Unlike other reagents that participate in the chemical reaction, a catalyst is not consumed by the reaction itself. The catalyst may participate in multiple chemical transformations.

The general feature of catalysis is that the catalytic reaction has a lower rate-limiting free energy change to the transition state than the corresponding uncatalysed reaction, resulting in a larger reaction rate at lower temperature. However, the mechanistic origin of catalysis is complex. Catalysts may affect the reaction environment favourably, e.g. acid catalysts for reactions of carbonyl compounds, form specific intermediates that are not produced naturally, such as osmate esters in osmium tetroxide-catalysed dihydroxylation of alkenes, or cause lysis of reagents to reactive forms, such as atomic hydrogen in catalytic hydrogenation. Kinetically, catalytic reactions behave like typical chemical reactions, i.e. the reaction rate depends on the frequency of contact of the reactants in the rate-determining step. Usually, the catalyst participates in this slow step, and rates are limited by amount of catalyst. In heterogeneous catalysis, the diffusion of reagents to the surface and diffusion of products from the surface can be rate determining. Analogous events associated with substrate binding and product dissociation apply to homogeneous catalysts. Although catalysts are not consumed by the reaction itself, they may be inhibited, deactivated or destroyed by secondary processes. In heterogeneous catalysis, typical secondary processes include coking where the catalyst becomes covered by polymeric side products. Similarly heterogeneous catalysts can dissolve or evaporate into the mobile medium, solution or gas-phase. These side reactions may even include following elementary reaction steps, as in alkali-catalysed hydrolysis of esters, thus requiring a stoichiometric amount of catalyst. A typical such catalyst consists of diatomaceous earth impregnated with upward of 7 per cent V_2O_5. Sometimes two grades are charged into the converter, a less active but harder type being used in the first pass of the converter, and a more active but softer type (increase of 30 per cent) in passes subsequent to the first. These catalysts are of long life, up to 20 years, and are not subject to poisoning, except from fluorine, which damages the siliceous carrier. Conversions are high, up to 99.8 per cent in double-absorber-type plants. Clean SO_2 must of course be supplied.

Recovery of Waste Sulphuric Acid

The recovery and reuse of sulphuric acid from inorganic and organic wastes are frequently not economical, nevertheless, they are usually necessary to avoid the pollution of streams. The more concentrated and cleaner the spent acid, the easier it is to recover the values. If sufficiently clean, these spent acids can be sent to a contact plant to be fortified, though this does not remove impurities. Some sulphuric acid is still used in the steel industry for pickling. However, because of the need to eliminate the discharge of spent liquors into streams, and because of the difficulty in treating such liquors for the recovery of acid values and to eliminate stream pollution, sulphuric acid is being replaced with hydrochloric acid for pickling. Spent liquor from the latter can be treated to recover acid values and to avoid stream pollution. A residual liquor analogous to steel-mill pickle liquor is obtained from titanium pigment plants which use sulphuric acid to produce titanium dioxide from ilmenite. These spent liquors contain large amounts of ferrous sulphate and free sulphuric acid. Because of increasing pressure to avoid the contamination of streams with ferrous sulphate, certain pickle liquors are concentrated and the ferrous sulphate recovered even though this is not profitable. Titanium dioxide pigment is made by the chloride route to avoid the disposal problems associated with the use of sulphuric acid.

SULPHUR DIOXIDE (SO_2)

Properties

It is a colourless gas or liquid with sharp pungent odour. Soluble in water, alcohol, and ether. Forms sulphurus acid, H_2SO_3. Sp. gr. 1.4337, liquid at 0°C; f.p. –76.1°C; b.p. –10°C; vapour pressure 3.2 at 20°C; refractive index (liquid) 1.410 (n 24/D). An outstanding oxidising and reducing agent. Noncombustible.

Derivation

1. By roasting pyrites in special furnaces. The gas is readily liquefied by cooling with ice and salt, or at a pressure of 3 atmosphere.
2. By purifying and compressing sulphur dioxide gas from smelting operation.
3. By burning sulphur.

Hazard

Toxic by inhalation; strong irritant to eyes and mucous membranes, especially under pressure. Dangerous air contaminant and constituent of smog. Tolerance, 2 ppm in air; US atmospheric standard, 0.140 ppm. Not permitted in meats and other sources of vitamin B_1.

Uses

Chemicals (sulphuric acid, salt cake, sulphites, hydrosulphites of potassium and sodium, thiosulphates, alum from shale, recovery of valatile substances); sulphite paper pulp; ore and metal refining; soyabean protein; intermediates; solvent extraction of lubricating oils; as a leaching agent for oils and starch; sulphonation of oils; disinfecting and fumigating; food additive (inhibition of browning of enzyme-catalysed reactions, bacterial growth); reducing agent, antioxidant.

SULPHUR TRIOXIDE

Sulphur trioxide (sulphuric anhydride) SO_3, $(SO_3)_n$.

Properties

Exists in three solid modifications; alpha m.p. 62°C, beta, m.p. 32.5°C; gamma, m.p. 16.8°C. The alpha form appears to be the stable form but the solid transitions are commonly slow; a given sample may be a mixture of the various forms, and its m.p. not constant. The solids sublime easily. All three forms boil at 45°C.

Derivation

Passing a mixture of SO_2 and O_2 over a heated catalyst such as platinum or vanadium pentoxide.

Hazard

Highly toxic, strong irritant to tissue. Fire risk in contact with organic materials. An explosive increase in vapour pressure occurs when the alpha form melts. The anhydride combines with water, forming sulphuric acid and evolving heat.

Uses

Sulphonation of organic compound, especially nonionic detergents; solar energy collectors. It is usually generated in the plant where it is to be used.

Chapter 19

Hydrochloric Acid and Miscellaneous Inorganic Chemicals

INTRODUCTION

Hydrochloric acid is the solution of hydrogen chloride (HCl) in water. It is a highly corrosive, strong mineral acid and has major industrial uses. It is found naturally in gastric acid. Historically called muriatic acid or spirits of salt, hydrochloric acid was produced from vitriol and common salt.

With major production starting in the industrial revolution, hydrochloric acid is used in the chemical industry as a chemical reagent in the large-scale production of vinyl chloride for PVC plastic, and for polyurethane. It has numerous smaller-scale applications, including household cleaning, production of gelatine and other food additives, descaling, and leather processing.

Hydrogen chloride (HCl) is a monoprotic acid, which means it can dissociate (i.e. ionise) only once to give up one H^+ ion (a single proton). In aqueous hydrochloric acid, the H^+ joins a water molecule to form a hydronium ion, H_3O^+:

$$HCl + H_2O \rightarrow H_3O^+ + Cl^-$$

The other ion formed is Cl^-, the chloride ion. Hydrochloric acid can therefore be used to prepare salts called chlorides, such as sodium chloride. Hydrochloric acid is a strong acid, since it is essentially completely dissociated in water.

Monoprotic acids have one acid dissociation constant, K_a, which indicates the level of dissociation in water. For a strong acid like HCl, the K_a is large. Theoretical attempts to assign a K_a to HCl have been made. When chloride salts such as NaCl are added to aqueous HCl they have practically no effect on pH, indicating that Cl^- is an exceedingly weak conjugate base and that HCl is fully dissociated in aqueous solution. For intermediate to strong solutions of hydrochloric acid, the assumption that H^+ molarity (a unit of concentration) equals HCl molarity is excellent, agreeing to four significant digits.

Of the seven common strong mineral acids in chemistry, hydrochloric acid is the monoprotic acid least likely to undergo an interfering oxidation-reduction reaction. It is one of the least hazardous strong acids to handle; despite its acidity, it consists of the nonreactive and nontoxic chloride ion. Intermediate strength hydrochloric acid solutions are quite stable upon storage, maintaining their concentrations over time. These attributes, plus the fact that it is available as a pure reagent, mean that hydrochloric acid makes an excellent acidifying reagent.

Hydrochloric acid is the preferred acid in titration for determining the amount of bases. Strong acid titrants give more precise results due to a more distinct endpoint. Azeotropic or constant-boiling hydrochloric acid (roughly 20.2 per cent) can be used as a primary standard in quantitative analysis, although its exact concentration depends on the atmospheric pressure when it is prepared. Hydrochloric acid is frequently used in chemical analysis to prepare (digest) samples for analysis. Concentrated hydrochloric acid dissolves many metals and forms oxidised metal chlorides and hydrogen gas, and it reacts with basic compounds such as calcium carbonate or copper(II) oxide, forming the dissolved chlorides that can be analysed.

MANUFACTURE OF HYDROCHLORIC ACID

Hydrochloric acid is obtained from four major sources: (i) as a by-product in the chlorination of both aromatic and aliphatic hydrocarbons, (ii) from reacting salt and sulphuric acid, (iii) from the combustion of hydrogen and chlorine, and (iv) from Hargreaves-type operations.

From Organic Chlorinations

Hydrochloric acid is manufactured when salt and sulphuric acid, sp. gr. 1.7 (slight excess), or salt and an equivalent amount of niter cake are charged to a furnace equipped with a rake agitator (Mannheim furnace), where the reacting mass is slowly heated to a temperature just below fusion (843°C). Hydrogen chloride is evolved and led through a cooling and condensing system to the absorbers. Salt cake (crude sodium sulphate) is continuously discharged from the periphery of the furnace and the following reaction takes place.

$$NaCl + H_2SO_4 \longrightarrow HCl + NaHSO_4$$

$$NaCl + NaHSO_4 \longrightarrow HCl + Na_2SO_4$$
$$98\% \text{ yield}$$

The combustion gases, containing hydrogen chloride (approximately 30 per cent HCl), leave the furnace at about 840°C, pass into silica bends (coolers) externally cooled with water, and leave the coolers at about 38°C (Fig. 19.1).

The cooled gases then pass through a coke-packed tower, in which sulphuric acid mist and any solid particles present are removed, and thence to another series of silica bends, where the hydrogen chloride is absorbed by water to produce hydrochloric acid. Exhaust gases from the absorber are scrubbed with water and discharged to the atmosphere. The more recently designed plants have made use of impervious graphite and structural plastics for absorbers and coolers.

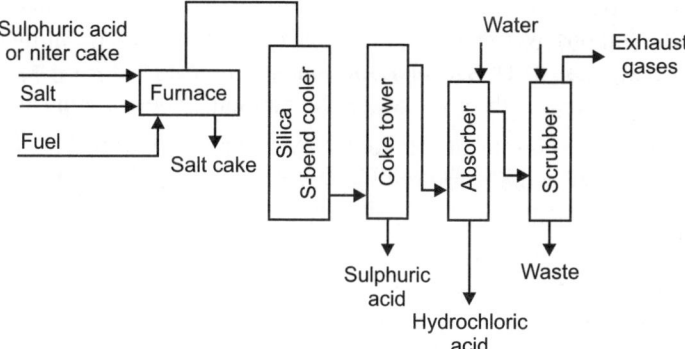

Fig. 19.1. Flow diagram for manufacture of hydrochloric acid from organic chlorinations.

Few plants also use the Laury furnace, a horizontal rotary kiln, in which the reacting mass comes into contact with fuel combustion gases. These gases, containing as little as 5 per cent hydrogen chloride, leave the furnace at 300°C, and are cooled and then recovered. In making potassium sulphate from potassium chloride, hydrochloric acid results as coproduct. The reactions are analogous, with potassium chloride in place of sodium chloride.

It is also manufactured when chlorine is burned in a slight excess of hydrogen to produce hydrogen chloride. Several types of burners are used: for example, the silica burner, the ceramic-lined burner, the graphite burner, and the water-jacketed steel burner. The last-mentioned burner cannot of course be used with wet gases (Fig. 19.2). The following reaction takes place.

$$H_2 + Cl_2 \longrightarrow 2HCl$$
90–99% yield

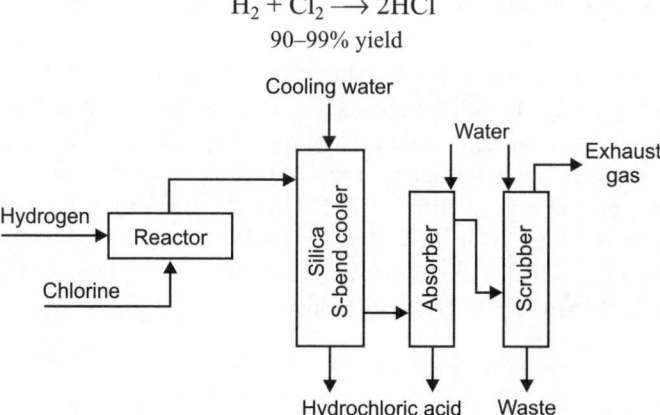

Fig. 19.2. Flow diagram for manufacture of hydrochloric acid from chlorine and hydrogen.

A submerged combustion burner, operating beneath a surface of muriatic acid, has been developed. The burner gases, practically pure hydrogen chloride, are cooled, absorbed, and scrubbed in a system essentially the same as that used in the salt process. The purifying coke tower, however, is omitted, and in some systems (where the gas concentration approaches 100 per cent HCl) the scrubber may be omitted. Strong hydrochloric acid (35 per cent) is removed directly from the bottom of the cooler by means of a trap, and weak acid (28 per cent) leaves the bottom of the absorber. Recently installed absorbers have been built of tantalum and structural carbon, as well as silica. This process readily gives a water-white product.

The concentration of hydrogen chloride in burner gases generally depends on the degree of chlorine utilisation. With 0.03 per cent or less chlorine in the product gases, hydrogen chloride concentration may be as high as 98.5 per cent, where 0.1 per cent chlorine is allowable, hydrogen chloride concentration may reach 99 to 99.5 per cent. Manufacturers often insist on 100 per cent chlorine utilisation in which case exit gases may be only 90 per cent hydrogen chloride because of the 5 to 10 per cent excess hydrogen used.

Hydrochloric acid is also made by burning chlorine in methane or water gas, according to the equation:

$$2Cl_2 + CH_4 \longrightarrow 4HCl + CO_2$$

The recovery system is the same as that used in the chlorine-hydrogen process.

Properties

The physical properties of hydrochloric acid, such as boiling and melting points, density, and pH depend on the concentration or molarity of HCl in the acid solution. They range from those of water at very low concentrations approaching zero per cent HCl to values for fuming hydrochloric acid at over 40 per cent HCl. Hydrochloric acid as the binary (two-component) mixture of HCl and H_2O has a constant-boiling azeotrope at 20.2 per cent HCl and 108.6°C (227°F). There are four constant-crystallisation eutectic points for hydrochloric acid, between the crystal form of $HCl \cdot H_2O$ (68 per cent HCl), $HCl \cdot 2H_2O$ (51 per cent HCl), $HCl \cdot 3H_2O$ (41 per cent HCl), $HCl \cdot 6H_2O$ (25 per cent HCl), and ice (0 per ent HCl). There is also a metastable eutectic point at 24.8 per cent between ice and the $HCl \cdot 3H_2O$ crystallisation.

Uses

Hydrochloric acid is a strong inorganic acid that is used in many industrial processes. The application often determines the required product quality: (i) pickling of steel, (ii) production of organic compounds, (iii) production of inorganic compounds, (iv) pH control and neutralisation, and (v) regeneration of ion exchangers.

Other uses

Leather processing, household cleaning, and building construction. Oil production may be stimulated by injecting hydrochloric acid into the rock formation of an oil well, dissolving a portion of the rock, and creating a large-pore structure.

Safety

Concentrated hydrochloric acid (fuming hydrochloric acid) forms acidic mists. Both the mist and the solution have a corrosive effect on human tissue, with the potential to damage respiratory organs, eyes, skin, and intestines. Upon mixing hydrochloric acid with common oxidising chemicals, such as sodium hypochlorite (bleach, $NaClO$) or permanganate ($KMnO_4$), the toxic gas chlorine is produced. Personal protective equipment such as rubber or PVC gloves, protective eye goggles, and chemical-resistant clothing and shoes are used to minimise risks when handling hydrochloric acid.

HYDROFLUORIC ACID

It is colourless, fuming, liquid which causes extremely serious burns on skin contact. It is prepared by dissolving hydrogen fluoride in water to various concentrates. It is also prepared from fluorspar and sulphuric acid.

By treating fluorspar (calcium fluoride) with concentrated sulphuric acid in a furnace, hydrogen fluoride gas is evolved, leaving a residue of calcium sulphate. After being cleaned of dust and sulphuric acid content, the gas is condensed as 98 to 98.5 per cent hydrogen fluoride. Distillation increases its strength to 99.9 to 99.95 per cent. Aqueous grades may be made by dilution with water (Fig. 19.3). The follwoingr reaction takes place.

$$CaF_2 + H_2SO_4 \longrightarrow 2HF + CaSO_4$$

85 to 95% yield

Crude fluorspar, fluorite, as it comes from domestic mines, varies in calcium fluoride content from about 50 to 90 per cent. To be suitable for acid production, the ore must be upgraded. This is usually done by flotation and results in an acid-grade fluorspar containing about 98 per cent calcium fluoride, 1 per cent silica, 0.03 per cent sulphur and less than 0.1 per cent moisture.

The finely powdered fluorspar is withdrawn from storage silos and transferred by air conveyors to steel conical-bottom hoppers. From here it is fed continuously along with concentrated (93 to 99 per cent) sulphuric acid into a rotating furnace in a refractory-lined shroud. The space between the shroud and the furnace is fired with oil or gas to provide heat for the endothermic reaction. The cylindrical steel furnaces have been constructed in a variety of dimensions, up to 3.7 m in diameter and 25 m long, having a metal thickness of 5 cm.

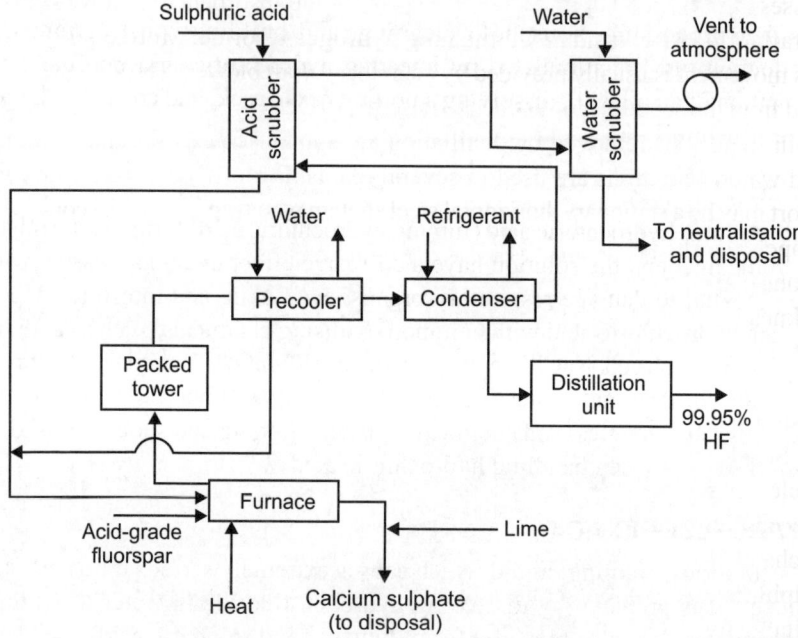

Fig. 19.3. Flow diagram for manufacture of hydrofluoric acid from fluorspar and sulphuric acid.

The raw materials are fed into the furnace at a ratio of 1 mole of fluorspar to 1.1 to 1.3 moles of acid. Several methods have been proposed to keep the reacting mass of fluorspar and sulphuric acid flowing freely enough to allow complete reaction. They include two-stage reaction system with kneading, fluidised beds, and stationary and rotary kilns. Temperatures of 250° to 300°C are commonly reached. Reaction is accomplished using oleum, sulphuric acid, or sulphuric acid-sulphur trioxide-steam mixtures. In the last case, the heat of reaction of sulphur trioxide and steam is used to maintain the temperature.

Calcium sulphate residue, containing 1 per cent or less unreacted fluorspar, is discharged continuously in amounts about 1.75 times the weight of the spar charge. From the feed end of the furnace, gaseous hydrogen fluoride (70 to 75 per cent HF) is withdrawn at a temperature of 120° to 175°C. The crude gases pass through a packed tower which provides surface area for the reflux of sulphuric acid and captures dust in the gas from the furnace. A precooler condenses a mixture of hydrogen fluoride, sulphuric acid, and water, which is combined with the feed acid and returned to the furnace. The cleaned gas is condensed in a heat exchanger using a refrigerant; hydrogen fluoride boils at 19°C at atmospheric pressure. Noncondensables pass through a packed tower, using the feed acid to absorb any uncondensed hydrogen fluoride. This gas passes through another tower in which water is the absorbent to remove silicon tetrafluoride (SiF_4) and any remaining hydrogen fluoride. Motive power for gas movement is usually provided by an induced draft blower. The 98.5 per cent acid from the condenser is distilled to raise its assay to 99.9 to 99.95 per cent. The main impurities removed by distillation are sulphuric acid, sulphur dioxide, and water. Variations are used in several plants. For instance, the furnace or retort may be a stationary, horizontal steel shell in which an alloy-steel conveyor propels the reacting fluorspar and product calcium sulphate through the unit. In one plant the fluorspar and 99 per cent sulphuric acid are mixed in a premixer or kneader, where the initial and highly corrosive stage of the reaction is carried out. The partially reacted mass is then discharged to the rotary kiln. It is claimed that the premixing is so thorough that excess acid is not required and capital costs are reduced.

Solid and liquid particulates may be removed by such techniques as cyclones, scrubbing with sulphuric acid, coke boxes, and demisters. The gases are cooled to 20° to 30°C by sulphuric acid scrubbing and indirect heat exchangers. The resultant acid can be 98 per cent pure, containing as impurities sulphuric acid, water, sulphur dioxide, and silicon tetrafluoride; this acid is suitable for cryolite manufacture without refining. The low boilers are readily fractionated as described. The vent gas from the last cooler will contain by-product silicon tetrafluoride and hydrogen fluoride in a mole ratio of 1:2. On absorption by water in a packed tower, fluosilicic acid is formed according to the following reaction:

$$SiF_4 + 2HF \longrightarrow H_2SiF_6$$

This by-product may be converted to the sodium salt for use in the fluoridation of water supplies; more often it is neutralised with lime and discarded with the calcium sulphate. Despite many efforts to utilise the calcium sulphate, it is normally dumped as landfill. It has b.p. (38 per cent solution) 112°C, will attack glass and any silica containing material, specific gravity 0.988 (13.6°C). It is highly corrosive to skin and mucous membranes, highly toxic by ingestion and inhalation.

Uses

It is used in the production of aluminium, fluorocarbons, pickling, stainless steel, etching glass, acidising oilwells, fluorides, gasoline production (alkylation), and processing uranium.

BROMINE

Bromine is a chemical element with the symbol Br and atomic number 35. A halogen element, bromine is a reddish-brown volatile liquid at standard room temperature that is intermediate in reactivity between chlorine and iodine. Bromine vapours are corrosive and toxic. The main applications for bromine are in fire retardants and fine chemicals.

Bromine is the only liquid nonmetallic element at room temperature, and one of only six elements on the periodic table that are liquid at or close to room temperature. The melting point of bromine is $-7.2°C$ and has the boiling point $58.8°C$. The pure chemical element has the physical form of a diatomic molecule, Br_2. It is a dense, mobile, reddish-brown liquid, that evaporates easily at standard temperature and pressures to give a red vapour (its colour resembles nitrogen dioxide) that has a strong disagreeable odour resembling that of chlorine. Bromine is a halogen, and is less reactive than chlorine and more reactive than iodine. Bromine is slightly soluble in water, and highly soluble in carbon disulphide, aliphatic alcohols (such as methanol), and acetic acid. It bonds easily with many elements and has a strong bleaching action. Bromine, like chlorine, is also used in maintenance of swimming pools.

Certain bromine-related compounds have been evaluated to have an ozone depletion potential or bioaccumulate in living organisms. As a result many industrial bromine compounds are no longer manufactured, are being restricted, or scheduled for phasing out. The Montreal protocol mentions several organobromine compounds for this phase out. Bromine is a powerful oxidising agent. It reacts vigorously with metals, especially in the presence of water, as well as most organic compounds, especially upon illumination.

The diatomic element Br_2 does not occur naturally. Instead, bromine exists exclusively as bromide salts in diffuse amounts in crustal rock. Due to leaching, bromide salts have accumulated in sea water (65 ppm), but at a lower concentration than chloride. Bromine may be economically recovered from bromide-rich brine wells and from the Dead Sea waters (up to 50,000 ppm).

The bromide-rich brines are treated with chlorine gas, flushing through with air. In this treatment, bromide anions are oxidised to bromine by the chlorine gas.

$$2Br^- + Cl_2 \rightarrow 2Cl^- + Br_2$$

Because of its commercial availability and long shelf-life, bromine is not typically prepared. Small amounts of bromine can, however, be generated through the reaction of solid sodium bromide with concentrated sulphuric acid (H_2SO_4). The first stage is formation of hydrogen bromide (HBr), which is a gas, but under the reaction conditions some of the HBr is oxidised further by the sulphuric acid to form bromine (Br_2) and sulphur dioxide (SO_2).

$$NaBr\ (s) + H_2SO_4\ (aq) \rightarrow HBr\ (aq) + NaHSO_4\ (aq)$$

$$2HBr\ (aq) + H_2SO_4\ (aq) \rightarrow Br_2\ (g) + SO_2\ (g) + 2\ H_2O\ (l)$$

Similar alternatives, such as the use of dilute hydrochloric acid with sodium hypochlorite, are also available. The most important thing is that the anion of the acid (in the above examples, sulphate and chloride, respectively) be more electronegative than bromine, allowing the substitution reaction to occur.

Organic compounds are brominated by either addition or substitution reactions. Bromine undergoes electrophilic addition to the double-bonds of alkenes, via a cyclic bromonium intermediate. In non-aqueous solvents such as carbon disulphide, this affords the dibromo product. For example, reaction with ethylene will produce 1,2-dibromoethane. Bromine also undergoes electrophilic addition to phenols and anilines. When used as bromine water, a small amount of the corresponding bromohydrin is formed as well as the dibromo compound.

So reliable is the reactivity of bromine that bromine water is employed as a reagent to test for the presence of alkenes, phenols, and anilines. Like the other halogens, bromine participates in free radical reactions. For example hydrocarbons are brominated upon treatment with bromine in the presence of light. Bromine will also oxidise metals and metalloids to the corresponding bromides. Anhydrous bromine is less reactive toward many metals than hydrated bromine, however, dry bromine reacts vigorously with aluminium, titanium, mercury as well as alkaline earths and alkali metals.

Uses

A wide variety of organobromine compounds are used in industry. Some are prepared from bromine and others are prepared from hydrogen bromide, which is obtained by burning hydrogen in bromine. Bromine is used as flame retardant, gasoline additive, in pesticides. It is also used for the production of brominated vegetable and as emulsifier in many citrus flavoured soft drinks.

Elemental bromine is toxic and causes burns. As an oxidising agent, it is incompatible with most organic and inorganic compounds. Care needs to taken when transporting bromine; it is commonly carried in steel tanks lined with lead, supported by strong metal frames.

When certain ionic compounds containing bromine are mixed with potassium permanganate ($KMnO_4$) and an acidic substance, they will form a pale brown cloud of bromine gas. This gas smells like bleach and is very irritating to the mucus membranes. Upon exposure, one should move to fresh air immediately. If symptoms arise, medical attention is needed.

IODINE

Iodine is a chemical element that has the symbol I and atomic number 53. Naturally-occurring iodine is a single isotope with 74 neutrons. Chemically, iodine is the second least reactive of the halogens, and the second most electropositive halogen; trailing behind astatine in both of these categories. However, the element does not occur in the free state in nature. Iodine and its compounds are primarily used in medicine, photography, and dyes. Although it is rare in the solar system and earth's crust, the iodides are very much soluble in water, and the element is concentrated in seawater. This mechanism helps to explain how the element came to be required in trace amounts by all animals and some plants, being by far the heaviest element known to be necessary to living organisms.

Characteristics

Iodine under standard conditions is a dark-purple/dark-brown solid. It can be seen apparently sublimating at standard temperatures into a violet-pink gas that has an irritating odour. This halogen forms compounds with many elements, but is less reactive than the other members of its Group VII (halogens) and has some metallic light reflectance.

Elemental iodine dissolves easily in chloroform and carbon tetrachloride. The solubility of elemental iodine in water can be vastly increased by the addition of potassium iodide. The molecular iodine reacts reversibly with the negative ion, creating the triiodide anion, I_3^-, which dissolves well in water. This is also the formulation of some types of medicinal (antiseptic) iodine, although tincture of iodine classically dissolves the element in alcohol. The deep blue colour of starch-iodine complexes is produced only by the free element. Iodine naturally occurs in the environment chiefly as a dissolved iodide in seawater, although it is also found in some minerals and soils. Organoiodine compounds are produced by marine life forms, the most notable of it is iodomethane (commonly called methyl iodide).

Manufacture

From the several places in which iodine occurs in nature only two are used as source for iodine: the caliche, found in Chile and the iodine containing brines of gas and oil fields, especially in Japan and the United States.

The caliche, found in Chile contains sodium nitrate, which is the main product of the mining activities and small amounts of sodium iodate and sodium iodide. During leaching and production of pure sodium nitrate the sodium iodate and iodide is extracted.

Most other producers use natural occurring brine for the production of iodine. The Japanese Minami Kanto gas field east of Tokio and the American Anadarko Basin gas field in northwest Oklahoma are the two largest sources for iodine from brine. The brine has a temperature of over 60°C due to the depth it came from. The brine is first purified and acidified using sulphuric acid, then the iodide present is oxidised to iodine with chlorine. An iodine solution is produced, but it is yet too dilute and has to be concentrated. Air is blown into the solution, causing the iodine to evaporate, then it is passed into an absorbing tower containing acid where sulphur dioxide is added to reduce the iodine. The hydrogen iodide (HI) is reacted with chlorine to precipitate the iodine. After filtering and purification the iodine is packed.

$$2HI + Cl_2 \rightarrow I_2\uparrow + 2HCl$$
$$I_2 + H_2O + SO_2 \rightarrow 2HI + H_2SO_4$$
$$2HI + Cl_2 \rightarrow I_2\downarrow + 2HCl$$

The production of iodine via electrolysis of seawater is not used due to the sufficient abundance of iodine rich brine. Another source of iodine was kelp, a kind of brown alga. This source was used in the 18th and 19th centuries but is no longer economically viable.

Commercial samples often contain a large amount of impurities; they may be removed by sublimation. The element may also be prepared in an ultrapure form through the reaction of potassium iodide with copper(II) sulphate, which gives copper(II) iodide initially. That decomposes spontaneously to copper(I) iodide and iodine:

$$Cu^{2+} + 2I^- \rightarrow CuI_2$$
$$2CuI_2 \quad \rightarrow 2CuI + I_2$$

There are also few other methods of isolating this element in the laboratory, for example the method used to isolate other halogens: oxidation of the iodide in hydroiodic acid (often made *in situ* with an iodide and sulphuric acid) by manganese dioxide.

Compounds

Iodine forms many compounds. Potassium iodide is the most commercially significant iodine compound. It is a convenient source of the iodide anion; it is easier to handle than sodium iodide because it is not hydroscopic. Sodium iodide is especially useful in the Finkelstein reaction, because it is soluble in acetone, while potassium iodide is poorly so. In this reaction, an alkyl chloride is converted to an alkyl iodide. This relies on the insolubility of sodium chloride in acetone to drive the reaction:

$$R\text{-}Cl \text{ (acetone)} + NaI \text{ (acetone)} \rightarrow R\text{-}I \text{ (acetone)} + NaCl(s)$$

Iodic acid (HIO_3) and its salts are strong oxidisers. Periodic acid (HIO_4) cleaves vicinal diols along the C-C bond to give aldehyde fragments. 2-Iodoxy-benzoic acid and Dess-Martin periodinane are hypervalent iodine oxidants used to specifically oxidise alcohols to ketones or aldehydes. Iodine pentoxide is a strong oxidant as well. Interhalogen compounds are well known; examples include iodine monochloride and trichloride; iodine pentafluoride and heptafluoride.

Uses

It is used as a disinfectant, in staining, as a radiocontrast agent, in radioiodine, and in organic synthesis.

Deficiency

In areas where there is little iodine in the diet, typically remote inland areas and semi-arid equatorial climates where no marine foods are eaten, iodine deficiency gives rise to hypothyroidism, symptoms of which are extreme fatigue, goitre, mental slowing, depression, weight gain, and low basal body temperatures. Iodine deficiency is the leading cause of preventable mental retardation, a result which occurs primarily when babies or small children are rendered hypothyroidic by a lack of the element. Iodine deficiency is also a problem in certain areas of Europe.

Toxicity

Excess iodine has symptoms similar to those of iodine deficiency. Commonly encountered symptoms are abnormal growth of the thyroid gland and disorders in functioning and growth of the organism as a whole. Elemental iodine, I_2, is a deadly poison if taken in larger amounts; if 2–3 grams of it are consumed, it is fatal to humans. Iodides are similar in toxicity to bromides.

FLUORINE

Elemental fluorine is the most chemically reactive and electronegative of all the elements. For example, it will readily burn hydrocarbons at room temperature, in contrast to the combustion of hydrocarbons by oxygen, which requires an input of energy with a spark. Therefore, molecular fluorine is highly dangerous, more so than other halogens such as the poisonous chlorine gas.

Fluorine (F_2) is a corrosive pale yellow or brown gas that is a powerful oxidising agent. It is the most reactive and most electronegative of all the elements, and readily forms compounds with most other elements. Fluorine even combines with the noble gases argon, krypton, xenon, and radon. Even in dark, cool conditions, fluorine reacts explosively with hydrogen. The reaction with hydrogen occurs even at extremely low temperatures, using liquid hydrogen and solid fluorine. It is so reactive that metals, and even water, as well as other substances, burn with a bright flame in a jet of fluorine gas. In moist air it reacts with water to form also dangerous hydrofluoric acid.

Fluorides are compounds that combine fluorine with some positively charged counterpart. They often consist of crystalline ionic salts. Fluorine compounds with metals are among the most stable of salts. Hydrogen fluoride is a weak acid when dissolved in water. Consequently, fluorides of alkali metals produce basic solutions. For example, a 1 M solution of NaF in water has a pH of 8.59 compared to a 1 M solution of NaOH, a strong base, which has a pH of 14.0.

Applications

Elemental fluorine, F_2, is mainly used for the production of two compounds of commercial interest, uranium hexafluoride and sulphur hexafluoride.

Manufacture

Industrial production of fluorine entails the electrolysis of hydrogen fluoride in the presence of potassium fluoride. This method is based on the pioneering studies by Moissan in the 1880s. Fluorine gas forms at the anode, and hydrogen gas at the cathode. Under these conditions, the potassium fluoride (KF) converts to potassium bifluoride (KHF_2), which is the actual electrolyte, This potassium bifluoride aids electrolysis by greatly increasing the electrical conductivity of the solution.

$$HF + KF \rightarrow KHF_2$$
$$2KHF_2 \rightarrow 2KF + H_2 + F_2$$

The HF required for the electrolysis is obtained as a by-product for the production of phosphoric acid. Phosphate-containing minerals contain significant amounts of calcium fluorides, such as fluorite. Upon treatment with sulphuric acid, these minerals release hydrogen fluoride:

$$CaF_2 + H_2SO_4 \rightarrow 2HF + CaSO_4$$

Precautions

Elemental fluorine (fluorine gas) is a highly toxic, corrosive oxidant, which can cause organic material, combustibles or other flammable materials to ignite. It must be handled with great care and any contact with skin and eyes should be strictly avoided. Fluorine gas has a characteristic pungent odour that is detectable in concentrations as low as 20 ppb. As it is so reactive, all materials of construction must be carefully selected. All metal surfaces must be passivated before exposure to fluorine.

Fluoride ions are also toxic and must also be handled with great care and any contact with skin and eyes should be strictly avoided.

FLUOROCARBON

Fluorocarbons, sometimes referred to as perfluorocarbons, are organofluorine compounds that contain only carbon and fluorine bonded together in strong carbon-fluorine bonds. The electron withdrawing nature, or electronegativity of fluorine results in many of the unique characteristics of fluorocarbons. For example, the electronegativity of fluorine makes single bonds to carbon remarkably strong. Resultingly, fluoroalkanes are more chemically and thermally stable than alkanes. However, fluorocarbons with double bonds (fluoroalkenes) and especially triple bonds (fluoroalkynes) are more reactive than corresponding hydrocarbons.

Also, the electronegativity of fluorine also reduces the cohesive intermolecular forces of fluorocarbons by mitigating the effect of the London dispersion force. Fluoroalkanes can serve as oil-repellant/water-repellant fluoropolymers, solvents, liquid breathing research agents, and powerful greenhouse gases. Unsaturated fluorocarbons tend to be used as reactants, as fluorocarbons with double and triple bonds are not as stable as fluorocarbons with single bonds.

Many chemical compounds are labelled as fluorocarbons, perfluorinated, or with the prefix perfluoro- despite containing atoms other than carbon or fluorine, such as chlorofluorocarbons or perfluoro-octanesulphonic acid (PFOS). These highly-fluorinated compounds are fluorocarbon derivatives, and not true fluorocarbons according to the IUPAC definition. Fluorocarbon derivatives share many of the properties of fluorocarbons, while also possessing new properties due to the inclusion of new atoms. For example, fluorocarbon derivatives can function as fluoropolymers, refrigerants, solvents, anesthetics, fluorosurfactants, and ozone depletors.

Properties of Fluorocarbons

Fluorocarbons with only single bonds are very stable because of the strength and nature of the carbon–fluorine bond. It is called the strongest bond in organic chemistry. Its strength is a result of the electronegativity of fluorine imparting partial ionic character through partial charges on the carbon and fluorine atoms. The partial charges shorten and strengthen the bond through favourable coulombic interactions. Additionally, multiple carbon–fluorine bonds increase the strength and stability of other nearby carbon–fluorine bonds on the same geminal carbon, as the carbon has a higher positive partial charge. Furthermore, multiple carbon–fluorine bonds also strengthen the 'skeletal' carbon–carbon bonds from the inductive effect.

Therefore, saturated fluorocarbons are much more chemically and thermally stable than their corresponding hydrocarbon counterparts. However, fluoroalkanes are not inert. They are particularly suceptible to reduction through the Birch reduction.

When fluorocarbons are unsaturated, they are less stable and more reactive than fluoroalkanes, or comparable hydrocarbons, due to the electronegativity of fluorine. The reactivity of the simplest fluoroalkyne, difluoroacetylene, is an example of this instability; difluoroacetylene easily polymerises. Another example is fluorofullerene, which has weaker and longer carbon–fluorine bonds than saturated fluorocarbons.

It is reactive towards nucleophiles and hydrolyses in solution. Additionally, the polymerisation of the fluoroalkene tetrafluoroethylene (which results in PTFE) is more energetically favourable than that of ethylene. Unsaturated fluorocarbons have a driving force towards sp^3 hybridisation due to the electronegative fluorine atoms seeking a greater share of bonding electrons with reduced s character in orbitals.

Examples of fluorocarbons

Fluoroalkanes

1. Polytetrafluoroethylene (Teflon).
2. Tetrafluoromethane.
3. Perfluorodecalin.

Fluoroalkenes

1. Tetrafluoroethylene.
2. Perfluoroisobutylene.

Fluoroalkynes

1. Difluoroacetylene.

Properties and examples of fluorocarbon derivatives

Fluorocarbon derivatives are highly fluorinated molecules that are commonly referred to as fluorocarbons. They are economically useful because they share part or nearly all of the properties of fluorocarbons. Some fluorocarbon derivatives have markedly different properties than fluorocarbons.

For example, fluorosurfactants powerfully reduce surface tension by concentrating at the liquid–air interface due to the lipophobicity of fluoro-carbons, due to the polar functional group added to the fluorocarbon chain. Other groups or atoms for fluorocarbon based compounds the oxygen atom incorporated into an ether group for anesthetics, and the chlorine atom for chlorofluorocarbons (CFCs). In a sharp contrast to true fluorocarbons, the chlorine atom produces a chlorine radical which degrades ozone.

Fluorosurfactants

1. Perfluorooctanesulphonic acid (PFOS).
2. Perfluorooctanoic acid (PFOA).
3. Perfluorononanoic acid.

Anesthetics
1. Methoxyflurane (contains chlorine).
2. Enflurane (contains chlorine).
3. Isoflurane (contains chlorine).
4. Sevoflurane.
5. Desflurane.

Halogenated derivatives
1. Polychlorotrifluoroethylene.
2. Perfluorooctyl bromide.
3. Dichlorodifluoromethane.
4. .Chlorodifluoromethane.

Hydrofluorocarbons
1. Polyvinylidene fluoride.
2. Tetrafluoroethane.

Environmental and Health Concerns

Despite the presence of some natural fluorocarbons and fluorocarbon-derivatives, such as tetrafluoro-methane and CFCs, which have been reported in igneous and metamorphic rock, man-made fluorocarbon based compounds are implicated in a variety of environmental and health related issues. For example, CFCs deplete the ozone layer while fluoroalkanes, commonly referred to as perfluorocarbons, are potent greenhouse gases. Also, the fluorosurfactants PFOS and PFOA, and other related chemicals, are persistent global contaminants. PFOS is a proposed persistent organic pollutant and may be currently harming the health of wildlife.

ALUMINIUM OXIDE/ALUMINA

Aluminium oxide is an amphoteric oxide of aluminium with the chemical formula Al_2O_3. It is also commonly referred to as alumina or aloxite in the mining, ceramic and materials science communities. It is produced by the Bayer process from bauxite. Its most significant use is in the production of aluminium metal, although it is also used as an abrasive due to its hardness and as a refractory material due to its high melting point. Corundum is the most common naturally-occurring crystalline form of aluminium oxide. Much less-common rubies and sapphires are gem-quality forms of corundum with their characteristic colours due to trace impurities in the corundum structure. Rubies are given their characteristic deep red colour and their laser qualities by traces of the metallic element chromium. Sapphires come in different colours given by various other impurities, such as iron and titanium.

Properties

Aluminium oxide is an electrical insulator but has a relatively high thermal conductivity (40 W/mK). In its most commonly occurring crystalline form, called corundum or α-aluminium oxide, its hardness makes it suitable for use as an abrasive and as a component in cutting tools.

Aluminium oxide is responsible for metallic aluminium's resistance to weathering. Metallic aluminium is very reactive with atmospheric oxygen, and a thin passivation layer of alumina quickly forms on any exposed aluminium surface. This layer protects the metal from further oxidation. The thickness and properties of this oxide layer can be enhanced using a process called anodising. A number of alloys, such as aluminium bronzes, exploit this property by including a proportion of aluminium in the alloy to enhance corrosion resistance. The alumina generated by anodising is typically amorphous, but discharge assisted oxidation processes such as plasma electrolytic oxidation result in a significant proportion of crystalline alumina in the coating, enhancing its hardness.

Manufacture

Aluminium hydroxide minerals are the main component of bauxite, the principal ore of aluminium. The bauxite ore is made up of a mixture of the minerals gibbsite [$Al(OH)_3$], boehmite [$AlO(OH)$], and diaspore [α-$AlO(OH)$] along with iron oxides and hydroxides, quartz and clay minerals.

Bauxite is purified by the Bayer process:

$$Al_2O_3 + 3H_2O + 2NaOH \rightarrow 2NaAl(OH)_4$$

The Fe_2O_3 does not dissolve in the base. The SiO_2 dissolves as silicate $Si(OH)_6$. Upon filtering, Fe_2O_3 is removed. When the Bayer liquor is cooled, $Al(OH)_3$ precipitates, leaving the silicates in solution. The mixture is then calcined (heated strongly) to give aluminium oxide:

$$2Al(OH)_3 + heat \rightarrow Al_2O_3 + 3H_2O$$

The formed Al_2O_3 is alumina. The alumina formed tends to be multi-phase, i.e. constituting several of the alumina phases rather than solely corundum. The production process can therefore be optimised to produce a tailored product. The type of phases present affects, for example, the solubility and pore structure of the alumina product which, in turn, affects the cost of aluminium production and pollution control.

Uses

The major uses of speciality aluminium oxides are in refractories, ceramics, and polishing and abrasive applications. Large tonnages are also used in the manufacture of zeolites, coating titania pigments, and as a fire retardant/smoke suppressant. Health and medical applications include it as a material in hip replacements. It is used in water filters (derived water treatment chemicals such as aluminium sulphate, aluminium chlorohydrate and sodium aluminate, are one of the few methods available to filter water-soluble fluorides out of water). It is also used in toothpaste formulations.

Aluminium oxide is used for its hardness and strength. Most pre-finished wood flooring now uses aluminium oxide as a hard protective coating. Alumina can be grown as a coating on aluminium by anodising or by plasma electrolytic oxidation. Both its strength and abrasive characteristics are due to aluminium oxide's great hardness (position 9 on the Mohs scale of mineral hardness).

It is widely used as a coarse or fine abrasive, including as a much less expensive substitute for industrial diamond. Many types of sandpaper use aluminium oxide crystals. In addition, its low heat retention and low specific heat make it widely used in grinding operations, particularly cut-off tools. As the powdery abrasive mineral aloxite, it is a major component, along with silica, of the cue tip chalk used in billiards. Aluminium oxide is widely used in the fabrication of superconducting devices, particularly single electron transistors and superconducting quantum interference devices (SQUID), where it is used to form highly resistive quantum tunnelling barriers.

ALUMINIUM SULPHATE

Aluminium sulphate is an industrial chemical used as a flocculating agent in the purification of drinking water and waste-water treatment plants, and also in paper manufacturing.

Aluminium sulphate is sometimes incorrectly referred to as alum but alums are closely related compounds typified by $KAl(SO_4)_2 \cdot 12H_2O$. The anhydrous form occurs naturally as a rare mineral millosevichite, found i.e. in volcanic environments and on burning coal-mining waste dumps. Aluminium sulphate is rarely, if ever, encountered as the anhydrous salt. It forms a number of different hydrates of which the hexadecahydrate $Al_2(SO_4)_3 \cdot 16H_2O$ and octadecahydrate $Al_2(SO_4)_3 \cdot 18H_2O$ are the most common. The heptadeca-hydrate, whose formula can be written as $[Al(H_2O)_6]_2(SO_4)_3 \cdot 5H_2O$, occurs naturally as the mineral alunogen. The crude bauxite ore is ground to a fine powder (80 per cent passes 200 mesh) and charged into open lead-lined steel reaction tanks. Sulphuric acid (sp. gr. 1.7) is added, and the raw materials are thoroughly agitated using paddle agitators, hot air, or live steam.

The reaction mixture is kept at a temperature of 105° to 110°C by the live steam or lead steam coils. An excess of bauxite is fed to the reactor, so that there is an excess of 0.1 to 0.2 per cent of soluble aluminium oxide. From 15 to 20 hrs are required to complete the reaction. At the end of this time, a reducing material is added to the reaction mixture to reduce the iron (ferric sulphate) to a colourless ferrous condition. Barium sulphide in the form of black ash is commonly utilised, although sodium sulphide, hydrogen sulphide, sodium bisulphite, or sulphur dioxide may also be used.

If the operation is performed in batches, the charge is allowed to settle in settling tanks. Flaked glue or some similar coagulable substance is generally added to remove the finely divided suspended material remaining in the supernatant liquid. This liquid is drawn off, and the residue is washed several times. The washings are combined with the decanted liquor, which is then sent to concentrators (Fig. 19.4).

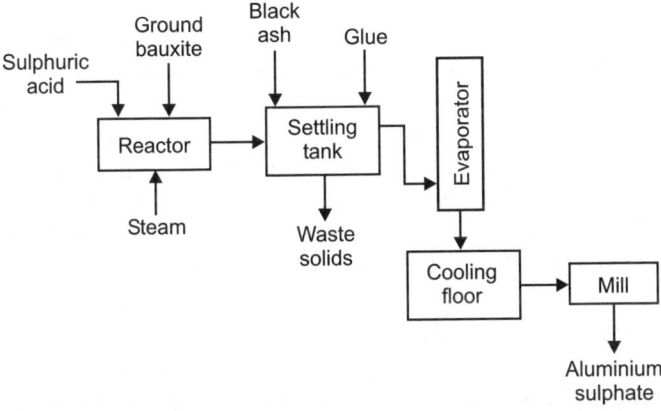

Fig. 19.4. Flow diagram for manufacturing aluminium sulphate from bauxite.

The process is generally operated in a continuous manner by using a battery of combined reaction and settling tanks. A common variation of this process is the Dorr procedure, which utilises reaction agitators in series. The reactants are thoroughly mixed and heated, using mechanical agitators, air, and live steam. Blach ash (barium sulphide) is added to the last reactor to reduce the ferric sulphate. The reaction mixture is sent through a series of thickeners, operating counter-currently, which remove the undissolved material. At the same time the waste is washed thoroughly, so that it contains practically no aluminium sulphate when discarded. Glue is generally added to the first thickener as a coagulant.

The clarified aluminium sulphate solution, from the counter-current decantation system is concentrated in open steam-coil-heated, lead-lined evaporators. Here, the specific gravity is increased from about 1.3 to about 1.7.

The concentrated solution is run into flat iron pans or onto a cooling table. The liquid quickly and completely solidifies and when cool is broken up and ground to a uniform powder for shipment. Commercial aluminium sulphate generally contains only about 13 or 14 moles of water instead of the theoretical 18 moles. Also, it is usually in the basic form contains excess alumina. Anhydrous aluminium sulphate may be obtained by dehydration. The yield of aluminium sulphate based on the amount of aluminium oxide in both the finished product and raw material is 90 to 95 per cent.

Preparation

Aluminium sulphate may be made by dissolving aluminium hydroxide, $Al(OH)_3$, in sulphuric acid:

$$2Al(OH)_3 + 3H_2SO_4 \rightarrow Al_2(SO_4)_3 \cdot 6H_2O$$

Uses

Aluminium sulphate is used in water purification and as a mordant in dyeing and printing textiles. In water purification, it causes impurities to coagulate which are removed as the particulate settles to the bottom of the container or more easily filtered. This process is called coagulation or flocculation.

When dissolved in a large amount of neutral or slightly-alkaline water, aluminium sulphate produces a gelatinous precipitate of aluminium hydroxide, $Al(OH)_3$. In dyeing and printing cloth, the gelatinous precipitate helps the dye adhere to the clothing fibres by rendering the pigment insoluble. Aluminium sulphate is sometimes used to reduce the pH of garden soil, as it hydrolyses to form the aluminium hydroxide precipitate and a dilute sulphuric acid solution.

Aluminium sulphate is usually found in baking powder, where there is controversy over its use due to concern regarding the safety of adding aluminium to the diet. In construction industry it is used as waterproofing agent and accelerator in concrete. Another use is a foaming agent in fire fighting foam.

It is also used in styptic pencils, and pain relief from stings and bites; it is the active ingredient in popular pain relief products such as Stingose. It can also be very effective as a molluscicide, killing Spanish slugs.

ALUM

Alum, refers to a specific chemical compound and a class of chemical compounds. The specific compound is the hydrated aluminium potassium sulphate with the formula $KAl(SO_4)_2 \cdot 12H_2O$.

Alums are useful for a range of industrial processes. They are soluble in water; have an astringent, acid, and sweetish taste; react acid to litmus; and crystallise in regular octahedra. When heated they liquefy; and if the heating is continued, the water of crystallisation is driven off, the salt froths and swells, and at last an amorphous powder remains. Potassium alum is the common alum of commerce, although soda alum, ferric alum, and ammonium alum are manufactured. Aluminium sulphate is sometimes called alum in informal contexts, but this usage is not regarded as technically correct. Its properties are quite different from those of the set of alums formally described above.

Uses

Alum is used in vaccines as an adjuvant. Alum is commonly used as a coagulant in water treatment.

Production

Alum from alunite

In order to obtain alum from alunite, it is calcined and then exposed to the action of air for a considerable time. During this exposure it is kept continually moistened with water, so that it ultimately falls to a very fine powder. This powder is then lixiviated with hot water, the liquor decanted, and the alum allowed to crystallise. The alum schists employed in the manufacture of alum are mixtures of iron pyrite, aluminium silicate and various bituminous substances, and are found in upper Bavaria, Bohemia, Belgium, and Scotland.

These are either roasted or exposed to the weathering action of the air. In the roasting process, sulphuric acid is formed and acts on the clay to form aluminium sulphate, a similar condition of affairs being produced during weathering. The mass is now systematically extracted with water, and a solution of aluminium sulphate of specific gravity 1.16 is prepared. This solution is allowed to stand for some time (in order that any calcium sulphate and basic ferric sulphate may separate), and is then evaporated until ferrous sulphate crystallises on cooling; it is then drawn off and evaporated until it attains a specific gravity of 1.40. It is now allowed to stand for some time, decanted from any sediment, and finally mixed with the calculated quantity of potassium sulphate (or if ammonium alum is required, with ammonium sulphate), well agitated, and the alum is thrown down as a finely-divided precipitate of alum meal. If much iron should be present in the shale then it is preferable to use potassium chloride in place of potassium sulphate.

Alum from clays or bauxite

In the preparation of alum from clays or from bauxite, the material is gently calcined, then mixed with sulphuric acid and heated gradually to boiling; it is allowed to stand for some time, the clear solution drawn off and mixed with acid potassium sulphate and allowed to crystallise. When cryolite is used for the preparation of alum, it is mixed with calcium carbonate and heated.

By this means, sodium aluminate is formed; it is then extracted with water and precipitated either by sodium bicarbonate or by passing a current of carbon dioxide through the solution. The precipitate is then dissolved in sulphuric acid, the requisite amount of potassium sulphate added and the solution allowed to crystallise.

Types of Alum

Soda alum

Sodium alum, $Na_2SO_4 \cdot Al_2(SO_4)_3 \cdot 24H_2O$, mainly occurs in nature as the mineral mendozite. It is very much soluble in water, and is extremely difficult to purify. In the preparation of this salt, it is preferable to mix the component solutions in the cold, and to evaporate them at a temperature not exceeding 60°C. 100 parts of water dissolve 110 parts of sodium alum at 0 °C, and 51 parts at 16°C. Soda alum is used in the acidulent of food as well as in the manufacture of baking powder.

Ammonium alum

Ammonia alum, $NH_4Al(SO_4)_2 \cdot 12H_2O$, a white crystalline double sulphate of aluminium, is used in water purification, in vegetable glues, in porcelain cements, in natural deodourants (though potassium alum is more commonly used), in tanning, dyeing and in fireproofing textiles.

Solubility

The solubility of the various alums in water varies greatly, sodium alum being readily soluble in water, while caesium and rubidium alums are only sparingly soluble.

Uses

It is used in cosmetic, medicines, and culinary. It is also used as a flame retardant and chemical flocculant.

ALUMINIUM CHLORIDE

Aluminium chloride ($AlCl_3$) is a compound of aluminium and chlorine. The solid has a low melting and boiling point, and is covalently bonded. It sublimes at 178°C. Molten $AlCl_3$ conducts electricity poorly, unlike more ionic halides such as sodium chloride. It exists in the solid state as a six-coordinate layer lattice.

Aluminium chloride is highly deliquescent, and can explode upon abrupt contact with water because of the high heat of hydration. Aqueous solutions of $AlCl_3$ are ionic and thus conduct electricity well. Such solutions are found to be acidic, indicative of partial hydrolysis of the Al^{3+} ion. The reactions can be described (simplified) as:

$$[Al(H_2O)_6]^{3+} + H_2O \rightarrow [Al(OH)(H_2O)_5]^{2+} + H_3O^+$$

$AlCl_3$ is probably the most commonly used Lewis acid and also one of the most powerful. It finds widespread application in the chemical industry as the classic catalyst for Friedel-Crafts reactions, both acylations and alkylations. It also finds use in polymerisation and isomerisation reactions of hydrocarbons.

Aluminium also forms a lower chloride, aluminium(I) chloride ($AlCl_3$), but this is very unstable and only known in the vapour phase.

Preparation

Aluminium chloride is manufactured on a large scale by the exothermic reaction of aluminium metal with chlorine or hydrogen chloride which are charged into a refractory crucible furnace in which it is melted at 660°C.

Dry chlorine is passed into the molten charge and forms aluminium chloride which vapourises and leaves the furnace through a vapour dust in the top. The vapours are then passed into air cooled iron condensers, where aluminium chloride sublimes below 178°C.

$$2Al + 3Cl_2 \rightarrow 2AlCl_3$$
$$2Al + 6HCl \rightarrow 2AlCl_3 + 3H_2$$

Hydrated forms are prepared by dissolving aluminium oxides with hydrochloric acid.

Uses

The Friedel-Crafts reaction is the major use for aluminium chloride, for example in the preparation of anthraquinone (for the dyestuffs industry) from benzene and phosgene. In the general Friedel-Crafts reaction, an acyl chloride or alkyl halide reacts with an aromatic system as given below:

With benzene derivatives, the major product is the *para* isomer. The alkylation reaction has many associated problems, such as in Friedel-Crafts, so it is less widely used than the acylation reaction. For both reactions, the aluminium chloride, as well as other materials and the equipment, must be moderately dry, although a trace of moisture is necessary for the reaction to proceed. A general problem with the Friedel-Crafts reaction is that the aluminium chloride catalyst needs to be present in full stoichiometric quantities in order for the reaction to go to completion, because it complexes strongly with the products.

This makes it very difficult to recycle, so it must be destroyed after use, generating a large amount of corrosive waste. For this reason chemists are examining the use of more environmentally benign catalysts such as ytterbium(III) triflate or dysprosium(III) triflate, which can be recycled.

Aluminium chloride can also be used to introduce aldehyde groups onto aromatic rings, for example via the Gattermann-Koch reaction which uses carbon monoxide, hydrogen chloride and a copper(I) chloride co-catalyst.

Aluminium chloride finds a wide variety of other applications in organic chemistry. For example, it can catalyse the 'ene reaction', such as the addition of 3-buten-2-one (methyl vinyl ketone) to carvone.

AlCl$_3$ is also widely used for polymerisation and isomerisation reactions of hydrocarbons. Important examples include the manufacture of ethylbenzene, which is used to make styrene and thus polystyrene, and also production of dodecylbenzene, which is used for making detergents. Aluminium chloride combined with aluminium in the presence of an arene can be used to synthesise *bis*(arene) metal complexes, e.g. *bis*(benzene)chromium, from certain metal halides via the so-called Fischer-Hafner synthesis. Aluminium chloride, often in the form of derivatives such as aluminium chlorohydrate, is a common component in antiperspirants at low concentrations.

Precautions

Anhydrous AlCl$_3$ reacts vigorously with water and bases, so suitable precautions are required. Hydrated salts are less problematic.

MOLYBDENUM

Molybdenum, is a group 6 chemical element with the symbol Mo and atomic number 42. It has the eighth-highest melting point of any element. It readily forms hard, stable carbides, and for this reason it is often used in high-strength steel alloys. Molybdenum is found in trace amounts in plants and animals, although excess molybdenum can be toxic in some animals. Molybdenum is a transition metal with an electronegativity of 1.8 on the Pauling scale and an atomic mass of 95.9 g/mole. It does not react with oxygen or water at room temperature. At elevated temperatures, molybdenum trioxide is formed:

$$2Mo + 3O_2 \rightarrow 2MoO_3.$$

In its pure metal form, molybdenum is silvery white with a Mohs hardness of 5.5, though it is somewhat more ductile than tungsten. It has a melting point of 2623°C, and of the naturally-occurring metals, only tantalum, osmium, rhenium, and tungsten have higher melting points. Molybdenum burns only at temperatures above 600°C. It also has the lowest heating expansion of any commercially used metal. Though molybdenum is found in such minerals as wulfenite (PbMoO$_4$) and powellite (CaMoO$_4$), the main commercial source of molybdenum is molybdenite (MoS$_2$). Molybdenum is mined as a principal ore, and is also recovered as a by-product of copper and tungsten mining.

A side product of molybdenum mining is rhenium. As it is always present in small varying quantities in molybdenite, the only commercial source for rhenium is molybdenum mines.

Production

The molybdenite is roasted at a temperature of 700°C and the sulphide is oxidised into molybdenum(IV) oxide by air.

$$2MoS_2 + 5O_2 \rightarrow 2MoO_3 + 2SO_2$$

The roasted ore is either heated to 1100°C to sublime the oxide or leached with ammonia, with which molybdenum(IV) oxide forms water soluble molybdates.

$$MoO_3 + NH_4OH \rightarrow (NH_4)_2(MoO_4) + H_2O$$

Copper is less soluble in ammonia, but to remove it from the solution the copper is precipitated with hydrogen sulphide. Pure molybdenum is produced by reduction of the oxide with hydrogen, while the molybdenum for steel production is reduced by the aluminothermic reaction with addition of iron to produce ferromolybdenum. Ferromolybdenum contains 60 per cent of molybdenum.

Compounds

Molybdenum has several common oxidation states, +2, +3, +4, +5 and +6.

The highest oxidation state is common in the molybdenum(VI) oxide MoO_3 while the normal sulphur compound is molybdenum disulphide MoS_2. The broad range of oxidation states shows up in the chlorides of molybdenum:
1. Molybdenum(II) chloride $MoCl_2$ (yellow solid).
2. Molybdenum(III) chloride $MoCl_3$ (dark red solid).
3. Molybdenum(V) chloride $MoCl_5$ (dark green solid).
4. Molybdenum(VI) chloride $MoCl_6$ (brown solid).

Like chromium and some other transition metals, molybdenum is able to form quadruple bonds. $Mo_2(CH_3COO)_4$ is an example for a quadruple bond. This compound can be transformed into the chlorine compound $Mo_2Cl_8^{4-}$.

Applications

The ability of molybdenum to withstand extreme temperatures without significantly expanding or softening makes it useful in applications that involve intense heat, including the manufacture of aircraft parts, electrical contacts, industrial motors, and filaments.

Molybdenum is also used in alloys for its high corrosion resistance and weldability. Moly contributes corrosion resistance to type 316 stainless steel by 'gettering' residual carbon, preventing the formation of chromium carbide at grain boundaries.

Because of its lower density and more stable price, molybdenum is implemented in the place of tungsten. Molybdenum can be implemented both as an alloying agent and as a flame-resistant coating for other metals. Although its melting point is 2623°C (4753°F), molybdenum rapidly oxidises at temperatures above 760°C (1400°F), making it better-suited for use in vacuum environments.

Molybdenum disulphide (MoS_2) is used as a solid lubricant and an extreme pressure (EP) antiwear agent. It forms strong films on metallic surfaces, and is highly resistant to both extreme temperatures and high pressure, and for this reason, it is a common additive to extreme pressure application greases; in case of a catastrophic failure, the thin layer of molybdenum prevents metal-on-metal contact. Molybdenum trioxide (MoO_3) is used as an adhesive between enamels and metals. Molybdenum powder is used as a fertiliser for some plants, such as cauliflower.

Lead molybdate (wulfenite) co-precipitated with lead chromate and lead sulphate is a bright-orange pigment used with ceramics and plastics. Also used in NO, NO_2, NO_x analysers in power plants for pollution controls. At 350°C (662°F) the element acts as a catalyst for NO_2/NO_x to form only NO molecules for consistent readings by infrared light.

STRONTIUM SALTS

Strontium is a chemical element with the symbol Sr and the atomic number 38. An alkaline earth metal, strontium is a soft silver-white or yellowish metallic element that is highly reactive chemically. The metal turns yellow when exposed to air. It occurs naturally in the minerals celestine and strontianite.

Uses of strontium salts are small in tonnage but important; they include red-flame pyrotechnic compositions, such as truck signal flares and railroad 'fusees', tracer bullets, miliatary singal flares, and ceramic magnets.

This ore (strontium sulphate) is finely ground and converted to the carbonate by boiling with 10 per cent sodium carbonate solution giving solution, giving almost a quantiative yield:

$$SrSO_4 + Na_2CO_3 \longrightarrow SrCO_3 + Na_2SO_4$$

From reaction of the strontium carbonate with appropriate acids, the various salts result.

LITHIUM SALTS

Some of the lithium salts are dicussed below.

Lithium Carbonate (Li_2CO_3)

It is a white powder; sp. gr. 2.111, m.p 735°C; b.p. decomposes at 1200°C. Slightly soluble in water, insoluble in alcohol; soluble in dilute acid.

It is prepared when finely ground ore is roasted with sulphuric acid at 250°C. Lithium sulphate is leached from the mass and converted to the carbonate by precipitation with soda ash. It is also prepared by reacting lithium oxide with carbon dioxide or ammonium carbonate solution.

Lithium Fluoride (LiF)

It is a fine white powder, sp. gr. 2.365 (20°C), m.p. 842°C, b.p. 1670°C, slightly soluble in water, does not react with water at red heat; soluble in acid; insoluble in alcohol. It is prepared by reacting of hydrofluoric acid with lithium carbonate.

Lithium Hydroxide (LiOH)

It is in the form of colourless crystals; sp. gr. 2.54; m.p. 470°C; b.p. 924°C, decomposes, slightly soluble in alcohol, soluble in water; absorbs CO_2 and water from air.

It is prepared by causticising of lithium carbonate or by action of water on metallic lithium, or by addition of Li_2O to water. The water solutions is strongly irritant and toxic. It is used in storage battery electrolyte as a carbon dioxide absorbent in space vehicles; lubricating greases; ceramics; catalyst; photographic developers; lithium soaps.

Lithium Chloride (LiCl)

It is in the form of white deliquescent crystals, sp. gr. 2.068, m.p. 614°C, b.p. 1360°C, very soluble in water alcohols, ether, pyridine, nitrobenzene. It is one of the most hygroscopic salts known. Low toxicity, but should not be used as dietary salt substitue.

It is prepared by reaction of lithium ores with chlorides; natural brines.

It is used in air conditioning, welding and soldering flux, dry batteries, heat-exchange media, salt baths; desiccant; production of lithium metal; soft drinks and mineral water to reduce escape of carbon dioxide.

Lithium Sulphate ($Li_2SO_4 \cdot H_2O$)

It is in the form of colourless crystals, sp. gr. 2.06, m.p. 130°C, soluble in water, insoluble in 80 per cent alcohol. Does not form alums.

It is preared by reaction of sulphuric acid with lithium carbonate or with spodumene ore. It is used in pharmaceutical products, and ceramics.

BORON

Boron is a chemical element with atomic number 5 and chemical symbol B. Boron is a trivalent metalloid element which occurs abundantly in the evaporite ores borax and ulexite. Boron is never found as a free element on earth.

Several allotropes of boron exist; amorphous boron is a brown powder, though crystalline boron is black, extremely hard (9.3 on Mohs' scale), and a weak conductor at room temperature (22°–28°C, 72°–82°F). Elemental boron is used as a dopant in the semiconductor industry, while boron compounds play important roles as light structural materials, nontoxic insecticides and preservatives, and reagents for chemical synthesis. Boron is an essential plant nutrient, although high soil concentrations of boron may also be toxic to plants. As an ultratrace element, boron is necessary for the optimal health of rats and presumably other mammals, though its physiological role in animals is poorly understood.

Brown amorphous boron is a product of certain chemical reactions. It contains boron atoms that are randomly bonded to each other without long range order.

Crystalline boron, a very hard, black material with a high melting point, exists in many polymorphs. Two rhombohedral forms, α-boron and β-boron containing 12 and 106.7 atoms in the rhombohedral unit cell respectively, and 192-atom tetragonal boron are the three most characterised crystalline forms.

Chemically boron is electron-deficient, possessing a vacant p-orbital. Boron is also similar to carbon with its capability to form stable covalently bonded molecular networks. Boron is also used for heat resistant alloys. Boron can form compounds whose formal oxidation state is not three, such as B(II) in B_2F_4.

Applications

The only major use of metallic boron is as boron fibre. The fibres are used to reinforce the fuselage of fighter aircraft, for example the B-1 bomber. The fibres are produced by vapour deposition of boron on a tungsten filament. (i) Glass and ceramics, (ii) soaps and detergents, and (iii) fire retardants.

Hardest Boron Compounds

The hardest boron compounds are created synthetically. Rhenium diboride (ReB_2) and cubic (or beta)-boron nitride can actually scratch diamond, but are still not as hard as diamond although rhenium diboride surpasses diamond in certain directions. Rhenium diboride has been reported to be nearly as hard as cubic boron nitride and boron suboxide, and much harder than osmium diboride (which was the first step towards rhenium diboride synthesis).

Boron Compounds

The boron oxygen compounds sodium tetraborate, boric acid and sodium perborate are the compounds with the largest production capacities. For some special applications the carbide and nitride boron carbide and boron nitride are produced. Sodium borohydride is one of the few hydrides produced on industrial scale as reduction reagent in chemical synthesis.

Boron does not appear on earth in elemental form but is found combined in borax, boric acid, colemanite, kernite, ulexite and borates. Boric acid is sometimes found in volcanic spring waters. Ulexite is a borate mineral that naturally has properties of fibre optics.

Pure elemental boron is not easy to prepare. The earliest methods used involve reduction of boric oxide with metals such as magnesium or aluminium. However, the product is almost always contaminated with metal borides. (The reaction is quite spectacular though.) Pure boron can be prepared by reducing volatile boron halogenides with hydrogen at high temperatures. The highly pure boron, for the use in semiconductor industry, is produced by the decomposition of diborane at high temperatures and then further purified with the Czochralski process.

RARE EARTH COMPOUND

The rare earths comprise a group of 17 elements in the periodic table and have similar properties in aqueous solutions. All are metals in the elemental state and all form salts that are strong electrolytes when dissolved in water. They ionise in this medium to give triply charged ions and, because of the high charge on these ions, react strongly with water dipoles to form a tight sheath of water molecules about them. Other ions in aqueous solutions only contact this sheath, giving rise to the similar properties of rare-earth cations in water.

The group consists of the following elements : scandium (^{21}Sc), yttrium (^{39}Y), lanthanum (^{57}La), all of which appear in the group IIIB of the periodic table, and cerium (^{58}Ce), praseodymium (^{59}Pr), neodymium (^{60}Nd), promethium (^{61}Pm), samarium (^{62}Sm), europium (^{63}Eu), gadolinium (^{64}Gd), terbium (^{65}Tb), dysprosium (^{66}Dy), holmium (^{6}Ho), erbium (^{68}Er), thulium (^{69}Tm), ytterbium (^{70}Yb), and lutetium (^{71}Lu). The rare earths are widely distributed in low concentrations throughout the earth's crust. They occur as mixtures in many massive rock formations, e.g. basalts, granites, gneisses, shales, and silicate rocks, in which they are present in amounts of 10–300 ppm. They also occur in ca. 160 discrete minerals, most of which are rare, but in which the rare-earth content, expressed as R_2O_3 or Ln_2O_3 (REO), can be as high as 60 per cent (REO is an abbreviation used in industry for rare-earth oxide content). Ln_2O_3 is used when Y_2O_3 is not present in the ore or mineral. Approximately ten of these occur in sufficient quantities that they may furnish some REO to commerce, but more than 95 per cent of the REO occurs in three minerals: monazite and bastnasite for the light rare earths, and xenotime for yttrium and the heavy rare earths. Xenotime occurs mixed with monazite in alluvial deposits.

Properties

Physical

The rare earth metals alloy with most metals to form intermetallic compounds and occasionally solid solution. In rare-earth-rich alloys, other elements can change the properties of the pure metal by drastically lowering (or, more rarely, raising) the melting point by 200°–300°C in some cases. Alloying with other elements can make the rare earth either pyrophoric or corrosion-resistant.

Chemical

The chlorides, bromides, nitrates, bromates, and perchlorate salts are soluble in water and, when their aqueous solutions evaporate, they precipitate as hydrated crystalline salts. The acetates, iodates, and iodides are somewhat less soluble. The sulphates are sparingly soluble and are unique in that they become less soluble with increasing temperature. The oxides, sulphides, fluorides, carbonates, oxalates, and phosphates are insoluble in water. The oxalate, which is important in the recovery of highly pure Ln, can be calcined directly to the oxide.

Anhydrous rare-earth salts usually cannot be prepared by evaporating the water of crystallisation. The lanthanides can form hydrides of any composition up to LnH_3. Small amounts of hydrogen dissolve interstitially but, with increasing amounts of hydrogen, a second phase (LnH_2) appears. These alloys are metallic in their properties and lose hydrogen at relatively high temperatures.

Uses

The rare earths are used in gasoline-cracking catalysts; in carbon arcs; as additives to steel and cast irons; as polishing compounds; to made optical glass; as both colourants and decolourants for glass, depending upon the rare earth element used; and in magnets. In the electronic area, the most important industrial rare-earth compounds are garnet-based materials, e.g. yttrium iron garnet and gadolium gallium garnet, used in microwave devices, memory storage, and oxygen sensors (Y_2O_3-stabilised ZrO_2). In nuclear reactors, the rare earths are used in the form of oxides as absorbers of neutrons. They are also used as additions to superalloys, as hydrogen-storing materials, and as synthetic gems.

SODIUM DICHROMATE

Sodium dichromate is the chemical compound with the formula $Na_2Cr_2O_7$. Usually, however, the salt is handled as its dihydrate $Na_2Cr_2O_7 \cdot 2H_2O$. Virtually all chromium ore is processed via conversion to sodium dichromate. In this way, many millions of kilograms of sodium dichromate are produced annually. In terms of reactivity and appearance, sodium dichromate and potassium dichromate are very similar. The sodium salt is, however, around twenty times more soluble in water than the potassium salt (49 g/l at 0°C) and its equivalent weight is also lower, which is often desirable.

Manufacture

Sodium dichromate is generated on a large scale from ores containing chromium(III) oxides. The ore is fused with base, typically sodium carbonate, at around 1000°C in the presence of air (source of oxygen):

$$Cr_2O_3 + 2Na_2CO_3 + 1.5O_2 \rightarrow 2Na_2CrO_4 + 2CO_2$$

This step solubilises the chromium and allows it to be extracted into hot water. At this stage, other components of the ore such as aluminium and iron compounds, are poorly soluble. Acidification of the resulting aqueous extract with sulphuric acid or carbon dioxide affords the dichromate, which is isolated at the dihydrate by crystallisation. Since chromium(VI) is toxic, especially as the dust, such factories are subject to stringent regulations. For example, effluent from such refineries is treated with reducing agents to return any chromium(VI) to chromium(III), which is less threatening to the environment.

Safety

Like all hexavalent chromium compounds, sodium dichromate is considered hazardous. It is also a known carcinogen.

HYDROGEN PEROXIDE

Hydrogen peroxide (H_2O_2) is a very pale blue liquid which appears colourless in a dilute solution, slightly more viscous than water. It is a weak acid. It has strong oxidising properties and is therefore a powerful bleaching agent that is mostly used for bleaching paper, but has also found use as a disinfectant, as an oxidiser, as an antiseptic, and in rocketry (particularly in high concentrations as high-test peroxide or HTP) as a monopropellant, and in bipropellant systems. The oxidising capacity of hydrogen peroxide is so strong that the chemical is considered a highly reactive oxygen species. Hydrogen peroxide is naturally produced as a by-product of oxygen metabolism, and virtually all organisms possess enzymes known as peroxidases, which harmlessly and catalytically decompose low concentrations of hydrogen peroxide to water and oxygen.

Uses

Industrial applications

About 50 per cent of production of hydrogen peroxide is used for pulp and paper-bleaching. Other bleaching applications are becoming more important as hydrogen peroxide is seen as an environmentally benign alternative to chlorine-based bleaches. It is highly corrosive to metal.

Other major industrial applications for hydrogen peroxide include the manufacture of sodium percarbonate and sodium perborate, used as mild bleaches in laundry detergents. It is used in the production of certain organic peroxides such as dibenzoyl peroxide, used in polymerisations and other chemical processes. Hydrogen peroxide is also used in the production of epoxides such as propylene oxide.

Reaction with carboxylic acids produces a corresponding peroxy acid. Peracetic acid and meta-chloroperoxybenzoic acid (commonly abbreviated mCPBA) are prepared from acetic acid and meta-chlorobenzoic acid, respectively. The latter is commonly reacted with alkenes to give the corresponding epoxide. In the PCB manufacturing process, hydrogen peroxide mixed with sulphuric acid was used as the microetch chemical for copper surface roughening preparation.

A combination of a powdered precious metal-based catalyst, hydrogen peroxide, methanol and water can produce superheated steam in one to two seconds, releasing only CO_2 and high temperature steam for a variety of purposes.

Recently, there has been increased use of vapourised hydrogen peroxide in the validation and bio-decontamination of half suit and glove port isolators in pharmaceutical production.

Hydrogen peroxide is also used in the oil and gas exploration industry to oxidise rock matrix in preparation for micro-fossil analysis.

H_2O_2 can be used either as a monopropellant (not mixed with fuel) or as the oxidiser component of a bipropellant rocket.

Manufacture

Hydrogen peroxide is manufactured by electrolytic and two organic oxidation processes. In the electrolytic methods the active oxygen is produced by the anodic oxidation of sulphate radicals to form peroxydisulphate intermediates. Sulphuric acid electrolyte has low current efficiency (70 to 75 per cent), but the use of ammonium sulphate causes crystallisation problems, so a mixture is used to obtain a current efficiency of 80 per cent or higher and yet not block the cell with crystal formation. The electrolyte is fed into a typical cell held at 35°C or below with platinum metal for the anode:

$$2H_2SO_4 \rightleftharpoons H_2S_2O_8 + H_2$$

The cell product, i.e. the persalt, or peracid, and sulphuric acid in water are subjected to hydrolyses at 60° to 100°C to yield hydrogen peroxide, as shown by the following simplified reactions:

$$H_2S_2O_8 + H_2O \rightleftharpoons H_2SO_4 + H_2SO_5$$

$$H_2SO_5 + H_2O \rightleftharpoons H_2SO_4 + H_2O_2$$

Physical Properties

As with all molecules, the physical properties of hydrogen peroxide are the result of its molecular mass, structure and distribution of atoms within the molecule.

Comparison with analogues

Analogues of hydrogen peroxide include the chemically identical deuterium peroxide and malodourous hydrogen disulphide. Hydrogen disulphide has a boiling point of only 70.7°C despite having a higher molecular weight, indicating that hydrogen bonding increases the boiling point of hydrogen peroxide.

Aqueous hydrogen peroxide solutions have specific properties that are different from those of the pure chemical due to hydrogen bonding between water and hydrogen peroxide molecules. Specifically, hydrogen peroxide and water form a eutectic mixture, exhibiting freezing-point depression. While pure water melts and freezes at approximately 273 K, and pure hydrogen peroxide just 0.4 K below that, a 50 per cent (by volume) solution melts and freezes at 221 K.

Storage

Hydrogen peroxide should be stored in a cool, dry, well-ventilated area and away from any flammable or combustible substances. It should be stored in a container composed of non-reactive materials such as stainless steel or glass (other materials including some plastics and aluminium alloys may also be suitable). Because it breaks down quickly when exposed to light, it should be stored in an opaque container, and pharmaceutical formulations typically come in brown bottles that filter out light.

References

Adams J.A., and Rogers D.F., *Computer-Aided Heat Transfer Analysis*, McGraw-Hill, New York.

Backhurst J.R., *Process Plant Design*, Elsevier, New York.

Camp, T.R., *Water and Its Impurities*, Reinhold, New York.

Chanlett E.T., *Environmental Protection*, McGraw-Hill, New York.

Cooper R.L., *Chemistry of Sugar and Starch*, Pergamon, UK

James A.N., *Industrial Chemistry*, Noyes, UK.

Jolles Z.E., *Bromine and its Compounds*, Noyes, UK.

Jurah J.M., *Quality Control Handbook,* McGraw-Hill, New York.

Kent J.A., *Riegel's Industrial Chemistry*, Reinhold, USA.

Kingzett, *Chemical Encyclopaedia*, Van Nostrand, USA.

Landau R., *Chemical Plant*, Reinhold, USA.

Morton M., *Rubber Technology*, Van Nostrand, Reinhold, USA.

Perry R.H., *Chemical Engineers' Handbook*, McGraw-Hill, New York.

R. Norris Shreve and Joseph A. Brink, *Chemical Process Industries*, McGraw-Hill, New York.

Ray W.H., *Process Optimisation*, Wiley-Interscience, New York.

Rudd D., *Process Synthesis,* Prentice-Hall, New York.

Sauchelli V., *Chemistry and Technology of Fertilisers*, Reinhold, UK.

Singer F.S., *Industrial Ceramics*, Chemical Publishing, UK.

Siting M., *Sulphuric Acid Manufacture and Effluent Control*, Noyes, UK.

Siting M., *Organic Chemical Process Encyclopaedia*, Noyes, UK.

Stephenson R.N., *Introduction to Chemical Process Industries*, Reinhold, USA.

Urbanski T., *Chemistry and Technology of Explosives*, Pergamon, UK.

Van Vlack L.H., *Physical Ceramics for Engineers*, Addison-Wesley, UK.

Yaffe L., *Nuclear Chemistry*, McGraw-Hill, New York.

Index